Technology, Engineering, and Economics

THE MIT PRESS Massachusetts Institute of Technology
Cambridge, Massachusetts, and London, England

PHILIP SPORN

Technology,
Engineering,
and Economics

Once again:
To Sadie

Acknowledgments

I am glad to acknowledge and express my appreciation for the help I received from Mr. Abraham Gerber in the editing of some of the material and from my secretary, Miss Dorothy Miesse, for her work in getting the manuscript ready for the printer.

PHILIP SPORN

Foreword

Philip Sporn's long and distinguished career has been at the interface of economics, engineering, and management. His accomplishments as an engineer, an engineering manager, and finally as an executive, have been marked by an innovative approach to problems and by an insight into the real needs of the corporate and, more broadly, the human environment. He was equally effective in leading the development of high-speed reclosing of circuit breakers to improve the reliability of power supply as he was in organizing the financing of major power projects. It is quite appropriate, then, that the lectures included in this volume should span the wide spectrum of interests from science and technology to financial management and public administration that was reflected by Mr. Sporn's experience in the power industry.

In a time when mastery of a single area of human knowledge is becoming a superhuman endeavor, Mr. Sporn has placed an additional challenge before us. The lectures move beyond the detailed history of a particular power project to the implication of engineering for a society characterized by rapid change and by a close interdependence of basic concepts, institutional structures, and ideas. His discussion of the conception and development of the OVEC power project demonstrates the need for a holistic approach to large-scale problem-solving. It shows that it is not enough to possess scientific and technical capability without the tools of economic analysis and without the depth of social perspective.

It would be fair to say that the effectiveness of his program depended, more than anything else, on successful engineering — the working amalgamation of all of these qualities and points of view. In this sense, engineering is far more than the mechanical marshaling of resources into working systems. It is an economic allocation which seeks the optimal arrangement of conditions within the human and historical context, as well as the purely technical.

Mr. Sporn has always had an optimistic faith in the future. Indeed, he feels strongly that we can and should "invent our future" by deciding where our society should go and then by exerting our best efforts to take it there. I believe that this point of view represents a special challenge for our universities. In today's systems society, education at the university has the distinctive task of reconciling the most ancient values of our culture with the most contemporary demands of our environment. The Sporn lectures address directly this match of expectations and capability; and, in this sense, they are a significant contribution to modern education.

We hope that more graduates of this and other universities will undertake useful roles of business leadership, while actively participating in governmental, social, and cultural activities. It would be wholly appropriate, then, in the context of this book, to say that our objective is to produce the Philip Sporns of the future.

Howard W. Johnson

Contents

Science, Technology, Engineering, and Economics: Capital as a Cost

Introduction

IT IS A universally recognized truism that science and technology in the two decades since the end of World War II have become interwoven with daily living on a scale far exceeding anything that prevailed previously. Throughout the world, science and technology have provided us a new primary source of energy, more rapid and reliable methods of transportation and communication, new methods of data processing, and a promise of new sources of food and new methods of, and mechanisms for, education.

Since the middle of 1966, for example, there have been placed on order in the United States some 51 nuclear power reactors, totaling approximately 40 million kw. These will be installed some time between 1968 and 1975 and will represent an investment cost of approximately $5.8 billion. In their active lifetime these reactors will generate about 8.5 trillion kwh, and displace over 3 billion tons of coal.

This is a technological-economic development of major significance. It already has had, and will continue to have for many years, an influence on coal mining, transportation, defense, and on international political relations. It is making it possible to look at the future with hope and confidence, and without fear for the possible exhaustion of our inanimate energy resources. While this fabulous aggregation of energy conversion equipment is not yet here, it is in the making and will continue to grow to still more vast proportions.

The foundation of this development is a scientific discovery — atomic fission. It all rests on one basic reaction — the splitting or fissioning of uranium-235, one of the lighter isotopes of the heavy element uranium. In that fissioning process, uranium-235 captures a neutron and for a brief period becomes uranium-236. It then splits into two nearly equal parts. These are lighter elements in the middle of the periodic table, but are not always the same. Altogether some 40 to 45 different fission products have been identified. The important fact is that the combined weight of the two atoms into which the uranium atom splits is less than that of the original. The lost mass of the annihilated matter, in accordance with Einstein's famous formula, has been converted into energy that appears as heat.

Einstein proposed his theoretical formulation some 60 years ago, and the phenomenon of fission was discovered by Hahn and Meitner some 30 years ago. The demonstration of a controlled fission reaction was provided in the famous experiment at Stagg Field in 1942 — 25 years ago — and at this point the main work of the scientists ended. The practical harnessing of that discovery has taken 25 to 30 years spent upon a great deal of developmental work on the part of engineers — and the engineers have just begun. Thus, technology and engineering based on scientific discovery have brought society a new source of energy — energy that is so essential for the continued development of an industrial society.

It is clear that science and technology together have played a major role in bringing this nation the highest standards of material welfare more broadly disseminated throughout its population than have ever been achieved by any society in the history of the world. Despite the troubled consciences on the part of a good many well-intentioned of our citizens, this scientific and technological progress has failed to bring about any revolutionary changes in our society, and the benefits it has brought have outweighed the problems and dislocations. And for the continuing welfare of our society science and technology need to be vigorously carried forward, perhaps even at a slightly accelerated pace, to enable us to solve the many economic and social problems confronting our world.

We need to continue to provide more goods and services, to raise living standards still further, and to disseminate these standards still more widely. We need greater productivity to give people more leisure and to provide many of our social-economic needs — housing, transportation, pollution control, continuing education, and beautification — all with a view of creating a better life and a better society. We need to help the underdeveloped world to lift itself from the slough of poverty to heights comparable with those our own society has managed to reach.[1]

To achieve these objectives, it is imperative that we understand better than we do the dual mechanism that is the principal tool — technology and engineering and the science on which they are frequently based, and particularly their interacting relationship with economics if these promising tools are to be properly developed for society's benefit, and if they are to be properly taught at such a great institution as M.I.T.

I shall attempt this task in the course of the four lectures I have undertaken to deliver at weekly intervals starting this afternoon — an undertaking I agreed to gladly even while entertaining some inner doubts as to my ability to carry it out with reasonable adequacy.

Science, Applied Science, Technology,
Engineering, and Economics

In connection with our Apollo program there is a well-worn witticism that if any particular operation goes off well

[1] National Commission on Technology, Automation, and Economic Progress, *Technology and the American Economy,* United States Government Printing Office, Washington, D.C., 1966.

it is reported as a scientific triumph, but if it fails it is an engineering failure. There is no question that since the end of World War II we have had an absolute rise in the enrollment of science students and an absolute drop in engineering students. For a country as technologically oriented as is the United States, this is a bad development and, because it is vitally related to the main theme of these lectures, I would like to clarify our coming discussions with a few brief explanatory definitions of the terms science, applied science, technology, and engineering, with a particular emphasis on their relationship to economics.

SCIENCE   Science and technology are both vitally important to our society. Science may be said to represent a body of systematic, experimentally verifiable knowledge regarding the relationships among the complex phenomena of the physical world. Scientists are concerned with improving man's understanding of his physical environment and with the expansion of the range of physical phenomena embraced by man's understanding. Science bears no direct relation to economics. In fact the attitude of many a scientist is that if what he is doing has any practical significance or application he is not interested in it.

APPLIED SCIENCE AND TECHNOLOGY   Applied science is science applied to the solution of a practical problem. Although in recent years many schools of engineering have changed their names to schools of applied science, the definition of the term remains vague. Too often it is simply poor technology because of the failure to consider economics adequately. Unfortunately, also, too frequently the general development of applied science is confined to areas of social technology — defense or space, for example, where the need for careful economic analysis is believed to be less urgent. The economics that enters into applied science, therefore, is frequently almost negligible. This may account for the high mortality rate among applied science organizations once they make an effort to transfer their activity to an atmosphere of civilian economics.

Technology is a better term for what is often referred to as applied science. It is a newly developed or an accumulated body of knowledge related to a specific area of economic activity, such as making steel or producing nylon or polyethylene or generating electric energy from fossil fuels. It is based upon scientific discovery or experimentation or merely successful practice over many years that makes possible the practicable production of a specific economic good or service.

Although I have specified practicable, I have not said economic. Essentially a technology can be developed in the sense that it provides a mechanism for achieving a stated purpose but still be economically infeasible. Thus, the technology is available for cleaning the SO_2 from the effluent of a power plant burning high-sulfur coal, but it cannot be done without imposing an extremely high-cost burden on energy production. Small nuclear reactors can be built to give every municipality, large or small, a supply of electric energy ash and sulfur-free, but the cost of a unit of energy so generated would be prohibitive. All transmission and distribution can be placed underground but at a cost in the case of 345,000-volt transmission of six to eight times the cost overhead. Technology recognizes economic limitations and is not satisfied with technical feasibility alone. It seeks to achieve — even though it sometimes does not find — economic feasibility as well.

7

ENGINEERING   In a lecture he delivered some five years ago,[2] Dean Gordon Brown of the M.I.T. School of Engineering stated, "engineering is practicing the art of the organized forcing of technological change," and ". . . when an engineer works at the frontier of his field his main function is to couple science . . . with his particular problem in order to build something and make it work."

This description, I believe, needs to be supplemented. The engineer must often go beyond the limits of science and question judgment based on alleged existing science. He must frequently exert his own overriding judgment and stake his reputation by going into areas beyond those which have been fully explored scientifically. After all, many advanced engineering structures were brought into being in the ancient world centuries before the existence of a scientific understanding of the interrelationship of the forces that made them work.

The engineer is the key figure in the material progress of the world. It is his engineering that makes a reality of the potential value of science by translating scientific knowledge to the extent it is available, filling in the scientific gaps with the help of experiment, past experience, and judgment — to marshal tools, resources, energy, and labor, and bring them into the service of man.

This need of the engineer to bring his work effectively into the service of society, involving as it does the need for society to judge his product good, i.e., serviceable, obviously recognizes the fact that the product must have economic validity.

Some six years ago a committee of the School of En-

[2] Brown, Gordon S., "The Engineer: Today's Pacesetter of Change," The Mendenhall Lecture, Harvey Mudd College, Claremont, California, May 7, 1962.

gineering at M.I.T. wrestling with this problem issued a statement[3] from which I quote:

Science . . . is a search for knowledge. The science of mathematics extends abstract knowledge. The science of physics extends organized knowledge of the physical world. In each of these consideration can be limited to a carefully isolated aspect of reality.

The engineer must deal with reality in all its aspects, he must not only be competent to use the most classical and the most modern parts of science, but he must be able to devise and make a product which will be used by people.

Thus, engineering goes well beyond technology — beyond putting together technical parts of systems that will work. Engineering must encompass the broad principle of economy of all resources — material, capital, labor — and their optimization at a given time in terms of the society of which they are a part. The principle of excellence is indispensable for engineering, but not in a limited sense of excellence, not merely in technical relationships. Excellence in engineering must include economic aspects — costs and values — and if it fails to include them it is not engineering.[4,5]

One would think this would be so patent as to be axiomatic. Yet only a few years ago the entire system of production in the USSR gave almost no recognition to the existence of economics as a part of engineering. A deputy

[3] Committee of the School of Engineering, M.I.T., Statement by Committee, Engineering and Education, March 1961.

[4] Chapters by I. I. Rabi, and Harvey Brooks in *The Impact of Science on Technology,* Edited by A. W. Warner, D. Morse, and A. S. Eichnar, Columbia University Press, New York, 1965.

[5] Sporn, Philip, *Foundations of Engineering,* Pergamon Press, Oxford, 1964.

premier of the USSR at a public luncheon given in his honor by the Secretary of the Interior of the United States, chided me for mixing, so he claimed, politics with engineering. When I replied that he was mistaken, that what he probably meant was that I was mixing economics with engineering, his reply was that it did not make any difference, economics too ought to be kept out of engineering. Surprised as I was at his statement, I was not too surprised to make the rejoinder that in our country we believed that engineering without economics is impossible and is no engineering. Indeed, the USSR itself is coming around to this view.

Science, Technology, and Engineering
as Agents for Advancing Human Welfare

Although science, technology, and engineering are commonly linked as the architects of the world of today and tomorrow, the mechanism by which they make their individual contributions is quite different and it stems from the differences between scientists and engineers in their training and basic approach to problems. They are, as Dr. Arthur M. Bueche pointed out at Cornell a little over a year ago, quite separate and distinct professions.[6]

There are, it is true, a number of common ties between them — the number who innovate, for example, in either group is relatively small and for most effective results they need to work together. But essentially science by itself and the work of the scientist is ineffective in advancing human welfare. For that technology and engineering are required. It is the engineer's function to bring together resources,

[6] Bueche, Arthur M., "A Look into the Future of Scientific Research in Industry," *Cornell Engineering Quarterly*, Vol. 1, Fall 1966.

tools (in the broadest sense), energy, and labor and to combine them in a productive entity to achieve or produce something wholly new or previously impossible, to achieve an improvement that yields a better product at the same cost or the same product at a lower cost, or even a better product at lower cost. Unless and until these factors are brought together in a productive combination, no social or economic benefit results.

The nature of the components of a productive entity is not new. They have always been the three elements: resources, tools, and labor. The big change that technology and engineering have brought about in our society today is to modify the relative contributions of resources, tools, and labor. Economics is particularly important in engineering because it provides the analytical mechanism for determining the relative participation of these three factors and the value of a new technology to society.

Thus, in its report "Technological Innovation: Its Environment and Management," the Panel on Invention and Innovation[7] set about to determine the costs in successful product innovation and particularly to examine the role of research and development in the total process of bringing a new product to market. Surprisingly it found that the step commonly called research and advanced development or basic invention accounts typically for less than 10% of the total innovative effort cost. The cost of engineering and designing the product, tooling, and manufacturing engineering amounts to more than eight times the cost of research and development, the approximate distribution being as shown in Table 1.

[7] Panel on Invention and Innovation, *Technological Innovation: Its Environment and Management*, United States Department of Commerce, January 1967.

TABLE I

*Typical Approximate Distribution in
Successful Product Innovations*

Operation	Approximate Percentage of Total Innovation Cost
Research-Advanced Development	8
Engineering and Designing the Product	15
Tooling-Manufacturing Engineering	50
Manufacturing Start-Up Expenses	10
Marketing and Start-Up Expenses	17
	100

The dominant item obviously is engineering in the several stages of bringing the product to market — engineering concerned with the design and organization of the tools of production. These tools, acquired by the investment capital, are a dominant item in bringing about innovation. If the economics of this important segment of the total cost is mishandled, the viability of the whole product is placed in jeopardy and scarce resources are wasted. And it is this segment of the total cost that is most frequently mishandled.

*The Seminal Effect of the Introduction
of Economics into Technology*

Proper economic analysis is indispensable to the design of productive technology and is independent of the particular form of social-economic organization. This sometimes takes a little while to find out by experience. Thus, the USSR discovered to its dismay that the lengthy construction time of some of its major hydroelectric projects involved costs and these costs, which in our economy are recorded as forthright interest during construction, but which they choose to label frozen costs — these costs were sufficient seriously to affect adversely the attractiveness of hydrogeneration. This discovery, interestingly enough, resulted in a change of policy and a shift of emphasis in their electric power program in favor of thermal generation.[8] And the dismay of the managers of Soviet manufacturing enterprises at having to give an account of the capital burdens of their plant in determining cost and profitability — a dismay that has only recently come to the foreground — is entirely understandable when many of them grew up in a society where the philosophic approach toward capital plant and equipment, even where the state was the capitalist, was heavily muddied by their ideological, anathematic view of a capitalistic society. In having all these years yielded to capitalistic society the exclusive utilization of economic evaluation, the USSR had, in fact, surrendered economics as a capitalistic monopoly.[9]

Economic forces not only provide a rational approach to the choice among many alternative routes in any complex technological development but serve to stimulate technological advance and the inventive process on which so much of our new technology is dependent.

It is interesting to examine a few examples that reflect the interplay of economic forces and the development and exploitation of technology. Agricultural employment in the United States declined between 1947 and 1964 from 14% to 6.3% of the total civilian employment, and the total employed in agriculture declined from 7.67 million to 4.4 million. Yet in that same period, with the total acreage in

[8] Sporn, Philip, and Abraham Gerber, "Soviets Find Capital Costs Make Hydro Less Economical," *Electrical World*, Vol. 158, No. 8, pp. 56–59, August 20, 1962.

[9] Schwartz, Harry, *Russia's Soviet Economy*, Prentice-Hall, Inc., New York, 1951.

the United States remaining practically unchanged at roughly 1.15 billion acres, output of wheat went up 17%, soya beans 264%, rye 41%, barley 51%, maize and corn 412%, and productivity rose some 161%.

The mechanization of American farms, shown by the increase of some 75% in available horsepower, the rapid increase in the use of fertilizer of 102% in the period, and the consolidation of farms into larger sizes, brought about the amazing improvement in farm productivity that made it possible for American agriculture to feed a growing population with a strikingly declining labor force. It represented a technological response to the changing relationships between supply and demand for the factors of production throughout the economy, especially the rapid growth in the demand for labor in the manufacturing and service industries. In turn this made it possible for our society to meet that demand with only relatively mild dislocation. Indeed it demonstrated the ability of a free economy in a free society to optimize the allocation of its resources. In contrast the failure of Soviet agriculture to respond similarly has limited the availability of labor and other resources for more intensive industrial development.

Next year will mark the second centenary of the patenting by Watt of his famous separate condenser steam engine, which is commonly accepted as marking the birthday of the industrial revolution, which in turn is the foundation on which the great industrial society that was England in the nineteenth century was built, and which served as the foundation of all other industrial societies, including ours.

Watt, with all his great genius, built on Newcomen and Newcomen built on Savery, and Savery, an inspired mechanic, was interested in solving the economic problem of keeping the mine pits in Cornwall free of water without incurring prohibitive costs. Neither science nor applied science entered into it, but economics did.

The story of the origin of the electric light and the electric power industry which came along to exploit the electric light is another example of economic motivation. Edison, having watched the development of the gas lighting industry to one of the leading industries in the United States (and this was matched in other industrially advanced countries) was prompted to start his work that was to lead to the electric light and the electric power industry of today by economic motivations. He recorded it in his notebook as follows:

Electricity versus gas as general illuminant. Object: Electricity to effect exact imitation of all done by gas, to replace lighting by gas by lighting by electricity, to improve the illumination to such an extent as to meet all requirements of natural, artificial and commercial conditions.

Economic incentives, or perhaps lack of economic incentives, are not always on the side of the angels — they do not always exert an influence to advance and improve technology, but frequently act as disincentives to technological progress. Here the examples are many. Let me cite a few.

The development of the great railroad systems in the United States left them at the turn of the century in the position of enjoying almost a complete monopoly on long distance passenger and freight traffic. The lack of an immediate economic incentive to improve facilities and service: road bed, locomotives, cars, speed, schedules, comfort, and tariffs, left them in the poorest sort of spirit to anticipate competition or to meet it, and so they eventually lost a good deal of their freight traffic to trucking and practically all their passenger traffic to the automobile, the bus, and air-plane. They were able for far too many years to live and live well without doing anything constructive or innovative. Today the railroad systems of the country are, with some exceptions here and there, completely decadent. As Alfred E. Perlman, then president, New York Central System, pointed out in *The Wall Street Journal* of July 14, 1967, "it is ironic that the United States, the most economically powerful nation in the world, has allowed its basic trans-portation facilities to nearly dry up."

The great coal industry and the energy base it provided the United States for many years made possible the in-dustrial development of the United States to the most productive nation of the world. The coal industry par-ticipated in that growth and experienced a period of 50 to 60 years of great prosperity. The most important single area of coal use was railroad transportation and it was also the most profitable. For many decades any coal mine locating on or near a railroad could be certain of having its share of that railroad's coal requirements for motive power at prices that were generous, if not too high. But the con-sequence of that was that the coal companies almost com-pletely eliminated themselves as parties at interest or critics in and of the cost of transportation. No incentive existed in the coal industry to develop any ideas to invent or con-ceive arrangements to cut transportation costs, even though these costs in many cases were equal to or greatly exceeded cost at the mine and even though in many cases it was the cost of transportation of coal which determined the eco-nomics of coal as a primary energy source to the user.

In the case of the electric utility industry we have the spectacle of the combination companies rendering impar-tially electric and gas service with full regulatory and stockholder approval, supremely unconscious of the deadening effect this has on management and operating personnel in their inability to exert effort to improve the condition of either service because of the economic loss this might lead to in the operations of the other.

Capital as a Cost in Economic Analysis

A little while ago in discussing the differences between science, applied science, technology, and engineering, I pointed out that until the engineer brings together resources, tools, energy, and labor and combines them in a productive entity to achieve or produce something wholly new or previously impossible no social benefit can result from either the science or technology. The largest of these components in our modern advanced technological societies is tools. Tools cost money. It is the tools — the complex plants and their equipment — that represent the capital cost of our modern capitalistic societies. There are few if any individuals who are in the least bit confused about the cost of smaller tools that they themselves have acquired. They found out the first time they tried to assemble an even modest and small group of such tools consisting of say a plane, a cross-cut saw, a soldering iron, two pliers, a set of screw drivers, a hammer or two, a few wrenches, and a few miscellaneous small tools that they had spent most of a $50 bill. But somehow, when confronted with the larger tools — the kind that cost $250 million, as is the case in an assembly to produce atomically 200 mw of electric capacity and 100 million gallons per day of desalinated sea water — doubts creep in as to whether these are costs of the same nature and need to be considered in determining the cost of the product they make possible.

In our times the leaders of Russian society throughout

the 50 years following the 1917 revolution have been the most remarkable followers of that folly. Of course, they had an excuse. They were Marxists — of a kind — and they were started on the wrong track by Marx himself in his theory of surplus value. But Russia has been taking a second look at costs. This started almost ten years ago. As early as 1959, Premier Khrushchev indicated that in Russia capital has a price. He justified a cutback in hydroelectric power construction on the grounds that such projects tie up great amounts of capital for long periods before making any return. Thus he admitted the cost of capital funds — in Soviet parlance a social cost, but in more pragmatic terminology a return on capital (interest, dividends, taxes and — once operation has started — depreciation).

It is so convenient to be able to utilize a flexible cost of capital, particularly when trying to promote a project or program that has a difficult time finding justification under a more rigid economic analysis. Thus, at the recent International Conference on Water for Peace, Mr. Gus Norwood in his paper, "Public Objectives in Water Resource Development," commends "the United States government policy of using low cost money to achieve optimum development of water resources."

There are, however, lower costs than those obtained by using low-cost money — at least they have been conceived and proposed. Some 17 years ago, the principal manager of the nationally owned water supply system in Israel expressed his judgment that the capital facilities of that system ought not to be burdened with any interest charges whatsoever in view of the great social importance of water to that society. Interestingly and not surprisingly the Minister of Finance disagreed completely with that view.

Another very low form of capital cost is what for want of

a better term I would call eleemosynary economics. In discussing with the secular managing head of a great United States church-supported university the relative economics of self-generation versus purchased electric energy, he rejected as an item of cost the carrying charges on the investment on the ground that the self-generating plant was a gift, rejected depreciation as a charge on the ground that when this plant was worn out some other donor would furnish a new plant, rejected cost of labor on the ground that the religious order administering the university had no difficulty in enrolling qualified members to its midst, and thus cost of energy to the university came down to fuel and some maintenance materials.

In the United States there is a pervasive body of opinion that costs vary with the form of social organization. The influential Federal Power Commission periodically prepares appraisal reports for licensing various potential hydroelectric projects. In one such report,[10] the Commission, using the exact same capital cost for the exact same project, arrives at the following annual costs:

By Private Financing	$4,442,000
Financing by the Federal Government	1,507,000
Financing by the Rural Electrification Administration	1,370,000

We note here three annual costs in descending order from private financing to federal financing to Rural Electrification Administration financing. The Rural Electrification Administration obtains its funds from the federal government at a fixed interest rate of 2%, irrespective of the cost to the

[10] Federal Power Commission, Bureau of Power: Appraisal Report, Chippewa River Basin, 1965.

federal government of borrowing such funds. In this particular case, to determine the cost to the federal government an interest rate of 3.125% was used. For private financing an interest rate of 6.25% was used. In the case of the federal government, no additional cost resulting from taxes forgone was included. In the case of the Rural Electrification Administration 0.5% was allowed for taxes or payments in lieu of taxes. For private financing 5.72% was included for federal and all other taxes. In all cases the annual benefits, based on the cost of an alternative privately financed thermal power plant, were exactly the same — $2,648,000. This yielded the following benefit cost ratios: for REA financing 1.93; for federal financing 1.76; and in the case of private financing 0.60 or less than 1.00.

This, therefore, allegedly demonstrated that on the basis of private financing the project was economically infeasible, while economically feasible under either of the other two alternatives. Thus, with exactly the same installation, the same use of real resources, and the same benefits, the economic desirability of the project varies not only in the degree of desirability or viability, but even to the extent of changing it to undesirable or nonviable, depending on what choice of ownership one selects. Furthermore, one is confronted with the absurd "logical" conclusion that an agency which, in the economic sense, is a ward of the federal government borrowing money from the federal government at what constitutes, in effect, negative interest, can do the job at a lower cost than the federal government itself. This obviously irrational conclusion, stemming from an Alice-in-Wonderland system of economics, did not in the least daunt the Commission. One can only conclude that this is another illustration of the failure to fully comprehend the real cost of a project to the economy.

*Sound Economic Evaluation Is Independent
of the Particular Economic System on
Which a Society Is Organized*

As you can see, taxes have a great deal to do with affecting the relative costs. Taxes are sometimes looked upon as an unfortunate economic necessity. But this is obviously short-sighted. In a modern socially advanced society government must carry out many activities that it alone can carry out on behalf of society as a whole. This covers such activities as defense, education, health, and many other broad scientific, technological, and social activities. None of these is possible without income to government. Since as a general rule these activities are carried out for the benefit of society as a whole, every branch of the national economy must contribute its share of what it costs to carry out these indispensable social activities. Taxes are the basic mechanism for making every member of that society contribute his share. When special groups are given special privileges and exempted from making their contribution, they create special burdens on the rest of the society which will have to be taxed for the benefit of the special or privileged group.

Taxes not paid or forgone represent a cost to the economy. It is part of the total return on investment. The fact that part of the return is distributed to the investor and part to the tax collector merely represents the allocation of the return on the investment between the investor and the government. Therefore, the return, including income taxes, represents the true investment opportunity cost to the economy. To put it another way, you cannot wash out the tax component, that part of the productive return on the investment that is allocated to support government functions, from consideration of alternative investment opportunities.

The Soviet economist, Z. F. Chukhanov, mentioned in

an earlier reference,[8] arrived at the same conclusion, although he expressed it slightly differently in a paper that he prepared a few years ago for *Teploenergetika*.[11] Of course, it was necessary for him to contrive an artificial substitute for a market rate of interest. However, he also allowed a rate equivalent to taxes, which he described as that income from investment necessary to meet the nation's "quite large nonproduction expenses for pensions, scientific research, defense, etc. . . . Since power generation, as most other branches of the national economy, must contribute its share to the national income, it is necessary to add another term to the (social cost) equation." Mr. Chukhanov, in his paper, by taking into account capital cost including the equivalent of taxes for support of the government, concludes that hydroelectric development in the Soviet Union has been wasteful when compared with the alternative of thermal power stations. Further, he concludes that in the period 1952 to 1958 no hydroelectric power capacity should have been built in the Soviet Union. The strong shift in emphasis in Soviet power development in recent years from hydroelectric to thermal indicates that this basic argument has had some effect on Soviet thinking regarding water resource development.

The elements of capital cost are rather simple and for any society are universal regardless of its social form of organization. They are the following:

1. Return on investment

This is the annual amount earned by an investment in any project or enterprise over and above its operating costs.
2. Depreciation

This is the annual charge against revenues used to repay

[11] Chukhanov, Z. F., "The Economic Effectiveness of Thermal and Hydroelectric Power Stations," *Teploenergetika*, No. 12, 1961.

the original cost of the investment over its productive life. The total depreciation charges over the life of the plant should add up to the total expended to provide the plant.

3. Taxes

Taxes are considered separately, although as was just pointed out they are really part of the return on investment. They are that part of the total return that by law must be allocated to the support of government services.

It is argued sometimes that capital costs would be lower if they were handled on a governmental basis. Loans for many classes of facilities could be obtained at lower interest charges, and taxes could be excluded from consideration.

This is, of course, true from the point of view of the money cost. This may or may not represent the true cost of capital in terms of the alternative opportunities available for capital expenditures for purposes of evaluating the desirability of a given project. For this purpose the total return on the investment must be considered, and this includes taxes and the full return available on alternative investment opportunities.

We also have the case of Great Britain, with a mixed economy, but one in which the national government owns all the electric power resources. A white paper published in 1961[12] pointed out:

> If the profitability of capital development is assessed on different (and easier) financial criteria from those adopted in industry generally, there is a risk that too much of the nation's savings will be diverted into the nationalized industries. Again, if the prices of the goods and services which the nationalized industries provide are uneconomically low, demand for them (and for in-

[12] Her Majesty's Stationery Office, *The Financial and Economic Obligations of the Nationalized Industries,* April 1961.

vestment to produce more of them) may be artificially stimulated. Thus the operation of the nationalized industries with an unduly low rate of return on capital is sooner or later damaging to the economy as a whole.

This white paper further pointed out that British industry as a whole is earning a gross rate of return, including taxes, of about 15%. However, it stated that because utility service involves somewhat less risk than other industry, the nationalized utilities can earn a somewhat lower rate, and a figure of about 12.5% for the return on Central Electricity Generating Board facilities was established.

Recently the Electricity Council was confronted with a decline in the rate of growth of demand on the nationalized power system. As a result it faced the prospect of an unexpectedly large reserve in generating capacity and therefore greater capital charges per unit of energy sold than anticipated. This would have reduced the net available for return on capital below 12.5%. When the appropriate ministry was asked what to do about this, the reply was a directive for an across-the-board rate increase of approximately 10%. This caused some dismay throughout the country, particularly since the entire British economy is operating under very rigid guidelines; so far, however, there has been no indication that the order will not be placed into effect.

From the point of view of optimum development and use of resources, the comparison of the costs of several alternatives needs to be based on true representations of the various alternative costs in terms of real resources rather than on distortions based on artificially contrived monetary arrangements.

Return includes an allowance to provide for risk. It is sometimes argued that where the full credit and faith of the government stands behind a loan, the rate of return can be lower because risk is eliminated. However, this refers only to the financial risk to the lender. The fact that the credit of the government stands behind an investment does not diminish the risk to the economy from such an investment. The same risks stemming from a variety of developments — technologic, economic, and social — will continue to influence every area of economic activity whether owned and operated by private capital enterprise or by government. Thus, in the case of the coal industry in Great Britain, which is nationalized, the fact of nationalization has in no way reduced the economic impact of a decline in coal production that has been taking place. The rate of obsolescence of many of the old mines has, if anything, been accelerated since nationalization, but only because of the competition of oil and, most recently, nuclear power and natural gas. The economic pressure of nuclear power is as great or greater than in the United States. And the social pressure on coal, owing to its greater proclivity to contribute to pollution, will be only slightly less, but may be more, as a result of coal's being a nationalized industry.

By and large, technological change is a factor that too often is underestimated or even ignored. Economic evaluations of water resource projects, for example, frequently are based on estimates of anticipated physical durability of the project rather than on its economic life, and the value of the output is projected without regard to the outlook for technological change that could affect this value. Economic obsolescence needs to be considered especially in the particular area of power generation, where technological progress has for some years now been proceeding at a very rapid rate. Thus, for example, regional shifts in growth and development can affect the demand for the output from a particular water project, or changes in technology can affect the competitive value of the output. For example, in connection with one proposal for a continent-wide multipurpose water development project in North America, it has been suggested that a large part of the cost of the project could be reimbursed by the sale of power from the project at 4.5 mills per kwh. Apart from many other considerations that might shed doubt on the economic feasibility of the project, this one basic assumption is open to grave doubt. Certainly the development of nuclear power technology over the next 20 to 25 years to the point where 3.5-mill or lower costs per kwh will be feasible, a more than fair probability, would make 4.5-mill power from a hydro project unmarketable.

In short, a sound economic evaluation of any major capital project is independent of the social, political, or other motivations, and is independent of the particular economic system on which a society is organized. If a given society is not to be led astray and if it is not to make a mess of the indispensable business in the proper allocation of its limited total resources, it is important that the proper — and this means total — costs be used in the evaluation. Having done this, the society is then in an excellent position to assign priorities. This does not prevent it from upgrading the priority of any socially desirable project at the expense of another less costly. But the intelligence and sound judgment with which this will finally be done will always be materially enhanced by having properly determined values and costs. Subsidies, desirable and granted, do not change cost.[13]

For economic studies and economic evaluations to make a contribution to the development of any complex technological project, it will be necessary to go beyond the critically important immediately relevant capital cost. It will always be necessary to make similar studies on and of operating and maintenance costs involving materials, transportation — of both the raw and finished product — operating labor cost,

[13] Engineers' Joint Council, "1957 Restatement of Principles of a Sound National Water Policy," *EJC Report No. 105*, May 1957.

cost of energy, water — including the obtaining and disposal — and the cost of maintenance and ecological measures. These need to be optimized not only vis-à-vis themselves but against variants in the main system plan, including its siting, the availability of labor, the effect of the labor and tax climates on the operation coming in, and finally the effect of such changes in the social-economic environment as inflation or deflation, in income tax structures, and likely changes in ecological regulations.

This involves in many cases side studies, looking far into the future, of the likely economic developments that will influence the availability and price at which raw materials entering importantly into the product will be available. Thus, in the case of an important thermal electric plant, the kind of plant, the terminal conditions of its thermodynamic cycle of conversion, the degree to which the efficiency of turbines, alternators, main transformers is pursued are all dependent in most cases on the present cost and the projected cost of fuel to be converted. It is the trend line of cost of the primary fuel that plays a dominant role in decision-making as to what is justifiable in capital expenditure in these areas. What is good economics for a given amount of dollars necessary to save 10 Btu per kilowatthour at 35¢ per million Btu may be wholly unjustified at 25¢ or 20¢ per million Btu.

If labor saving opportunities present themselves, the expenditure one can afford with good economic result is totally different if an inflationary trend — slow, medium, or rapid — is anticipated than if stability is to be anticipated.[14]

These so obvious observations would be wholly unjustified if industrial records were not replete with scores of

[14] Sporn, Philip, and Abraham Gerber, "The Social-Economic Evaluation of Water Resource Projects," International Conference on Water for Peace, Washington, D.C., May 1967.

examples where poor engineering and poor economics — or more precisely, poor economics and, therefore, poor engineering — overlooked these main bases for decision but concentrated fully and, almost microscopically, on technological details.

Many a brilliant project — brilliant technologically — has been wrecked on the shoals of unsound economic evaluation. To the extent these shoals were not charted it represented bad economics and bad engineering. And this in large measure distinguishes engineering from science and technology. It requires a broad understanding of the technological and economic environment and the ability to organize resources to achieve economically and socially desirable objectives.

The application of these ideas to the problem of primary fuels for electric energy generation will be the subject of my next lecture.

Engineering and Economics in Primary Fuel for Electric Energy Generation

Introduction

THE PRIMARY FUEL sources for electric generation have, with few exceptions, been overlooked for far too long by electric utility management, although they are a major item of expense and an essential ingredient of successful utility operation.

This neglect can be attributed to several interesting factors. In the early days of the industry the rates at which lighting was sold ranged from 10¢ to 20¢ per kwh. Several decades after the birth of the industry, 33 kwh per month, that is 400 kwh per year, was considered the height of the saturation in the residential electric market. Even with coal at $4 per ton, representing a cost of 16¢ per million Btu, and with thermal efficiency of only 8% to 9%, coal represented a cost of only 6.4 mills per kwh — hardly noticeable in a selling price of 10¢ to 20¢ per kwh. Capital cost was the major cost item demanding attention.

The idea of 4000 kwh per year domestic usage, when suggested some decades later as a goal for the utility industry, caused questioning of the sanity of the speaker. Had the speaker dared to project utilization of 25,000 to 30,000 kwh, as can now be found in electrically heated homes, he most certainly would have been rushed to the nearest mental ward.

This low horizon projection of the possibilities for expanding electric energy use residentially, commercially, and industrially particularly overlooked the potentials of expanding the industrial market by reducing cost, and especially the very energy intensive industries in which electric energy in effect becomes a raw material.

The possibility of generating more kilowatthours for each kilowatt of capacity and thus sharply reducing the capital cost per kilowatthour was given no serious thought. Consequently the importance of fuel cost and the opportunities for technology to achieve the dual function of lowering capital cost and reducing fuel cost through improved efficiencies were overlooked. Furthermore the need to emphasize the importance of obtaining primary fuel at the lowest possible costs was neglected. In short, the prospect of combining advanced engineering and intelligent fuel economics to encourage major growth trends in the use of electric energy was, for the most part, overlooked.

The History of Growth of Electric
Energy and Its Indicated Future

Since its beginning almost 85 years ago, the electric utility industry has been one of the most dynamic growth industries in the United States. Over this entire period the output of the industry has grown at a rate of about doubling every ten years. Between 1920 and 1966 electric generation grew almost 30-fold from 39.4 billion kwh to 1144 billion kwh. In the period 1920 to 1946 the growth rate was at about 7% annually despite almost a decade of one of the most severe depressions ever experienced in modern economic history. In the 20 years since 1946 the growth rate of the electric utility industry has maintained its dynamic character, and growth has been at a rate of 8.5% annually.

This growth has been the result of two complementary forces: first, the application of electric energy in new uses for which inanimate energy had theretofore not been used and, second, through the substitution of energy in the electrical form for the direct use of fossil fuels. Both of these forces continue to be important at the present time. The substitution effects are illustrated by the substitution of electric energy for oil or gas in open-hearth manufacture of steel by the oxygen process and the expanding use of electric energy in place of other fuels for space heating. Still in its early stages of development, this last source of electric energy growth is expected to become increasingly important in the next several decades. A new area of substitution which has the potential of becoming significant, although it is presently in its very earliest stages, is the substitution of electric energy for liquid fuels in transportation. The rapid acceptance and growth of space cooling illustrate the potential effect of new uses of energy on electric energy growth.

For the next 20 years there is likely to be little diminution in the historical growth rate of the electric utility industry, and by 1987 electric utility generation is expected to reach 4 trillion kwh. Beyond that there can be visualized factors of saturation in energy use moderating the growth rate of electric energy use. Nevertheless, total electric power generation by the utility industry is expected to approach some 6.8 trillion kwh by the end of this century. These projections do not include any substantial quantities of electric energy use for transportation. Although by 1987 significant numbers of electric vehicles are likely to be in use, they are unlikely to have a major impact on the total pattern of energy use. At the most, the additional electric energy may be of the order of 60 billion kwh per year or 1.5% of the total that would prevail in the absence of electric transportation. Yet, by the end of the century electric vehicles may well grow to the point where they will be adding approximately 125 to 150 billion kwh to annual electric energy consumption, so that the 6.8 trillion kwh noted above may well be raised to about 7 trillion. Electric vehicles by that time could possibly offset the tendency for saturation to slow down the rate of growth.

The growth in electric energy consumption has for a long time been more than double the rate of growth in total energy use in the United States. Total energy use has grown at a rate of about 3% a year from about 817 million tons of bituminous coal equivalent in 1920 to some 2000 million tons in 1966. By 1987 this figure should rise to slightly over 3000 million tons and reach 4250 million tons by the year 2000. Between 1920 and 1966 the share of total energy use consumed for the purpose of electric power generation has more than doubled from about 10% to close to 23%, and it is expected to increase to over 40% by 1987, and reach 50% by the end of this century.[1]

In the period since 1920 the structure of energy use in the United States has changed considerably. In 1920 coal accounted for more than three-quarters of total energy consumption, petroleum constituted only 13.3%, natural gas 4.4%, and hydro 3.9%. By 1966 coal consumption was slightly below the 1920 level in terms of absolute tonnage, but in terms of per cent of total it had declined to less than one-quarter. Petroleum had risen to some 40% and natural gas to more than 30%, while hydro had declined to slightly over 3%. For the first time nuclear power generation began to appear in a noticeable quantity.

Over the next two decades the relationships among the several sources of energy are expected to change, but only moderately. Coal, because of its importance as a source of electric energy and coke for steelmaking, is expected to remain at about 23% of the total. Petroleum is likely to decline to about 35%, and natural gas should decline to about 25%. Hydro will continue to be negligible, declining to under 3%, but nuclear power can be expected to become an important factor in our energy resources by 1987, accounting for over 15% of our total energy use.

[1] Sporn, Philip, Energy Resources and Technology, Hearings before the Subcommittee on Automation and Energy Resources of the Joint Economic Committee, Congress of the United States, 86th Congress, October 13, 1959.

The importance of nuclear power will continue to grow so that by the end of the century it will account for close to one-quarter of total energy use while the share of all other sources will continue to decline by from 10% to 20%. Coal is likely to decline slightly to about 20%, and except for an insignificant percentage for hydro — under 3% — petroleum and natural gas will account for the remaining 53%. The importance of nuclear power will stem from its application in electric energy generation and the growing share of electric energy in total energy use.

In 1920 and 1960 coal accounted for slightly under 56% of the total electric energy generation and by 1966 it still accounted for about that share. Fuel oil, which accounted for a little over 4% in 1920 had, by 1966, grown to approximately 6%. There was a major increase in the importance of natural gas, however. Starting with only 1% of the total generation in 1920, natural gas grew to almost one-quarter of total generation by 1966. Hydroelectric power, however, showed a very sharp decline from over 40% in 1920 to only about 15% in 1966.

In the next 20 years nuclear power's importance will grow in electric energy generation, rising to about one-third of the total, while coal will decline to under 50%. Oil will tend to diminish in importance, settling back toward the 1920 level, and natural gas is also likely to decline to well below 20%. By the year 2000 nuclear power can be expected to account for about 50% of total electric generation while coal declines to about 35%. Oil and gas will be negligible, with gas at about 8% and oil at less than 2%, and hydro power declining to where it will represent less than 5% of the total electric energy.[2]

[2] Sporn, Philip, *Energy, Its Production, Conversion and Use in the Service of Man*, Pergamon Press, Oxford, 1963.

I have summarized this in two tables (Tables 1 and 2) showing 1960, 1966, and projections for 1987, two decades hence, and for the year 2000.

TABLE I

*Total Energy Consumption in the United States
for Selected Years*
(Million Tons of Bituminous Coal Equivalent)

Year	Total	Coal	Petroleum	Natural Gas	Hydro	Nuclear
1960	1714	23.2%	41.7%	31.6%	3.5%	0
1966	2000	24.9%	40.9%	30.9%	3.15%	.15%
1987	3000	23.0%	34.0%	25.0%	2.6%	15.4%
2000	4250	20.0%	30.5%	22.2%	2.3%	25.0%

Total primary energy consumption in the United States has grown at a relatively moderate rate — less than 3% compounded. For the next two decades total energy use is projected to rise to 3000 million tons of bituminous coal equivalent and by the year 2000 to 4250 million tons. Also shown is the dominance of petroleum in the entire four-decade period, 1960–2000, the beginning of the relative decline of natural gas, and the relatively rapid rise of nuclear energy from zero in 1960 to 25% in the year 2000. It is interesting to note the figures in the total energy picture for controversial hydro. It is clear that hydro is not now much of a factor in the total energy picture, and its future shows a still further relative decline to less than 2.5% by the year 2000.

Even in the electric energy projections shown in Table 2, hydro does not show up as much of a factor. By the year 2000 it will account for less than 5% of the total electric energy generated. Looked at in this perspective, it would

TABLE 2

Electric Utility Generation and Fuel Consumption for Selected Years
(Millions of Tons of Bituminous Coal Equivalent)

Year	Generation Kwh $\times 10^{12}$	Generation Electric	Fuel Consumed Total	Coal	Petroleum	Natural Gas	Hydro	Nuclear
1960	0.75	18.8%	322	55.5%	5.6%	20.5%	18.4%	0
1966	1.14	23.0%	460	55.3%	5.5%	24.9%	13.7%	0.6%
1987	4.00	42.4%	1270	44.0%	3.0%	13.5%	5.5%	34.0%
2000	7.00	50.0%	2125	35.6%	1.8%	8.1%	4.5%	50.0%

appear hardly worth the heated controversy which discussions of hydroelectric development have invariably brought forth in the past.

Table 2 also clearly emphasizes that the two fuels that will account for the largest share of the electric energy generation in 2000 are coal and nuclear fuel, with nuclear fuel dominant in a ratio of 50:35.[3]

The Relative Importance of Fuel Cost

Whether fuel costs are given a position of major importance in the foundation of the utility structure depends a great deal upon the development of a basic philosophy founded upon a thorough understanding of the economics of fuel. If fuel economics are misunderstood and emphasis is placed solely on capital cost of facilities, then obviously fuel costs tend to be disregarded — they can be almost anything. But if it is understood that capital costs and operating costs together — including fuel — are importantly interrelated, then the fuel situation appears in a totally different light.

Fuel cost can be disregarded only if it is assumed that sales will be sales and returns on capital will be returns no matter what. It involves an attitude that there is just so much business that can be obtained and if only capacity is provided the sales will come. If it is assumed that rates can be adjusted to take care of all your capital and operating costs, then fuel costs and indeed other costs are unimportant. Another way of stating this is, you have a monopoly position and you are going to do the job and your price structure is set so that you recover cost, what-

[3] Sporn, Philip, "Atomic Energy — The Coming Great Rival of Coal," AAAS Symposium on Coal in the United States — Problems and Promises, Philadelphia, Pa., December 27, 1962.

ever it is, plus the necessary return on capital. It sounds like a very simple series of relationships.

But in fact the relationships are not so simple, because costs depend on technology and on management and on sales. It is sales that determine the demand on your system. Technology, however, which generally can be applied effectively only on something you build brand-new to take care of increased demand, cannot be more effective in helping your operations if you cannot apply it and to apply it you need sales, and sales do depend on economic cost. Furthermore, the monopoly position was only an illusion. Under these conditions the cost of the primary energy could determine whether there is or there is not a basis for sales.

For example, while sales programs and selling have a great deal to do with sales in the domestic and commercial markets, electric cooking, electric water heating, electric laundry drying, and electric space heating and cooling, all are materially influenced by prices that are charged for the energy involved.

In the industrial field there are a host of applications where the cost of electric energy determines whether the application will be made electrically at all or whether, even though it is made electrically, it will be made in a specific service area. Among these items are electric heating, electric welding and forging, metallurgical operations such as production of electric steel, ferroalloys, the melting of brass and titanium, iron, and many of the nonferrous metals, all of which are determined in large measure by the cost of electric energy. In the areas of heavy application of electric energy, there is not only competition as between adjoining or even nonadjoining utility systems, that is between companies in the Ohio Valley, the Southeast, and the Northwest, but there is also competition between the investor-owned and governmentally owned and/or financed utilities. For example, aluminum reduction was for many years located primarily where government-financed hydroelectric power was available in the Southeast and Northwest. It was not until a great deal of engineering and economic effort had been made to reduce the cost of fuel per kilowatthour and the total cost of thermally generated electric energy that it became possible to bring the aluminum industry based upon coal generated electric energy into the Ohio Valley.

But, of course, the competition can go beyond national boundaries, and indeed has done so, so that foreign hydro power has competed as locations for aluminum reduction operations based upon heavy economic support of the foreign hydrogenerated energy from the United States itself.

It is perfectly astonishing, and yet should not be, how small a differential in energy cost can determine the location of an operation such as a large aluminum reduction plant. The determining factor can, and has in the past, been as little as 1¢ per million Btu in the cost of the raw fuel. To the question, is it conceivable, can this actually be so, the answer, based upon experience, has to be yes. And why not?

A five potline aluminum operation will have a demand from 400 to 485 mw and an annual use of some 3.8 billion kwh. A 1¢ per million Btu differential can make a difference in cost of 0.09 mill per kwh. At 3.8 billion kwh, this represents an annual cost differential of close to $350,000. In one specific case, an offer was made to locate a plant if a reduction could be made in the price quoted from 4.39 mills to 4.3 mills per kwh. Since neither prospective fuel economies or other operating economies gave any basis for this reduction, the suggested 2% reduction in the price of energy had to be rejected and the load was lost to the system.

The great effect of fuel and fuel conversion cost on growth, where a realization of its significance and possibility of exploiting it to stimulate growth exists, cannot be over-emphasized. When all these factors are considered, and particularly the influence in a competitive economy of small differentials in cost, the question of cost of primary fuels, the possibility of reducing their mined cost, their transportation costs, and the possibility of increasing the thermal efficiency of conversion of these primary fuels into electric energy — all these fall into place as does the possibility of combining and linking these with improvements in a program of system development. Fuel cost, in other words, can become a most important tool in system building.

Cost of Present Energy Sources
in Electric Energy Generation

In the light of the 40 million kw of nuclear reactor capacity placed on order in the last 12 months — a block of nuclear capacity which will displace some 105 million tons of coal a year when its installation is completed some time in 1975–1976 — is cost of *present* energy sources significant? If we are coming into the nuclear age, why study present energy sources?

There are two answers to this question. The first is a reference to Table 2. Note that by the year 2000 coal will still account for over 35% of the primary energy involved in the generation of some 7 trillion kwh and this will require some 750 million tons of coal (0.356×2125 million). The second is an examination of Table 3. This is not a projection, but what I have chosen to call an economic reverie regarding the electric energy to be generated in the years 2015, 2035, and 2065. But while this is not a pro-

TABLE 3

Electric Generation — A Century from Now — An Economic Reverie

Year	Electric Generation Kwh × 10¹²	Total Primary Energy Consumed — Millions of Tons of Bituminous Coal Equivalent	Nuclear Generation Per Cent	Per Cent Petroleum, Gas, Hydro	Coal Consumed Millions Tons	Coal Consumed %
2000	7.0	2125	50	14.4	755	35.6
2015	12.0	3100	70	8.0	680	22.0
2035	21.0	4600	85	4.6	478	10.4
2065	40.0	8000	95	2.5	200	2.5

jection, I visualize a gradual switching over from fossil primary fuels to nuclear, but the retention by coal of some of the market even as late as 2065. Thus, with 70% of the total electric energy being nuclear in 2015, 680 million tons of coal will still be required to supply the portion of the remaining 30% not suppliable by petroleum, natural gas, and hydro. In 2065, with an energy generation of 40 trillion kwh and a total primary energy requirement of 8000 million tons bituminous coal equivalent, of the 5% that will still be non-nuclear 2.5% will have to be carried by coal and this will require 200 million tons of coal and that is 80% of the total coal burned in 1966.

In light of all the above, it appears to me imperative to take a good look at cost of present energy sources in electric energy generation.

Since the end of World War II the cost of fuel for electric generation in the United States has declined moderately from 26.6¢ per million Btu in 1948 to a low of 24.3¢ in 1955. Since 1955, however, costs have fluctuated from a peak of 27.1¢ during the Suez crisis in 1956 to a low of 25.2¢ in 1965. These averages for the United States, however, obscure significant regional differences.

In New England fuel costs per million Btu were 38.6¢ in 1948 and declined until 1965 when they were 33.9¢, except for the period of the Suez crisis when they rose temporarily to a peak of 43.1¢. Similar declines took place in the Middle Atlantic region where costs were 29.8¢ in 1948 and only 27.7¢ in 1965. In the areas closer to the coal fields, that is, the East North Central, West North Central, and South Atlantic states, fuel costs also declined, but much more modestly, over this period. In the East North Central area they declined from 26.9¢ in 1948 to 24.2¢ in 1965; in the West North Central from 25.6¢ to

25.3¢; in the South Atlantic states from 29.7¢ to 26.8¢. In the East South Central states the decline in magnitude was similar to that which took place in the New England and Middle Atlantic states, falling from 23.1¢ in 1948 to 19.4¢ in 1965.

In the remainder of the country, that is, the West South Central, Mountain, and Pacific states, where the major fuel has been natural gas, costs have risen markedly. In the West South Central states, where in the immediate postwar period abundant gas supplies were available and the inter-state pipeline system was in the very earliest stages of development, natural gas was available in 1948 for electric generation at an average cost of 7¢ per million Btu. As older contracts for fuel were terminated and new contracts were made at higher prices, the average began to climb steadily and by 1965 the cost per million Btu had reached 19.5¢. Similarly, on the Pacific Coast, average fuel costs rose from 28.2¢ per million Btu to a peak of 34.5¢ in 1962, and since then have declined moderately to 31.8¢ in 1965. In the Mountain States costs rose from 21.4¢ to a peak of 26.2¢ in 1962, and have since declined to 22.2¢. In the latter two areas the recent decline in fuel costs reflects moderate declines in the cost of fuel oil and the introduction of additional quantities of lower cost gas, especially imported Canadian gas. In the areas east of the Mississippi River the decline in the average cost of fuel for electric generation has been the result largely of reductions in the delivered cost of coal. This has come about as a consequence both of moderate reductions in the mine price of coal and in significant declines in the cost of rail transportation of coal.[3]

In facing the competitive threat of imported oil, the expansion of the natural gas pipeline to the East Coast, and the developing nuclear power technology, the coal industry, together with the railroads, undertook a vigorous competitive effort to maintain coal's markets in the Eastern areas. Particularly in the period since the peak of prices during the first Suez crisis, the coal industry lowered costs and prices through increases in productivity that reduced the labor cost per ton of coal despite rising wages, and at the same time the railroads introduced the unit train idea and made available trainload rates which reduced significantly coal's freight cost burden. In the past year, however, there appears to have developed a decided change in this situation. The railroads have obtained a rate increase on coal shipments to Eastern markets and recently received approval for a general freight rate increase that will further affect coal. The coal industry, operating very close to capacity and with new coal-fired electric generating plants now being put on the line offering the prospect of continued high levels of production for a number of years into the future, and with wage increases continuing and the increase in productivity slowing down noticeably, has begun to raise prices. Thus 1966 witnessed an increase in coal prices that is being continued in 1967.

While there is some economic basis for this development, the noneconomic contributions were in all probability much more significant. These came about as a result of two possible developments:

1. Tiredness, and this is a pity. Certainly a study of Tables 1 and 2 and even of 3 would indicate that the future still to be developed and the work ahead should not lead to any such resigned acceptance of the adequacy of the effort by the coal suppliers, the railroads, and other transportation agencies, the utilities, the conversion equipment manufac-

turers, when contemplating the loss of over 100 million tons of coal per year markets to the nuclear fuel industry in one 12-month period.

2. A resignation to the inevitable, a sort of attitude that says let us get while we can the prices we want and hang the future. Right now the market for coal could not be better.

Surely that is no solid foundation on which to build an industry. This is no way to bring the market battle home to an overwhelmed nuclear industry, bemused by its apparently great achievement in making such, it has been told, phenomenal progress in the application of atomic power as, to quote some of its scientific enthusiasts, to constitute an atomic revolution.[4] Unfortunately, the revolution not only has never truly materialized but to the extent it did make a feeble effort at showing its head it stopped right there and never finished the process. The revolution, in short, became aborted.

Similarly, the price of oil has firmed and further declines in the price of oil for electric generation are unlikely. In the case of gas, finding new reserves is becoming more costly and demand for uses other than electric generation continue to rise. As a result the price of gas on a firm delivery basis has hardened and further declines in the near future are unlikely. Indeed, the price of fossil fuels, whether coal, oil, or gas, east of the Mississippi River can be expected to remain firm at present levels or tend to rise, with the limitation being the competitive cost of nuclear power generation. West of the Mississippi further declines

[4] Weinberg, Alvin M., and Gale Young, "The Nuclear Energy Revolution — 1966," *Proceedings of the National Academy of Sciences,* Vol. 57, No. 1, p. 1, January 15, 1967.

in the price of oil or gas are unlikely. In the gas-producing areas of the Southwest gas prices may have little room to increase before running into the competitive opposition of nuclear power plants. There is already some indication of this in the announcement by Middle South Utilities, Inc., that it is ordering a nuclear plant to be installed at an as yet undetermined site in Arkansas where gas costs are slightly above 25¢ per million Btu. In areas further removed from the gas fields and on the West Coast where oil is also an important factor, the price of oil and gas is likely to remain around present levels, largely because of the competitive alternative of nuclear power, although in this case the ability to introduce nuclear power as a competitive force in some areas is being limited by licensing difficulties.

However, nuclear power economics, instead of developing by a series of rational steps from technological innovation to innovation stands today no farther advanced in its progress toward lower costs than it was two years ago. Today, with increases in unit size of almost 40%, atomic energy based upon the installation of two 1100-mw units is competitive with coal at 22.0¢ ±1¢, contrasted with 24¢ ±2¢ on the basis of a post-Dresden No. 2, 800-mw nuclear unit in December 1965. And conventional fuel, coal to be specific, instead of taking courage at the clear indication of higher capital cost even in these stretched-out units, larger than can be supported by any but a very few power systems in the United States, stands relatively immobilized without any significant challenge or response while losing in one year a market for coal equal almost to what for many decades of the past was its most important market — the railroads.

How this came into being deserves more attention.

*Development of the Nuclear Market
and the Reasons for the Takeoff*

Until 1965 the commercial development of nuclear power proceeded at a very modest rate. All the plants in operation were prototypes and few commercial plants were being ordered. In 1964 the manufacturers undertook a major effort to develop the nuclear market. Although 200 mw was the largest size plant that had until then been constructed and operated, size was stepped up and the manufacturers began to offer reactors in ratings of 400 and 500 mw to take advantage of the substantial economies of scale available in this size range. At the same time electric utility growth had made it possible for an increasing number of utilities to undertake conventional generating units of that size, and these size developments in conventional plants also helped make larger size nuclear units technically feasible. In addition, the manufacturers undertook to provide nuclear power units on a turn-key basis at fixed prices, thus assuming most of the financial risk in the construction of these units. Furthermore, the sharp competition among the manufacturers to gain position in the market resulted in significant price reduction. As a result, in 1964 General Public Utilities ordered a 500-mw nuclear generating unit for its Oyster Creek site and Niagara Mohawk ordered a similar type unit for its system to be located at Nine-Mile Point. These units were widely publicized as being competitive with alternative fossil fuel plants at these same sites. The true significance of the Oyster Creek and Nine-Mile Point plants I discussed in my 1964 report to the Joint Committee on Atomic Energy.

At the same time electric utility use of coal for electric generation had risen to well over 200 million tons and total coal production exceeded 500 million tons. The coal industry had reached a level of production and prosperity that it had not known since the end of World War II, with the possible exception of the Suez crisis year. Coal production began to press against capacity and the construction of additional coal-fired generating units promised a continuing rise in the use of coal for electric generation for several years ahead. This change in coal's markets resulted in a considerable firming of coal prices. This was coupled with a growing concern about the problems of pollution resulting from the combustion of fossil fuels and the additional costs of pollution control equipment, together with the sharply rising costs of fossil-fuel-fired equipment generally, which followed the price readjustments of the antitrust episode and which favored the nuclear turbines. As a consequence both the absolute and relative costs of conventional generating plants were raised.

These developments favorable to the nuclear plant's competitive position were further compounded by the effects of increasing size. The increase in unit size of nuclear plants resulted in significant reductions in costs, although not to the extent that was indicated by the pricing of these plants which largely reflected efforts by the manufacturers to establish their position in the nuclear market.

In 1965 Commonwealth Edison ordered an 800-mw nuclear generating unit. Thus, although 200 mw still was the largest operating nuclear unit, size was extrapolated still further. However, this was not reflected in any reduction in the cost of nuclear power, but it did hold these costs constant while conventional power costs — both fuel and capital equipment — continued to rise. This stimulated what might be termed a bandwagon effect, with many utilities rushing ahead to order nuclear power plants, often on the basis of only nebulous analysis and frequently because of a desire to at least get started in the nuclear business. At this point the manufacturers began to abandon the turn-key offer and published price lists for the nuclear steam supply systems that they were prepared to offer, leaving it to the utility and its consultants to design the balance of plant.

The final push to the nuclear power boom that developed in 1966 and has carried over into 1967 came with the purchase by TVA of two 1100-mw units, to which a third unit was added quite a bit later. This brought nuclear power into the heart of the coal country, and TVA's analysis of alternative costs indicated that on their system nuclear power was competitive with 13¢ per million Btu coal. An analysis of costs based on investor-owned utility financing indicated that the TVA job was competitive with about 16¢ per million Btu coal. With fuel available in only very few locations in the United States at prices below 16¢, and with escalation provisions in fuel contracts, especially coal contracts, tending to increase fuel costs over the operating life of the generating unit, further impetus was given to the rush to nuclear power. But as will be shown later, TVA Browns Ferry Units 1 and 2, whatever the reasons for the particular bid from a standpoint of market development and whether intended so or not, did in fact become a come-on bid. It was immediately withdrawn, even for additional units for TVA.

This impetus coincided with the period during which the electric utility industry, having experienced two years of very rapid growth, began to accelerate sharply its orders for generating equipment, so that total generating capacity on order increased rapidly at the same time that nuclear power's percentage of the total also increased dramatically. The result was a stretch-out in delivery time and the ordering of nuclear equipment for delivery six years, and in one recent case as much as eight years, ahead. The increase in lead time for nuclear orders resulted in the ordering of over 17 million kw of nuclear capacity by the end of 1966 or about 51% of the total generating capacity then on order, and exaggerated nuclear power's apparent share of the market beyond its actual position.

Where did this leave nuclear power economics? Let us take a look.

Status of Nuclear Power Economics — 1967

The economic competitiveness of nuclear power at Oyster Creek reflected four distinct but closely related factors:

1. The unusually intense competitive pressures among the manufacturers at the time of the bidding for the Oyster Creek job.

2. The culmination of vigorous research efforts by the manufacturers which resulted in simplification of the nuclear steam supply equipment and the improvement in the quality, reliability, performance, and cost of nuclear fuel elements.

3. The lower conventional equipment prices against which the nuclear manufacturers had to bid.

4. The overoptimism with which the need for contingencies, particularly those stemming from siting problems, capacity stretch-out, and the like were evaluated.

Beginning in 1962 there had been intense competition among at least two and sometimes three nuclear manufacturers. These manufacturers, committed to a nuclear future, were determined to maintain strong and closely competitive market positions. Any pulling ahead or advantage gained by one organization called forth a challenging competitive response from the others, and these were very strong at the time of the Oyster Creek bidding. The result was the incorporation of every available technological development and the most optimistic market projections based on new pricing that these developments were thought to make possible. None of these developments, of course, had yet passed the test of experience, although they appeared to have sound technical foundations.

General Electric, the manufacturer that finally won the bid, priced the Oyster Creek plant on the assumption that at least three very similar units could be sold to minimize the financial risk of this plant. The competitive cost level of the Oyster Creek plant was the result of the combined impact of the technical and market forces at play at this time. However, it seemed at the time, and has since become increasingly clear, that the manufacturer risked somewhat greater uncertainty in his turn-key price than was tolerable on a repeated basis, and it became clear, on the basis of pricing put into effect shortly after the Oyster Creek contract was let, that prices similar to those provided at Oyster Creek would thereafter no longer be available and that nuclear power could not achieve the cost levels it seemed to achieve at Oyster Creek.[5] A carefully studied analysis of the Oyster Creek costs indicated that the Oyster Creek plant and similar plants following might actually be competitive with coal prices at about 27¢ per million Btu ±2¢ and would yield total costs of about 4.43 mills per kwh with capital costs at about $139 per kw.[6] While the analysis suggested cost and competitive cost levels considerably above those that were then being claimed for the plant, it has

[5] General Electric Company, "Annual Report to Shareowners for 1966," Schenectady, N.Y.

[6] Sporn, Philip, "Nuclear Power Economics — Analysis and Comments," prepared for Joint Committee on Atomic Energy, Congress of the United States, October 1964.

since become clear that actual costs are going to be much higher still.

Thus neither Oyster Creek nor the Dresden units that followed were the indices of progress they first appeared to be. Dresden was a unit with a capability of close to 800 mw. Analysis of these costs indicated that the increase in size would make possible a reduction in capital costs to $123 a kilowatt, but an increase in estimated fuel cycle costs raised the total costs to 4.42 mills per kwh or about the same level of costs as this author estimated would be achieved at Oyster Creek.[7]

In the meanwhile the base of conventional costs shifted upward. This brought the competitive cost level of nuclear power to about 24¢ per million Btu ±2¢. The nuclear power cost per kilowatthour remained virtually the same between Oyster Creek and Dresden but the competitive cost level declined largely as the result of an increase in capital cost of coal-fired units. The 600-mw coal-fired unit that American Electric Power constructed at Cardinal at an estimated cost of $107 per kw against which the Oyster Creek nuclear unit was compared to determine the competitive cost level should have yielded a substantially lower figure for an 800-mw unit. Instead, the figure for an even more advanced 800-mw unit analyzed by AEP against which to compare the Dresden unit rose to $112 per kw. Furthermore, to achieve even this cost it was necessary to design the unit with an increase in heat rate or a reduction in thermal efficiency from 8650 Btu per kwh for the Cardinal Plant to 8775 Btu per kwh for the more advanced unit.

The TVA order of two nuclear units for a new plant at

[7] Sporn, Philip, "Nuclear Power Economics — An Appraisal of the Current Technical-Economic Position of Nuclear and Conventional Generation," Morgan-Guaranty Hall, March 17, 1966.

Browns Ferry, Alabama, gave the final fillip to the rush toward nuclear power. TVA actually found that the nuclear unit had a capital cost slightly lower than an available coal-fired unit, or $116 a kw compared with $117 a kw, and with the nuclear unit TVA received bids for firm, guaranteed fuel costs considerably below the cost of coal. The bids for coal were 18.6¢ per million Btu delivered with escalation, and at the attainable heat rates this would have given a fuel cost of about 1.6 mills per kwh. The nuclear fuel costs were guaranteed firm for ten years at 1.5 mills per kwh. However, the TVA bids with which General Electric won the contract are irrelevant to evaluations of the current competitive status of nuclear power.

At the most charitable interpretation, the TVA bids were obviously designed to introduce nuclear power to the largest coal-burning utility in the United States and have not been repeated. Specific requests made of General Electric for a duplicate of the TVA quotation yielded the reply that it was no longer available. As a matter of fact, TVA itself has found, in seeking options for a third nuclear unit, that prices have risen considerably. The manufacturers have abandoned the turn-key concept and capital costs have risen to the range of $140 to $150 a kw and more for units in the size range similar to the 1100-mw TVA units. Furthermore, the costs of nuclear fuel as offered by the manufacturers have risen and the manufacturer is no longer willing to offer firm contracts for nuclear fuel.

At the present time (October 1967) the costs of nuclear plants of 1100-mw size range between $140 and $150 a kw and, despite considerable improvement in the fuel performance reflected in the TVA bids, nuclear fuel costs are about 1.5 mills per kwh with allowance for escalation. Costs per kilowatthour of a nuclear plant ordered at the present time are about 4.25 to 4.35 mills.

TVA Browns Ferry Unit 3 will range in incremental cost around $132 per kw and on 12.5% fixed charges will result in an energy cost at 7000 hours of 4.04 mills per kwh. It will thus be competitive with coal at 19¢ per million Btu.

The announcement of the AEP 2.2-million-kw plant to be located at Bridgman, Michigan, and consisting of two Westinghouse P.W. 1100-mw reactors gave the cost of the plant minus substations as $300 million or $137 per kw. Correcting for this and for delayed interest charges results in an additional $2.46 per kw or $139.50. On this basis energy costs at 12.5% fixed charges and 7000 hours are

Capital Costs	$2.49
Fuel	1.54
Operating and Maintenance	.20
Insurance	.07
	$4.30

This is competitive with coal at 21.95¢ per million Btu, a close enough agreement with the announcement range of 21¢ to 22¢ per million Btu.

Further, interesting and substantiating figures are given in *Nucleonics Week* of July 20, 1967. For the two-unit Vepco Surry Station consisting of 815.5 mw Westinghouse units, costs including substation costs are given at $152.55 per kw, while costs for the single-unit 1060-mw Diablo Canyon Plant of PGE are given at $144.94 per kw. For a single-unit plant, if substation costs are included, the figure of $144.94 would apear somewhat low.

However, the cost of an 800-mw conventional plant has also risen and is likely to be about $117 a kw. The net result is a further reduction in the competitive cost levels of nuclear power from the post-Dresden level of 24¢ to about

22¢ to 23¢ ±1¢ per million Btu. This brings nuclear power into competitive contention in all areas of the United States except those relatively close to the centers of coal production.

There are, however, several reservations that need to be considered. Nuclear power costs will remain uncertain until some of the plants already on order have been built. The capital costs indicated earlier are subject to escalation which is uncertain at the present time. Furthermore, the longer lead time required for construction increases the escalation risk. The ability to bring nuclear plants on the line in accordance with construction schedules remains to be demonstrated, and several of the earlier plants, such as Oyster Creek, will exceed schedules by a considerable time period. Plant licensability remains a major uncertainty, and there has been some increase in the time required to obtain construction permits. At the same time, with each new plant the AEC Advisory Committee on Reactor Safeguards has tended to impose additional safety requirements that have tended to increase the cost of plant. The ACRS letter granting a construction permit for the TVA Browns Ferry job raised questions related to the size of plant which will have to be answered prior to receipt of operating license and thus could create some additional difficulties. Fuel performance remains to be demonstrated in operating plants.

On the coal side, uncertainties also remain. The areas in which coal can be delivered at 20¢ per million Btu or less have narrowed, and escalation provisions, both in the price of coal and in the cost of transportation, are likely to have an impact on coal prices almost immediately after any coal contract is negotiated. The major impact of unit train movements of coal on costs has already been reflected in the averages and little further effects are anticipated. Indeed, it appears that rather than further cost reductions, increases in rail transportation costs, even for unit train movements, are likely over the next several years. And the effect of antipollution regulations will continue to be a handicap to coal in the electric energy generation market.

Yet the net result of these developments in coal and in nuclear power indicates that, while nuclear power has slightly widened the area of competitiveness in the United States, it has not achieved dominance to any significant degree despite the overwhelming rush of orders, so that there will continue to be vigorous competition between nuclear and conventional fuels in the market for new power plant construction. The initial bandwagon rush to nuclear units can be expected to subside in due course and new plant decisions are likely to be subjected to more stringent analysis. The present high level of nuclear plant orders includes capacity for as much as eight years ahead. This unusually long lead time is not likely to be sustained so that, to a considerable extent, present orders have been borrowed from future years. Furthermore, total orders for new capacity can be expected to decline somewhat, especially if the current slower rate of expansion in economic activity continues for several more months.

Therefore, given a little more time the result to be expected is a slowing down in the rate of growth of nuclear orders to a more sustainable level with somewhat less ebullient nuclear capacity totals than some of the more recently revised projections for 1980 would indicate. The earlier 115,000-mw projected by the AEC until their recent upward revision of the figure to 150,000 mw and later still to 180,000 mw is probably more realistic than the recently revised figures. In my judgment both revised AEC figures are too high and I believe the 115,000-mw figure will prove to be closer to what will be realized.

What does all this add up to?

It seems to me that the following is quite clear:

1. As far as the economics of competitive atomic power are concerned, we have made remarkably little progress, as far as any absolute reduction in cost per kilowatthour is concerned, with some question as to whether there has been any progress at all in the two-year interval 1965–1967. Post-Dresden 2, with its competitive level of 24¢ per million Btu, is superior in performance to Surry, with its 24.8¢. The Browns Ferry 1 and 2 performance is in a class by itself not explainable at all by conventional economics. The 1100-mw Bridgman units of AEP at 21.95¢ are so little improvement over Dresden 2 as barely to show any influence at all of the almost 40% increase in size.

2. No examination of the history of the past two years as summarized in Table 4 can leave one with anything but the inescapable feeling that, as stated by the French proverb, "Plus ça change, plus c'est la même chose."

The latter observation is particularly pertinent in the light of the apparent progress made in a joint effort by the author and a number of his colleagues and the nuclear organization of one of our great manufacturing companies to meet the challenge of 18¢ conventional fuel through a program to develop the bases for improving factors of specific performance, eliminating stretch-out, and yet carrying nothing beyond a safe point. This very stimulating effort, initiated in January 1965, had by May 1966 produced a design with a performance arrived at by joint evaluation to yield an energy cost of 4.11 mills, competitive with conventional fuel at 20.4¢. Being shy some 13.5% of the goal sought, it was not accepted, although I believe could

TABLE 4

Some of the Erratic Progress on the Road to Competitive Atomic Power

Year	Atomic Unit	Size Mw	Plant Capital Cost $/kw	Total Production Cost Mills/kwh	Competitive Point ¢/M Btu
1964	Post Oyster Creek	605	139	4.48	26.3
1965	Post Dresden 2	800	123	4.42	24.0
1966	Browns Ferry 1, 2	1100	115	3.78	16.0
1967	Browns Ferry 3	1100	132	4.04	19.0
1966 (Apr)	Project 202*	800	124	4.11	20.4
1966 (Sept)	Project 202A*	1121	118.10	3.93	18.3
1967	Bridgman	1100	139.50	4.30	21.95
1967	Surry	815.5	152.55	4.55	24.83

* A bilateral ambitious and expensive dream project — unilaterally treated seriously until it evaporated.

59

have been found available on a contractual basis. But after a presentation of the whole project was made to the Atomic Energy Commission a decision was made to make one more effort to explore the economics of size, with a view of seeing how close to the 18¢ goal it would be possible to come. This gave the results shown in Table 4 under Project 202A. It is unfortunate that, by the time this was put into a formal request for a proposal and bid, figures entirely different by more than 5¢ per million Btu of competitiveness had to be confronted. By then, also, Browns Ferry Units 1 and 2 identical in their basic design parameters to Project 202A had been negotiated and the order placed and its price structure, having apparently served its purpose of stimulating the desired level of interest in nuclear orders, had become a thing of the past.

3. Under such conditions the failure of the coal industry to strengthen its position vis-à-vis nuclear fuel is a measure of the coal industry's failure to respond adequately. The 105 million tons of coal per year lost by virtue of the 40,000 mw of nuclear reactor capacity ordered in the last 12 months will not yield to recapture by any currently conceivable technological-economic development in conventional fuels over the next 25 years.

4. Except for minor details, the current technology has concentrated on two reactor designs with one only a slight variant of the other and the sponsor of each having undertaken the delivery of not less than five reactors of over 1000-mw size rating, with experience in back of them confined to a single reactor of approximately 200-mw rating. The risk in such a major extrapolation is not inconsiderable based upon experience in other technological areas and in conventional power.

5. In the light of manufacturer defense of and satisfac-

tion with 30% to 50% price increase[8] in nuclear steam supply systems, it is hard to see how a continuation of the present nuclear trend will add any economic support to the need of the electric power industry to continue to reduce its cost of generation in order to maintain the dynamic character of growth which has so prominently supported it these many years.

6. Finally, what seems to have fallen victim to this semi-competitive atmosphere in which this transition from fossil fuel to nuclear fuel is being fought out is the loss of the grand concept: coal costs lowering atomic costs and atomic costs setting a ceiling on coal. This would still be a sound idea if it could be brought back to life. It can be revived, but not without a mighty effort by the nuclear manufacturers, the manufacturers of conventional conversion equipment, the coal industry, the uranium mining industry, and the utilities. In such an effort there are bound to arise some internal conflicts of interest, but if these could be met and hurdled this prospect could still be brought into reality.

How to Make Coal and Nuclear Fuel More
Constructive Elements in the Economic
Development of the Country

Before we yield without a struggle to the prospect of coal and nuclear fuel failing to become more constructive competitive elements in our economic system, it needs to be pointed out that without question one of the great contributors to nuclear energy having reached the position it occupies today is the acceptance by the nuclear industry some ten years ago of the fact that in coal-based generation it was facing a moving target. When the indications showed up on the horizon that nuclear energy was likely to reach that target,

[8] *Nuclear Industry,* Vol. 14, No. 7, p. 3, July 1967.

coal itself at first reacted very responsibly and very encouragingly to the nuclear threat. The present becalming of the sea of struggle between these two fuels certainly need not be accepted by coal, because coal has any number of directions in which to go to renew its vigor as a competitor of atomic power. It is a better competitor even now than its sponsors give it credit for. It is, to use a race track metaphor, a better horse than its clocking would indicate. It apparently needs a better jockey to help improve its performance.

The automation of a modern coal mine has not reached its ultimate, and in the sizes that are inevitable in the coal industry of the future more automation can certainly be installed.

More effective use of its large capital investment needs to be brought about. This means around-the-clock operation of mines with three shifts versus one shift or the current occasional two-shift operation. In this regard the coal industry has myopically overlooked the valuable lesson in capital economics it could have learned — but it may not be too late even now — from the AEC in operating its three great diffusion plants at load factors of 99.99%. The capital cost per ton per year of a modern underground mine is too high to be wastefully used by single- or even two-shift operations. In this effort the mine workers themselves need to be challenged and, properly challenged, I believe will respond.

Transportation improvement needs to be pursued with renewed vigor and the concepts of integral trains and quick turnarounds developed to their utmost. The possibilities of transportation of coal by pipeline need to be pursued as a competitor of rail transportation. Nor can water transportation be neglected on the assumption that it has already achieved its optimum. Better barges and barges that will stay seaworthy, around-the-clock loading, and complete coordination of every step between arrival of the barge at a plant, its unloading, and its turnaround need to be optimized so as to reduce the capital cost component of this form of transportation.

Finally, electric transportation, particularly at voltages as high as 765 kv, ac, the highest thus far developed to the stage of practice, needs to be explored to its limits.

As already pointed out earlier, coal is dependent for its place in the fuel economy upon the efficiency with which it is converted into electric energy. Thus more attention needs to be given to improving the basic conversion cycle, the design, purchase, construction, and operation of modern coal-fueled electric plants. Perhaps here the element of competition needs to be strengthened. And there are a number of significant ways of doing that which have been tried but never heretofore fully exploited.

There is in the future, a great prospect of a 30% improvement in the thermal efficiency of conversion of coal from the current limit of 40% to a figure of 52% by the introduction of the magnetohydrodynamic (MHD) cycle. This has been explored by a very small utility group with some sympathetic support from the Office of Coal Research, but up to the present the support has not taken more tangible form than expressions of good will. The attitude of the utility group toward continuing its efforts to build an experimental plant is now uncertain, so that currently MHD research in the United States is becalmed. However, the prospect of the significant improvement in thermal efficiency that MHD offers is so enticing and so promising that its possible contribution to making both coal and nuclear power more constructive and economically contributory to our society cannot be overlooked without a much more serious effort to bring it about.

Better technology, better engineering, economics-oriented

both in conventional fuel conversion and in atomics, carrying through the MHD program,[9] more true competition in atomic power and between atomic power and fossil-fueled power, and a much greater effort by the coal industry to build its future to levels of coal utilization much higher than any reached in its past history are what we need to maintain the dynamism of our industry and of our energy industries. It is to be hoped that these can be brought about. If the atomic industry and the coal industry properly react to such a dynamic challenge, the American energy economy will indeed be well served.

[9] Sporn, Philip: *Research in Electric Power,* Pergamon Press, Oxford, 1966.

The Dynamics of a Specific Power System: Sales, Production Costs, and Underlying Technology: Their Interlocked Relationships

THE THESIS of these lectures — the linkage of technology, engineering, and economics and their interaction one on the other in a feedback relationship — cannot be considered to any better advantage than in an analysis and discussion of the development of a specific power system. Specifically, I should like to discuss the American Electric Power System, with which I have been associated for over 45 years.

The AEP System, from the standpoint of electric energy generation or sales, is the largest investor-owned power system in the United States. But it was not always that.

The territory and the principal communities are served by distribution facilities encompassing approximately 45,000 square miles. The perimeter of its main transmission lines encloses an area of close to 100,000 square miles. This area includes parts of seven states: Ohio, a small portion of southwestern Michigan, the central portion of Indiana, the western portion of West Virginia, the western portion of Virginia, the eastern portion of Kentucky, and a small area in the northeast of Tennessee. The total communities served as of the end of 1966 numbered 2567 and the average population of these communities was less than 2500. Total population served is some 5,700,000, representing 2.86% of the population of the United States. In 1966 they were supplied with some 48.75 billion kwh, or 4.3% of the total electric energy supplied to the entire country. The people of the territory served by the AEP System, therefore, used about 1.5 times the electric energy per capita as did the total United States.

Almost all this electric energy was generated in thermal electric plants at an average thermal efficiency of close to 36%, the highest efficiency of any power system in the world. The average price received for that energy was 1.08¢

per kwh, the lowest of any investor-owned power system in the United States.

Over the past 30 years the AEP System has had an average growth of 7.9%, a compound rate of growth that has resulted in doubling every nine years. The corresponding growth rate of the total electric power industry was 7.04%, or a doubling every 10.25 years.

The territory served by the system is not outstandingly dynamic. It is characterized by many small towns. In the entire area there are only three cities with a population over 100,000 compared with 130 such cities in the United States. On the basis of its total population, its proportionate share of cities of this size would be four. Furthermore, the combined population of the three cities with population of 100,000 is only slightly above 400,000, compared with the average population of the 130 largest cities in the United States of close to 400,000.

Much of the system service territory is rural. Yet, although important parts of three such great agricultural states as Michigan, Indiana, and Ohio are served by the system, the average income from farm marketings in the year 1965 for all six states is quite a bit less than the average for the United States. Parts of the system are close to being depressed, and some areas are severely depressed. This is especially true of certain parts of West Virginia and Kentucky. There are areas served in southwestern Ohio where the population today is less than it was at the time of the Civil War. Thus, although the territory contains many communities that have had a healthy growth, as a whole it cannot be called outstandingly dynamic in the same sense that Florida, Texas, and California have been dynamic over the past decade. Nevertheless, if the territory in terms of over-all economic development cannot be called dynamic, from the standpoint of growth of energy-intensive industry it has enjoyed dynamic and significant growth.

It is, therefore, interesting to ask how such dynamic energy growth has come about in such an area. This growth, I believe, has been the result of a conscious policy of power development that has sought to exploit the region's strong points, such as its highly adaptive labor force, its centralized location, good transportation, and the existence of some of the best bituminous coal deposits in the world widely distributed geographically. These advantages have been combined with imaginative development of some of the most advanced concepts in energy conversion to achieve a superior system of power supply that made it possible to attract to the area those industrial operations for which an economic supply of energy is critically important. These economic developments in turn made possible still further technologic advances.

It is no accident that the electric power system serving this territory is the most advanced integrated power system within the United States (and the world for that matter); that it has the benefit of the technologically and economically most advanced transmission network developed over a period of over a third of a century; that this has produced the highest quality of service known anywhere; and that has progressively reduced system transmission losses from high levels running in excess of 12% of the energy carried over this transmission system in the early 1930s to between 7.1% and 7.4% in the last three years. It is no accident either that the average price per kilowatthour, which in the 1929–1931 period ran around 19 mills, has dropped in the subsequent three decades to something under 11 mills, a reduction of 43%. Nor is it chance that the system is in the process of the most intensive development of the residential electric heating market.

These developments have been unique and in many respects contrary to what one might generally expect. It could be argued with considerable persuasiveness that the low endemic fuel cost warrants no development of advanced conversion devices to achieve fuel efficiency. The availability of plentiful sites with access to water and fuel suitable for generation of electric energy did not indicate the need for solid electrical integration of the system, and with the most advanced transmission facilities. The large territorial spread of the system, with a total transmission and distribution mileage very close to 83,000 miles, each foot of which is responsible for some energy loss, it could be rationalized is the last system in the world one would expect to operate with the lowest transmission losses. Because of the low population density of much of the territory, one might expect the price per kilowatthour to be among the highest.

Yet none of these debilitating characteristics is part of the system. The accomplishments of the system, confronted by seemingly insuperable obstacles, illustrate the interplay of engineering and economics. It illustrates what I have chosen to define as engineering — the integration of broad social objectives, economic understanding and analysis, and an advanced and imaginative spectrum of technology into a system for the attainment of a set economic objective.

Advanced Conversion Technology

At the outset it was necessary to reject conventional thinking. Rather than concentrating on minimizing capital cost at the expense of thermal efficiency, the availability of low-cost fuel was exploited more intensively through the application of the most advanced thermodynamic cycles coupled with imagination. The result was advanced conversion technology yielding not only the lowest practical fuel cost but also the lowest capital cost. In addition this combination also made possible lower unit labor costs. The result of successfully pursuing advanced technology and optimum thermal efficiency has been very low total cost per kilowatthour that has made possible the attraction of such energy cost sensitive industries as electric steel, ferroalloys, aluminum reduction, and ammonia and chlorine production. The residential markets for hot water, cooking, and, over the past decade, home heating and air conditioning were also stimulated in their development.[1,2]

At the same time the development of a high quality of service, in spite of the apparent handicaps associated with a geographically widespread system, has made possible the attraction of industry for which the highest reliability of electric supply is critical to their production processes — such industries as glassmaking and glass products, textile spinning, and high-compression compressor loads.

As a consequence, about one-third of the total load now being served by the system was successfully attracted only because either the low price or high quality of electric supply, or both, were of such importance that the location of this particular system was most advantageous. In the absence of these considerations, these loads may well have located elsewhere. However, these loads alone are enough to account for the $1:1.5$ ratio between average United States and the system use of electric energy per capita mentioned earlier. It also explains the rapid growth of electric use in a relatively depressed service area that includes

[1] Sporn, Philip, *The Integrated Power System*, McGraw-Hill Book Company, Inc., New York, 1950.

[2] Sporn, Philip, H. A. Kammer, and S. N. Fiala, "The Development and Implementation of a Generating Program on the American Gas and Electric Company System," *Transactions ASME*, Vol. 74, pp. 603–619, 1952.

much of Appalachia. While some labor-intensive industry is located in the area, much of the industry in the area is capital and energy-intensive rather than labor-intensive.

These developments had feedback effects on system performance which in turn led to still further economic and technological advance. They made possible a system load factor of about 71% compared with a present industry average of about 60%. Such a high load factor further reinforces the economic validity of the investment to achieve higher thermal efficiency. Although similar efficiencies were developed in transmission, in interconnection, and in distribution, I should like to confine the discussion in fairly full detail to the work that was done to combine engineering and economics to achieve the striking advances in reducing the costs of electric power generation.

THE AMERICAN ELECTRIC POWER SYSTEM: PEAK LOAD AND GENERATION   The system performance in peak demand and kilowatthours generated is shown in Figure 1. As a point of reference for the rapid growth rate, a trend line

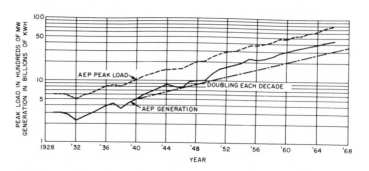

FIGURE I

AEP system annual peak load and generation

showing a doubling in growth every ten years has been drawn starting with 1940. The chart clearly shows that AEP System growth has considerably exceeded that rate. As mentioned earlier, the long-term growth trend rate of the system is 7.9% as against an industry figure of 7.03%.

Electric generation of the AEP System in 1966 was 45.1 billion kwh with a peak load of 7843 mw in 1966. System-wide annual load factor for 1966 was 71%. Of the energy generated on the system, 99% is supplied by coal-burning steam electric plants located mainly on the Ohio River or its various tributaries.

One major plant of 675,000-kw capacity is located on a small river and is served by a large cooling tower installation. This plant location was selected contrary to generally accepted criteria of soundness. The site was cramped, condensing water almost nonexistent. But its location, from considerations of system and transmission, was ideal. And it offered an unusually attractive supply of coal. The latest plant to be brought under construction is a mouth-of-mine plant, and will consist of two 800-mw units as the initial installation and, although located on the Ohio River, will also be served by cooling towers to avoid thermal pollution of the river.

Centralized engineering for all phases of system and plant planning, design, construction, and supervision of all phases of operation and maintenance is carried out by the American Electric Power Service Corporation, a nonprofit-making subsidiary of American Electric Power Company. This organization supplies the continuity and provides the very important experience-integrating or time-binding function for the system. It (or its predecessor) also supplies — and has supplied for a period of 40 years from 1927 to 1967 — the locus of discontent of the system; that is of particular importance in bringing about change, advances, and improvement.

RESULTS OF DEVELOPMENT   Cost of equipment, materials, and labor used in the generation of thermal electric energy has increased sharply over the last 37 years. The striking increase is shown in Figure 2, which shows the Handy-

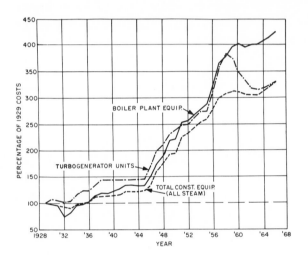

FIGURE 2

U.S. utility construction costs

Whitman Index of Public Utility Construction Costs in the United States over the 37-year period 1929–1966, using 1929 as a base of 100. As can be seen, over-all construction costs have risen more than 230%. Turbogenerator prices had by 1959 risen over 275%, then dropped following the antitrust litigation, but since 1962 have been rising again.

In the face of these large increases in the cost of the

components that are required in the generation of electric energy, the low actual production cost that has been achieved testifies to the success of the technological advances resulting from continuous and unremitting efforts both in research and in development of new ideas in engineering design and construction and in operation and maintenance of new facilities, all stemming from an attitude of the indispensability of an unremitting search for means to nullify rising cost trends.

One of the most significant of these developments in the past 36 years has been the improvement in thermal efficiency of power plants. This too had behind it a philosophy: If higher capital costs have to be encountered, let us find and bring into being a compensating benefit. The resultant development was in a particularly dynamic state for the first 30 years. Since 1960, however, there has been a considerable slowdown in the rate of improvement.

Table 1 shows average heat rates for the AEP System and for the United States utility industry as a whole from 1930 through 1966, as well as the performance of the best AEP plant in the same period. In these 36 years, the best plant has improved from a heat rate of 13,710 Btu per kwh (24.9% over-all efficiency) to 8842 Btu per kwh (38.7% over-all thermal efficiency), or an improvement of 55% in thermal efficiency. Note the consistent AEP lead in performance over the industry of between 1000 and 2000 Btu per kwh for a period of over a third of a century. This can, perhaps, be seen better from the graph, Figure 3.

The most important manifestation of this improved efficiency is shown in Figure 4, where the cost of coal on the AEP System is compared to the fuel cost per kilowatthour generated. Whereas the price paid for coal has risen 103% in the past 37 years, because of reduced heat rate the fuel

TABLE I

Heat Rates Btu Net Kwh

Year	AEP System		U.S.A. National Average
	Best Plant	Average System	
1930	13,710	17,740	19,800
1935	13,750	16,450	17,850
1940	13,370	16,270	16,400
1945	12,000	13,960	15,800
1949	11,200	13,450	15,030
1954	9,110	10,720	12,180
1959	9,011	9,725	10,879
1960	8,975	9,619	10,701
1961	8,819	9,363	10,552
1962	8,842	9,420	10,497
1963	8,898	9,457	10,438
1964	8,971	9,478	10,407
1965	8,935	9,476	10,384
1966	8,942	9,543	10,396 (Est.)

cost per kilowatthour has increased only 11%. Note here too the clear indication of a determination, following the inflationary flare-up of 1946, to get operating costs under control and the remarkable success in achieving this.

Other components of production cost have been similarly controlled so that the over-all production cost has been reasonably stable.

Figure 5 shows the variation in another production cost component over the same interval of 37 years, in one important rate and in one index. Figure 6 shows the variation in other costs — other than fuel and maintenance costs — over the same period. Finally Figure 7 shows the variations in total production costs. This total production cost is made up of the fuel cost shown in Figure 4, the maintenance costs shown in Figure 5, and all other operating costs shown in Figure 6.

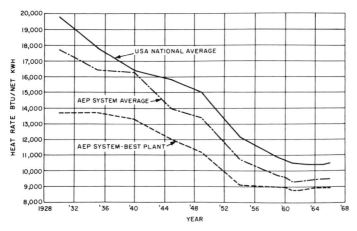

FIGURE 3

Heat rates

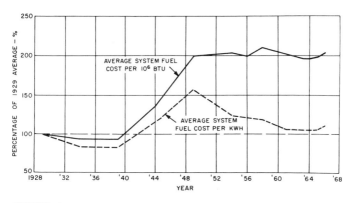

FIGURE 4

AEP system fuel costs

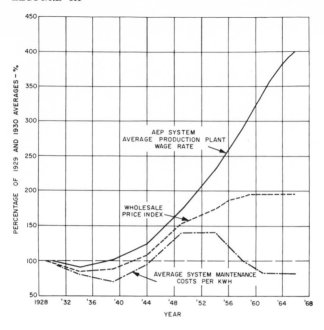

FIGURE 5

AEP system maintenance costs

In the face of the rise in fuel cost (Figure 4), average production plant wage rate (Figure 5), and wholesale price index (Figure 5), each one of which shows rises of from 100% to 300% in the interval covered, the achievement of average maintenance costs per kilowatthour in 1966 below the 1929 level and the average of other operating costs 54% below 1929 (Figure 6) have made possible an average production cost per kilowatthour 1963–1966 at 8% below 1929 (Figure 7). This is a truly remarkable performance. It is a particularly graphic illustration of the interaction of engineering, technology, and economics to obtain an over-all result contrary to what might have been expected. But

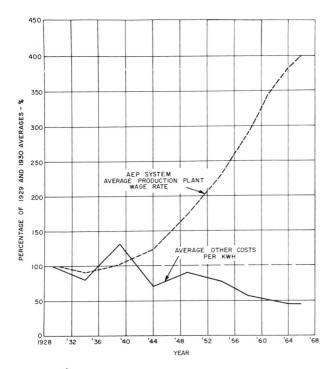

FIGURE 6

AEP system other costs

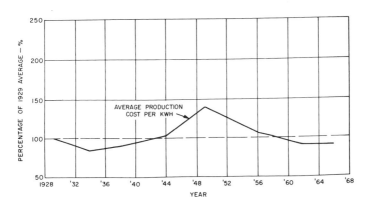

FIGURE 7

AEP system production costs

then it is no particular achievement to get merely the expected.

Maintenance cost is composed mainly of labor and material. The effectiveness of maintenance cost control can be seen from a comparison of the variation in the average manpower wage rate and the wholesale price index, which reflects the cost of manufactured articles, with the unit maintenance costs. In this period, even though wage rates have risen 300% and the wholesale price index has risen 98%, maintenance cost per kilowatthour has not increased at all — in fact it shows a 6% drop.

Many of the reductions in operating costs shown in integrated form in Figure 7 are subject primarily to changes in average wage rates. These costs per kilowatthour have also declined. This was brought about partially by larger sized units, but to a much greater extent by more skillful manning and effective use of automation. It all was in the main the result of sharply directed conscious planning.

KEEPING THE CAPITAL COST COMPONENT OF ENERGY PRODUCTION EXPENSES UNDER CONTROL The plant capital cost component of energy production has been one of the most difficult aspects of cost to control over the past 37 years. Figure 2, referred to earlier, shows the variation of plant construction costs in the area in which the AEP System operates. Using these construction cost variations as a basis for correction to a consistent base, Figure 8 has been developed to show the cost per kilowatt of generating capacity additions to the AEP System in terms of 1929 dollars over the past 37 years. This, therefore, shows what improved design techniques, particularly with larger units, higher pressures, higher temperatures, better utilization of area and cubage, simplified concepts, and better construc-

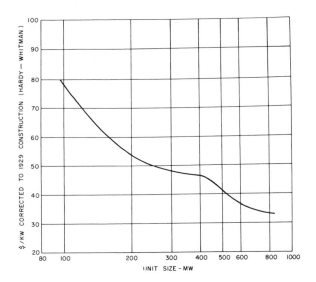

FIGURE 8

AEP system unit size versus corrected cost

tion planning and execution, have been able to bring about in substantial reductions in cost per kilowatt.

As a measure of the value of this decrease in cost of capacity, it might be noted that in the absence of these savings the annual capital requirements of the system for plant expansion alone would today be approximately $40 million greater than they are. Here the driving force followed this rationalization: Since the capital cost is the predominant element in cost of generation, reduce the capital cost per kilowatt to reduce the generating cost per kilowatthour.

As already noted, one of the significant ways in which capital cost has been controlled has been the reduction in cubic content of plants by use of larger unit sizes, the unit type arrangement of components, and by the closest kind of attention to actual area and volume needs and the elimination of the unnecessary. Figure 9 shows how the cubic content per kilowatt of net output capacity has been reduced over the past 29 years. The trend line shown ap-

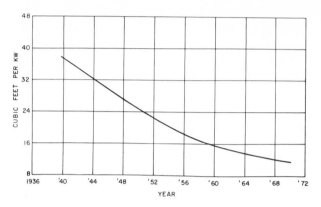

FIGURE 9

AEP system plant cubage

plies to plants on the Ohio River where deep basement structures are required to accommodate the large rise and fall of the river (reaching 73 feet in some locations). The reduction in cubage brought about in the 29-year period involved a decline from approximately 40 cubic feet to 11.5 cubic feet per kilowatt — a reduction of 71.5%. This is another striking example of technological development to bring about a much-needed economic result.

TECHNIQUES OF DEVELOPMENT   Improvements in the generation of power came into being as new ideas, supported by past experience, were first tested and then, when proved, fully exploited. While throughout this period considerable laboratory and factory development was almost always going on, it was necessary to make and then to execute plans to use the utility system itself as the laboratory proving ground for new designs. Experience, punctuated by hard work and reinforced by numerous perils not always ridden through painlessly, taught an appreciation of the fact that size extrapolation must be quite conservative, the ratio not being permitted very frequently to be greater than two to one. As the technology developed, developmental installations came to involve fairly large generating installations with a single unit representing a large investment on the part of the utility. To protect its availability required a program of very close understanding and cooperation with the participating manufacturers.

The improved efficiencies, the lower relative operating costs, and the lower relative capital costs which have been achieved on the AEP System over the past three and a half decades have followed this pattern. Forward-looking development type units have been carefully studied, designed, constructed, and operated using the best judgment available from all responsible sources. The results of these developments have been integrated and then used as the basis for design of more advanced, higher efficiency but still high reliability production units.

Table 2 shows the capacity additions to the AEP System over the past 34 years and includes capacity projected and under actual construction to be completed by 1972. (This list excludes the Deepwater and Missouri Avenue Plants of Atlantic City Electric Company which added materially to the AEP System's technological progress but which are no longer part of the system.) By the end of 1972, 47 units will have been built with a total net generating capacity of 14,579,000 kw. In addition, AEP Service Corporation

TABLE 2

Capacity Additions to AEP System

	Date	No. of Units	Size Mw	Steam Conditions		Design Heat Rate Btu/kwh	Notable Features
				psi	°F		
DEVELOPMENT UNITS							
Logan A	1937	1	90	1250	925	12,000	High-pressure, high-temperature first 1,000,000 lb/hr, single boiler; topping installation.
Twin Branch 3	1940	1	72	2400	940/900	10,200	High-pressure, high-temperature, first single-boiler reheat installation, more centralized controls.
Philo 6	1957	1	107	4500	1150/1050/1000	8,650	Supercritical pressure, once-through boiler, 1150°F temp., double reheat, cyclone firing, hollow cooled generators bars.
PRODUCTION UNITS							
Philo 3	1929	1	160	600	720/720	12,000	Large triple compound, reheat 60-mw topping unit over 3 30-mw units.
Windsor 7	1939	1	150	1250	925	12,000	Advance in steam temperature, large size, cross compound.
Philo 4 etc.	1941	5	95	1300	950	10,850	

Station	Year	Units					Remarks
Glen Lyn 5, etc.	1944	3	100	1300	925	10,900	Single-cylinder 1800 rpm, one installation with single boiler.
Philip Sporn 1, etc.	1949	7	150	2000	1050/1000	9,200	Largest sized units, high-pressure, high-temp., modern reheat, single-boiler, centralized controls.
Kanawha 1, etc.	1953	5	215	2000	1050/1050	9,070	Largest sized units, high reheat temperature.
Glen Lyn 6, etc.	1957	9	225	2000	1050/1050	8,920	Adopted single-turbine-driven boiler feed pump.
Breed and Philip Sporn 5	1960	2	475	3500	1050/1050/1050	8,700	Largest sized, supercritical pressure, once-through boiler, double reheat, single-boiler feed pump.
Big Sandy 1	1963	1	280	2400	1050/1050	8,900	High-pressure natural circulation boiler, natural draft cooling tower, single-boiler feed pump.
Tanners Creek 4	1964	1	600	3500	1000/1025/1050	8,450	Supercritical pressure once-through boiler, double reheat, single-boiler feed pump.
Cardinal 1	1966	3	615	3500	1000/1025/1050	8,580	Largest sized tandem turbine-generator, supercritical pressure once-through boiler, double reheat, single-boiler feed pump.

TABLE 2 (*continued*)

	Date	No. of Units	Size Mw	Steam Conditions		Design Heat Rate Btu/kwh	Notable Features
				psi	°F		
Big Sandy 2	1969	4	800	3500	1000/1025/1050	8,660	Largest sized tandem turbine-generator, supercritical pressure once-through boiler, double reheat, single-boiler feed pump Units 2 and 3 use a single 1200' stack for diffusion of SO_2 in stack effluent and natural draft cooling towers to avoid heating the Ohio River.
Nuclear Plant Lake Michigan	1972	2	1,100	675	501	10,650	Pressurized water reactor, largest sized 1800-rpm tandem turbine-generator.
TOTALS		47	14,579				

designed and constructed and continues to advise on maintenance and operation of 11 identical 215,000-kw units which are installed in the two plants of Ohio Valley Electric Corporation which supply electricity to the Atomic Energy Commission's installation at Portsmouth, Ohio. AEP is one of the joint owners of OVEC. If these units are included, the total number of units designed and built is raised to 58 and the total generating capacity to 16,294,000 kw.

Three highly significant developmental units have been listed at the top of the table, and the remaining 44 have been grouped essentially in accordance with the similarities in their particular design parameters. A list of the significant features of each of these groups of units is also shown in the table.

IMPROVED EFFICIENCY   The development of higher thermal efficiencies can be traced in Table 2. Deepwater — a plant currently outside the AEP System — in 1930 utilized 1200-psi steam pressure and 725°F steam temperature at a time when 600 psi and 650°F were considered above-average conditions. Operation of Deepwater proved conclusively the sound economics of higher steam pressures at a time when most people were debating the validity of 600 to 800 psi.

In the mid-1930s, as load on the AEP System began to increase with improved economic conditions in the United States, studies dictated topping installations at two older plants. Based on Deepwater experience, 1250 psi was chosen as the steam pressure for these topping units. Metallurgical progress made available superheater tubing and alloys for turbine parts suitable for raising the steam temperature to 925°F. Using these conditions, Logan A and a similar unit at Windsor were designed and put in operation, bringing the over-all plant heat rate to the 12,000 Btu per kwh level (28.4% efficiency).[3,4]

The limited application of topping units was recognized at the start, and efforts were concentrated on the next development of the condensing plant. These resulted in a decision to take full advantage of all previous improvements and to establish, in a single design, an installation which would not only represent the most advanced conditions that could then be incorporated in the immediate design, but also would offer no foreseeable limitations to betterment and refinement in design in future years.

The resulting unit, a 67,500-kw (operated subsequently at ratings close to 75,000 kw), 2500-psi, 940°F initial steam temperature, 900°F reheat temperature, single-boiler, single-turbine generating unit, operating on the reheat cycle, was projected for the Twin Branch Plant and was placed in operation in March 1941. This was the forerunner of the postwar single-turbine–boiler reheat cycle unit development now reaching commercial sizes of 800,000 kw in single-shaft units.

The Twin Branch development can be traced in some measure to the large-scale reheat adaptation at Philo Plant in Units 1 and 2 prior to 1929 and in Unit 3 in 1929. But to a much larger extent, Twin Branch represented a triumph of clarification of the true principles of circulation in a natural circulation boiler which led to the simplified circulating circuits and so completely eliminated the circulation problem at any pressure up to a range of 2500–

[3] Sporn, Philip, "The Logan Steam Station of Appalachian Electric Power Company," *Southern Power Journal,* Vol. 56, No. 6, June 1938.

[4] Sporn, Philip, and F. M. Porter, "Application of and Operating Experience with Hydrogen Cooled Synchronous Condensers and Alternators," AIEE Winter Convention, New York, January 22–26, 1940.

2750 psi. This had been the bête noire of boiler design at pressures one-sixth that of Twin Branch, that is, 400 psi. The heat rate of the Twin Branch unit of 10,200 Btu per kwh (33.4% over-all efficiency) established a new record of thermal efficiency.[5,6]

To maintain this trend, it was necessary to develop higher operating temperatures, and this step was accomplished by the installation of the first steam electric plant to use 1000°F at 1300 psi on a unit of 25,000 kw at the Missouri Avenue Plant of Atlantic City Electric Company.[7]

Wartime material limitations in the period 1941–1945 precluded further advancement in technology. However, in 1946, as soon as conditions permitted, plans were formulated for the next developmental stage of the postwar series of installations. Full exploitation of a single-boiler, single-turbine, high-pressure, high-temperature reheat type unit, designed in the maximum size which could be accepted by the system and could be expected to be designed safely by the manufacturer, was projected for the new Philip Sporn Plant and other plants of the system. To proceed along this route it was necessary to fight the temptation to use standard designs — particularly attractive when speed of construction was needed as it was in the postwar boom.

The new units were rated at 150,000-kw net output at 2000 psi, 1050°F, with reheat to 1000°F. All austenitic material was used in the high-temperature superheater tubes

[5] Sporn, Philip, et al., "Twin Branch Plant of I&ME Company, A Pioneering Development in High-Pressure Steam-Electric Power Generation," Electrical World, Vol. 116, No. 16, pp. 80–82, 104–106, October 18, 1941.

[6] Sporn, Philip, and E. G. Bailey, "Operating History of the 2500 Psi Twin Branch Plant," Transactions ASME, pamphlet in Vol. 66, 1944.

[7] Sporn, Philip, "One Thousand F — A New High in Steam Temperature," Electrical World, Vol. 126, No. 7, pp. 60–67, August 17, 1946.

and the high-temperature parts of the turbine on the first six of the seven identical units. The last unit of this series received the benefit of the metallurgical improvements in the use of ferritic materials, and it was decided to construct the high-pressure turbine inner shell of ferritic material to check the suitability of the least expensive type of construction capable of operating at 1050°F. These units were designed and are operated today at a heat rate of 9200 Btu per kwh (37.1% efficiency).[8]

The next series of units advanced the reheat temperature 50° to 1050°F and witnessed another projection in size to 215,000-kw net output.

In 1953 it became quite obvious that unit size increase alone would not achieve the badly needed improvement in thermal efficiency, and it was concluded that pressure, temperature, and reheat should be extended to what appeared to be their practical limits to obtain a base for a series of new unit sizes with improved thermal economy. The result was the placing in service in 1957 of the Philo 6 supercritical unit with steam conditions of 4500 psi, 1150°F, with two stages of reheat: one to 1050°F and the second to 1000°F. Successful operation of this installation at Philo demonstrated the feasibility of the elevated steam conditions and proved, on a relatively large scale, the once-through type boiler operating at supercritical steam pressure.[9]

However, metallurgical research has not yet produced a low-priced, high-temperature material suitable for a large-scale installation at temperatures above 1050°F. The units

[8] Sporn, Philip, "The 2000 Psi, 1050F and 1000F Reheat Cycle at the Philip Sporn and Twin Branch Steam-Electric Stations," *Transactions ASME*, Vol. 70, pp. 287–294, May 1948.

[9] Sporn, Philip, and S. N. Fiala, "The Development of the Supercritical 4500 Psi Multiple Reheat Generating Unit for Philo," Fifth World Power Conference, 1956.

based on the success of the Philo development have consequently been designed at slightly lower temperature and pressure to give the most economical combination. In 1960, two units, each rated at 475,000 kw, were placed in operation at 3500 psi, 1050°F, with two stages of reheat, each to 1050°F conditions. These units have an over-all net heat rate of 8700 Btu per kwh (39.3% over-all efficiency).

The development of higher initial steam conditions was accompanied by advances within the heat cycle and the supporting auxiliary equipment. More stages of feedwater heating, with the 475,000-kw unit designs now employing nine stages, better distribution of bleed points, use of desuperheating zones in heaters, and use of heater designs having approach temperatures down to zero, are some of the examples. Boiler efficiencies have been radically increased through the extensive use of large air heaters, improved combustion techniques, and pressurized firing. Parenthetically, pressurized firing was first introduced on the Philip Sporn series of units in 1949. Forced draft fans handling cool combustion air were used rather than the previous practice of using forced draft and induced draft fans in series. The Philip Sporn units and the Kanawha units were provided with induced draft fans for emergency operation only. The reliability of pressurized operation was gradually improved and was brought to the point that, with the Glen Lyn 6 series of units, I.D. fans were omitted.

Improved over-all cycle efficiency has been aided by the use of steam heating of combustion air which, although extracting steam from the turbine cycle, allows boiler operation with lower outlet gas temperatures for a given critical air heater metal temperature. Critical air heater metal temperatures are determined by the sulfur content of the coal burned, and the resulting air heater pluggage which must be held at a tolerable level even with newly developed in-service air heater washing techniques.

Auxiliary power requirements in the generating units have been reduced by improved efficiency of components. Much of this has been the result of increased size. Two F.D. fans, for instance, serve the 475,000-kw Philip Sporn 5 unit, whereas 16 F.D. fans were used in the design of the 160,000-kw Philo 3 unit in 1929. Boiler feed pumps have also been radically reduced in number. Steam-turbine-driven boiler feed pumps were used in the 1920s, often three per unit. These were gradually displaced by squirrel-cage induction-motor-driven feed pumps with generally no less than three one-half capacity pumps being provided per unit. In 1955, AEP staff studies indicated that significant thermal efficiency and capital economies could be achieved by reliance on one full-size steam-turbine-driven feed pump. The efficiency of this important auxiliary was not only increased by its larger size but the over-all cycle efficiency was improved by the strategic location of the feed pump drive turbine in the heat cycle. In the first of these applications on the nine 225,000-kw units, beginning with Glen Lyn 6, the pumps are driven by an extraction turbine which serves three feedwater heaters below the reheat point with only moderately superheated steam rather than the highly super-heated steam that would have been extracted from the main turbine. The 475,000-kw units utilize a single condensing turbine-driven pump, which improved cycle efficiency by lowering turbine exhaust losses, and so does the Big Sandy 800-mw unit.[10]

To understand the full impact of this drive on reliability

[10] Tillinghast, J. A., "Application of Single Turbine Driven Boiler Feed Pump to 225,000 Kw Generating Units," Proceedings of the American Power Conference, 1956.

and capital economy, it is well to recall that the Missouri Avenue 1000°F, 25-mw unit utilized three boiler feed pumps, whereas Big Sandy No. 2 will utilize a single pump and turbine for its 800-mw boiler — a step-up of 96:1.

Substantial improvements have been made in turbine and generator efficiencies during this period. The major improvement in generator efficiency has been the development of hydrogen cooling, described later.

More detailed discussions of a number of the technologic-economic developments discussed earlier relating to the Twin Branch 2500-psi unit, the Philip Sporn Station, the Philo 4500-psi supercritical installations, and the use of single-turbine-driven boiler feed pumps are given in footnotes 2, 5, 6, 9, and 10.

REDUCED LABOR AND MAINTENANCE   The lowered labor and maintenance costs shown in Figures 6 and 7 have been achieved by developments in several directions. Important among these has been the use of larger integral blocks of generating capacity. The practice of the 1920s was to use several boilers serving one or more turbines. Philo 3 is a good example of this: here six evaporating and two reheating boilers serve three turbine elements, a single high-pressure element and two parallel low-pressure elements. The largest amount of capacity that could be forced out of service for an extended period was, therefore, one-half the unit capacity of 160,000 kw. For the determination of reserve requirements this unit could be classified as an 80,000-kw unit. Similarly, Logan A and Windsor 7 could be classed as 40,000- and 60,000-kw units, respectively, since these were the largest blocks of generating capacity which could be forced out of service for an extended period. The units in the Philo 4 series have two boilers and high- and low-pressure turbines so that the largest amount of capacity that can be forced out by a single failure is 50,000 kw. All subsequent units, however, have a single-turbine, single-boiler design so that the full unit can be forced out of service.

But this interesting mathematical evaluation of the economic costs of an outage for maintenance purposes was a mere beginning. More fundamental questions of an engineering nature were asked and answered. Among them were: Is it necessary to open a steam electric turbine every year? How about every two years? Five? Ten? The unprecedented answers of pushing the period between openings up to ten years, as long as critical internal and external control point readings gave no warrant for concern, contributed heavily to reduction of maintenance costs.

Figure 10 shows the rapidly increasing size of units by the percentage relationship between the unit at its time of initial operation and the total load on the AEP System as the largest size on the system increased from 95,000 kw in 1940 by more than eightfold to 800,000 kw projected for 1969 and 1100 mw for 1972. But Figure 10 also shows the trend, as the system grew, toward greater conservatism in the per cent of total system capacity exposed to outage as a result of a mishap to, or failure of, a single boiler, turbine, or any other significant element in its cycle of operation.

Of vital importance, not only to reducing labor and maintenance costs but to the safer operation of the large and costly pieces of equipment in a modern power plant, has been the development of centralized automatic control. Philo 3 unit in 1929, with eight boilers and three turbines, had control stations located at each pair of boilers, the turbines, the condenser pit, and the electrical station's switching dispatcher. In all, 11 operators were required per

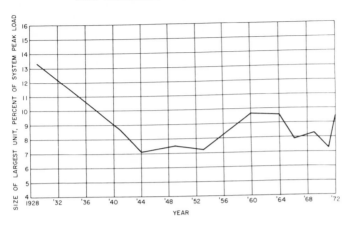

FIGURE 10

AEP system largest unit size

shift. In 1940 Twin Branch 3, with a single reheat boiler and reheat turbine, was put in operation with control centralized at a boiler control panel, a turbine panel, and at the dispatcher's panel. Tidd Units 1 and 2 of the Glen Lyn 5 series were the first completely centralized control units. All functions of unit control from admission of coal to the boiler to switching of high-voltage output were brought into a single control room. Both of these 100,000-kw units were operated with seven men per shift. Today these controls have been brought to the point where on the 475,000-kw Breed unit four operators per shift control all functions of the generating plant, and in the case of the new Cardinal Plant these are down to three per 615-mw unit.

Another important feature contributing to the reduction of maintenance and also tending to increase the safe effective output of units has been the extensive instrumentation of critical areas of major equipment which has been devel-

95

oped over this period of almost four decades. This instrumentation allows operation of equipment at the true design limits of the components rather than at some artificially created limit which does not bear any rational relationship to the properties of the constituent material of the equipment in question. Examples of these measurements are boiler drum metal temperatures which determine safe rate of pressure rise in the boiler, turbine stage pressure measurements which indicate internal condition of the turbine, and turbine metal temperature measurements in shells and bearings. Particularly important are the turbine supervisory instruments which record vibration, eccentricity, and differential expansion to give a continuous record of the condition of the machine. These turbine supervisory instruments were first used on the Windsor topping unit of 1939 and have come to be completely standard on all units throughout the industry.

A particularly important type of instrumentation which has been used and which is presently being developed to a higher reliability is the direct measurement of superheater and reheater tube metal temperatures. As the cycle reaches out for higher steam temperatures, the margins which the boiler designer puts in for unbalance firing, slagging, or other phenomena that affect tube metal temperatures become more onerous. Direct measurement of tube metal temperatures allows the reduction of these uncertainty factors to a minimum and the exploitation of a given tube material to its safe ultimate.

Another practice which has been developed over the past 36 years on the AEP System which has contributed substantially to the reduction of maintenance cost has been the special design provisions for maximum interchangeability of spare parts. Spare rotating parts are provided for all major sections of turbogenerators, as well as other components, such as feed pump motors, fan rotors, transformers, high-voltage bushing, and other critical, longer delivery items. Details are engineered so that provisions are built into the plant to permit ready replacement of certain pieces of equipment by spare units, in some cases even those of another manufacturer.

REDUCED CAPITAL COSTS   Costs of all components that go into the design of a thermoelectric power plant have risen sharply over this long period in spite of many cost reduction efforts throughout the industry. These trends are illustrated in Figure 2.

The reductions in capital cost in the face of these increases are shown in Figure 8. They have been importantly the result of increased size of unit and the development of the unit-type single-boiler, single-turbine, single-auxiliary concept. No small contribution, however, has been made by better utilization of area and cubage, simplified concepts, and better construction planning, supervision, and execution. These cost reductions are all the more significant when consideration is given to the fact that greater use is being made of the more expensive materials needed for operation at the higher steam conditions. Philo 3, first operated in 1929, utilizes 600 psi, 720°F, whereas the Breed–Philip Sporn units of 1959 utilize 3500 psi and 1050°F with their associated high-cost alloys. Yet the capital cost of new generating plant, when adjusted by the index of national average cost, has been reduced about 70% in 36 years and at the same time thermal efficiency has been improved some 57%.

A particularly significant contribution to the capital cost reduction per kilowatt has been the abandonment of the multiple system of auxiliaries; in this process individual auxiliaries have been greatly increased in size as illustrated by the employment of a single boiler feed pump serving the 475,000-kw Sporn series of units. The driver of this pump is a 20,000-hp condensing turbine. Where the cycle demands variable speed, it is made possible by a process of two energy conversions in series as against seven where a hydraulic coupling is employed.

Switching and transformer arrangements have been simplified. In units designed during the early 1920s, low- and high-voltage switching were required. Low-voltage buses and switching were utilized for plant synchronization and to supply low-voltage feeders and auxiliaries. High-voltage switching was utilized for transmission system switching of transformers and transmission lines. Even when low-voltage buses were abandoned, low-voltage switching was still utilized to obtain fast closing time to synchronize units, and high-voltage switching for fault protection. Philo 3 was the first unit to utilize high-speed, high-voltage outdoor switching, operating at 138 kv without low-voltage switches. This practice became standard.

Main step-up transformers have undergone an evolutionary simplification also. Practice in 1929 called for three single-phase step-up transformers with a spare single-phase unit. First, multiple, three-phase units were utilized, three per unit, on the 150,000-kw units of the Philip Sporn series; but later, starting with the 225,000-kw Glen Lyn 6 series, a single three-phase step-up transformer was introduced to handle the output of the entire unit. In the case of Muskingum 5 this has become a single three-phase transformer stepping up from 23 kv to 345 kv with a rating of 725 mva, and in the case of Mitchell this rating has been increased to 950 mva.

A most significant generator development over the past 30 years has been the hydrogen cooling system, which has permitted efficient high-speed machines in sizes which could not otherwise have been designed, shipped, or economically installed. The first hydrogen cooling installation in the world was made on the AEP System in 1928 on a 20,000-kva synchronous condenser. This provided the operating and design information for the construction of the first central station generator, placed in service in 1937 at the Logan Plant — a 50,000-kva, hydrogen-cooled generator connected to a topping turbine. The efficiency of this machine was 98.4% compared to 97.2% for a similarly rated air-cooled machine. At first, hydrogen was operated at one-half psi pressure, but in subsequent designs this was raised first to 15 pounds and then to 30 and 45 psi. The 45-psi pressure has been associated with the conductor type cooling in which the hydrogen is circulated directly through small passages inside the hollow conductors of the stator and field coils of the generator. This arrangement was first used on the Philo 6, 156,000-kva generator in March 1957.

Hollow cooling of rotor conductors was used on the Breed 475,000-kw unit. Stator windings, however, were cooled by direct circulation of oil through the hollow stator winding conductors. As a further development of this technique, because of its improved heat-removing capabilities, water cooling was used on the stator bars of the otherwise duplicate Philip Sporn 5, 475,000-kw machine.

With the exception of Big Sandy 1 (1953), rated at only 280 mw, this combination — hydrogen cooling of rotors and water cooling of stators — has been utilized on all subsequent turbine alternators on the system.

The net result of this consistently followed practice is that there are now in service on the AEP System (or will be upon completion of the presently initiated additional capacity in 1971) a total of 6493 mva in hydrogen-cooled units, both in rotor and stator, and another group consisting of 10,257 mva in rotor hydrogen-cooled and stator liquid-cooled units.

The universal acceptance of hydrogen cooling, in which the AEP System carried out the world pioneering, led to improvements in design and progressive reduction in size per kilovolt-ampere, and to expansion in size of units. Upon completion of the present program there will be operating on the AEP System 45 machines comprising 14,330,000 kw of turbogenerator capacity cooled in one form or another by hydrogen; this development, extending back almost 40 years, has made a significant contribution to very much increased output through compact designs.

Present State of Development
and Future Prospects

At this writing, the construction of the two largest and most advanced type units in the AEP System has been started. The first should be in commercial operation in 1970 and the second will follow in 1971. These are 800-mw net output units designed to operate at 3500 psi and 1000°F with double reheat at 1025 and 1050°F. It will be recalled that these are single-shaft, tandem compound units. They represent the very best that the art and technology in thermal electric energy conversion can currently do to give an optimum economic result.

But progress in thermal electric generation cannot stop with these 800-mw supercritical units. Further development of thermal electric generation must continue in the future along many lines.

More work needs to be done on still larger sized units, particularly on single-shaft designs and in a range of 1050 to 1350 mw. These increases in size, up to 1350 mw, are particularly important to be able to provide a solid design base for the 4000-mw plant that is so necessary to exploit fully any reasonably good plant site, and which in addition is the mechanism for keeping the per kilowatt cost under control in the face of continuing inflationary pressures.

Improved thermal efficiency is badly needed, first to provide the continuity in reducing the operating expenses in the advanced plants and, second, to keep the problem of size under control, since with every per cent improvement in thermal efficiency the increased amount of kilowatts that can be obtained from a given size plant is in direct ratio to the improvement in efficiency. Unfortunately, improved efficiency through the use of higher steam conditions awaits a breakthrough in metallurgy. Moderately priced materials for service in the 1100 to 1300°F range are badly needed. But we do not have them today. Thus, while there is nothing major that can be anticipated in the way of efficiency improvement, there are many small items that collectively can make a significant contribution. It needs to be remembered that in a 4000-mw plant an improvement in thermal efficiency of 1 Btu per kwh represents an annual saving at a 20¢ per million Btu energy cost level of $6,000 or a capital saving of $50,000. Then, too, studies and research on combined coal-burning–gas-turbine–steam-turbine cycles continue, with some progress even though the development work has so far not reached a point where a pilot plant appears to be near realization.

And, of course, in the distance there looms the possibility of successful development of MHD with a 30% jump in thermal efficiency of a combined cycle to 52%.[11] It is a

[11] Sporn, Philip, and Arthur Kantrowitz, "Magnetohydrodynamics — Future Power Process?" *Power,* Vol. 103, No. 11, pp. 62–65, November 1959.

great pity that the utility group which jointly with Avco has been doing pioneering work in this area, having reached the stage where an experimental unit was called for, currently finds itself in a state of indecision as to whether or not to continue its research efforts.

Beyond that we have nuclear generation which currently, in very large sizes in the neighborhood of 1100 mw, is competitive with coal at a Btu cost of between 21.5¢ and 22.5¢. This has given atomic power admission, so to speak, into many energy areas of the United States. Whether such admission was earned is a matter of some doubt considering that absolute progress in improving the competitive position of atomic power has been relatively negligible in the last year. Yet, for a number of reasons, most of which I covered in my previous lecture, nuclear power has made remarkable progress in breaking into markets that coal might have continued to hold. Yet nuclear power has found admittance into some of the very low cost coal areas, such as TVA. It is now pushing into, or is being pulled to enter, other areas. In July it was, as a matter of fact, admitted into one such highly important additional area.

The working out of this competitive struggle between nuclear and more conventional fuels needs very badly the dedicated efforts, a reoriented philosophy and approach, and a strongly cooperative effort on the part of the reactor manufacturers, the AEC, the utilities, and the coal mining and uranium mining industries in which technology, engineering, economics, and social considerations are each given their full measure of consideration. This, too, I discussed more fully in my previous lecture.

An attitude toward research and development that is summed up in the policy, "We will carry on research if you will pay us to do so," does not bode well for the proper working out of this so significant and important struggle.

It is obvious that a complete interlinkage of technology, engineering, and economics and their mutual interaction, which I set out to illustrate, cannot be covered completely merely by the development of the history of the advance in power generation both economically and technologically. After all, it takes more than generation to develop a system.

Specifically, it takes transmission and distribution and system itself and financing and commercial policy, particularly a farseeing commercial policy that can sense the potentialities of using a new area of service to develop expansion that can make possible economic service to this particular area, and imagination to visualize serving any load that the economy of the area could possibly bring into being or that can be helped to come into being without setting any limitations on the quantity of power, on its quality (no matter how precise the power that is required), or the economics within the realm of the attainable. It takes very farseeing management to realize that this kind of policy is not a one-time thing but that resources, technology, and engineering, coupled with a forward-looking program of research and development aimed at and supported by economics, so as to improve the economics of the supply, are factors that have to be studied and permitted to interact continuously and over a long period to yield a successful result.

In the case of the AEP System, equally fascinating discussions, but going beyond the scope and particularly the time we have available for this lecture, can be brought forth to trace the development of transmission, distribution system, and commercial activities. If this were done, I would hope the exercise would be fully as interesting as the discussion of generation. All of these were, as a matter of fact successfully integrated into a single operation and its great success is the result of their fullest coordination and interaction.

The Systems Approach, Operations Research, Engineering, and Economics

Introduction

THE SYSTEMS APPROACH, sometimes termed operations research, is commonly thought to have been introduced, indeed made possible, by the introduction and development of the digital computer. Without even remotely intending to downgrade the value of the digital computer as an enormously useful tool in modern technological operations to reduce the cost and improve the thoroughness of economic evaluation of complex technological operations, it is still important to point out that systems analysis and the systems approach are at least a half century old and were for decades practiced effectively in our society, and particularly in the communications and electric power fields[1,2] well before the development of the modern computer.

This is important, because the basics — the areas of concern and study — antedate the computer and are not computer determined — they are determined, instead, by the project or system analyst, the system engineer in most cases. The engineer's analysis necessarily involves economics. In that process the computer is an enormously effective tool. But the degree of vision, sophistication, and judgment regarding the elements that are vital and paramount in a system analysis are not determined by the computer. The results of extracomputer judgments are fed into and then processed by the computer. These can, however, be processed and handled without the computer, but not as quickly, not as economically, and therefore not quite as thoroughly.

[1] Sporn, Philip, "Planning and Carrying Out of a Program to Provide Adequate Generating Facilities on an Extensive Integrated Power System," *Power Plant Engineering*, Vol. 42, No. 5, p. 304, April 1938.

[2] Sporn, Philip, *The Integrated Power System*, McGraw-Hill Book Company, Inc., New York, 1950.

As an effective case study to illustrate how such an analysis was handled in a situation where technological and economic boldness were called for, yet where assumptions and perceptions difficult to grasp and evaluate were handled in a manner to lead to a successful consummation of a major project, I should like to discuss the OVEC project — a project that I had a major hand in conceiving, analyzing, organizing, and constructing and then supervising in all its activities for a period of 15 years following its corporate creation.

The History of the Project[3,4]

On October 15, 1952, the Ohio Valley Electric Corporation and the Atomic Energy Commission executed a 25-year power agreement calling for the supply of 1.8 million kw of capacity and 15 billion kwh per year to the gaseous diffusion plant known as the Portsmouth Area. This was the largest single power contract ever entered into by a single customer in the 85-year history of the electric utility industry. To fulfill its commitments, OVEC, and its subsidiary Indiana-Kentucky Electric Corporation, constructed two large steam electric plants, one with a capacity of 1.2 million kw and the other of 1 million kw. An expenditure of close to $400 million was required to construct the two generating stations with the necessary 330,000-volt transmission lines to deliver the power to AEC and to provide the required interconnections with the participating power companies.

[3] U.S. Atomic Energy Commission, 12th Semiannual Report, Washington, D.C., July 1952.

[4] Sporn, Philip, and V. M. Marquis, "The OVEC Project: Economic, Engineering and Financing Problems of the 2,200,000 Kw and 18,000,000,000 Kwh Power Project of the Ohio Valley Electric Corporation," AIEE Winter General Meeting, New York, January 1954.

Following a first meeting on January 23, 1952, and a second meeting on February 1, 1952, representatives of almost a score of electric utilities in the North Central Area of the United States met with representatives of AEC to discuss the Commission's part of the proposed expansion of its diffusion program tentatively selected for location in the Ohio Valley at an undetermined site between Parkersburg, West Virginia, and Evansville, Indiana, which would require meeting the electric demand and energy use earlier described. The utility group agreed to undertake a study of their ability to supply the necessary power requirements and to report promptly their preliminary findings.

On February 18 a preliminary report was made to AEC, and later confirmed, indicating that power costs were about even as between any one of three preferred locations. This was followed by a second report less than a month later which confirmed the conclusions reached in the first report.

This second report also emphasized a fundamental conclusion previously expressed that it was the group's judgment that power to be supplied to AEC should be furnished from two generating stations, one to be located in Ohio and the other in Indiana, but connected together to operate as a single project, and interconnected and also coordinated with the systems of the participating companies. Four reasons were especially important in reaching such a conclusion:

1. The undertaking required the participation of a large number of systems in the North Central Area.

2. The power supply to AEC had to have a high degree of reliability.

3. Large fuel requirements pointed up the advisability of tapping two major coal fields — the southern Indiana–west Kentucky and the western Pennsylvania–eastern Ohio–

northern West Virginia fields — and thus reducing the impact on the marketing area of a single coal field.

4. The problem of supplying interim power, supplemental or emergency back-up, and the absorption of the generating capacity in case of discontinuance of the diffusion operation all had an unavoidable bearing on the basic design of solid interconnection with the systems of all the sponsors as a basis for furnishing this power.

The group quickly organized itself to carry on the detailed studies of the economic and engineering phases and of the many financing, regulatory, and legal problems. Study groups, operating under the direction of selected chairmen, and employing consulting service when necessary, were set up and immediately undertook the basic studies required for making a definite proposal to AEC.

By May 12 a definite proposal was submitted, in considerable detail, to the AEC on behalf of 15 participating power companies. This proposal was on the basis of an assumed diffusion plant site, since AEC had as yet not selected the site. As a matter of fact, the ultimate site did not prove to be the site assumed in the May 12 proposal. Fortunately, the analysis of the basic problems that had been made and the solutions adopted were such that no material change in the plan was called for when the final site north of Portsmouth was selected.

In the May 12 proposal to AEC, the 15 participating electric utilities proposed that a new corporation or corporations be formed by the group. The new enterprise would build, operate, and maintain the necessary generation and transmission facilities to supply the power required. It would obtain the required capital, up to an estimated amount of $440 million. This would be accomplished, as the proposal fully developed, by borrowing up to $360 million from a group of insurance companies, pension funds, and savings banks on 3.75% mortgage bonds; obtaining $60 million from a group of banks and pension funds on 14-year unsecured notes; and obtaining $20 million by selling common stock to the participating companies.

Shortly after the submission of the May 12 proposal, AEC accepted the proposal as a basis for further negotiations and these were promptly started.

Because of defense urgency, negotiations were carried out with all possible expedition and a final power agreement executed between OVEC and AEC on October 15, 1952.[5] OVEC and its Indiana subsidiary, IKEC, had been organized some two weeks previously. Immediately upon execution of the October 15 agreement, the necessary clearances were obtained from the Securities and Exchange Commission (SEC), and Federal Power Commission (FPC), and the involved state regulatory commissions to permit raising equity capital to make possible an early start on construction.

The October 15 contract was based on the location of the diffusion plant on the Scioto River between Portsmouth and Chillicothe, Ohio, which the AEC had announced on August 12. Promptly following the execution of the contract preliminary commitment discussions were carried out with the coal suppliers and the work of optioning the controlling acreage for the two plant sites was commenced. On October 29, an announcement was made of the definite location of the Clifty Creek station near Madison, Indiana, and of the Kyger Creek station near Gallipolis, Ohio.

Contracts and orders for all major equipment were

[5] U.S. Atomic Energy Commission; Power Agreement between Ohio Valley Electric Corporation and United States of America, October 15, 1952.

awarded within the first several months after October 15, 1952. The engineering and drafting work on such a large project, brought into being under circumstances such as prevailed here, can economically be moved along at a rate to keep plans only moderately ahead of actual construction. Yet, as of October 15, 1953, this work was more than 35% completed.

The original tight schedule was not only met in full but many of the units were completed months ahead of time. The first of the 11 turbogenerator units was brought on the line in December 1954, a bare two years after the start of construction. The eleventh and final unit went into full commercial operation in March 1956 and established a unique record for speed in the planning, design, and construction of power facilities.

I would now like to discuss the major areas of activity and technologic and economic involvement to indicate how each was resolved while in each case the economic factor was given full evaluation and introduced into the equation and how they were integrated into the whole technological-economic concept to reach a hard-set goal.

The Basic Economic Problems in Launching OVEC;
Their Economic Involvement; Their Satisfactory Solution

The basic problems in launching OVEC stemmed from the economic size of the project; from the character of the principal recipient of the product of the project — the government of the United States; and from the need for the utmost flexibility in the effective term originally contracted for, that is, the need to be able to cancel at any time — even before, or on the very eve of, the project's coming into effective being. These conditions created numerous difficult and special problems — all, or most of them, heavily charged with difficult but important economic aspects. To expand briefly:

PROBLEMS STEMMING FROM SIZE

1. These created a problem in size of group to combine in a joint venture that could command confidence so as to be able to finance the project. It was evident that to supply a load of 1800 mw the additional capacity needed to supply losses plus reserve would be 300 mw, thus a total of 2100 mw. At approximately $150 per kilowatt the total capital required to provide plant, plus transmission, plus working capital, would be $400 million. A group with a combined peak of about 8 million kw was estimated as about the level to aim for.

2. With a capacity required of 2100 mw, the question arose of whether one or two or three plants needed to be built. This was a very complex problem involving balance and interaction among cost per kilowatt of capacity, cost of fuel, effect on economic dislocation, and thus on cost of construction, effect on air and water pollution, effect on transportation costs of fuel from mine to generating plant, and optimization among rail, barge, and electric transportation. The final decision: two plants, one containing six units and the other five, all of the same 200-mw nominal size, the plants interconnected at 330 kv and delivering the power at the same voltage.

3. The necessary reserve to protect the continuity of supply and the proper amount to be furnished from the facilities reserved and from the interconnected system.

4. The provisions for cancellation: These were particularly difficult matters to negotiate since they involved projections of system conditions as well as general economic conditions under a complex variety of circumstances and there existed great need to handle this contingency with fairness to AEC but also to the banks and insurance institutions furnishing the capital and to the sponsor companies.

The final solution was based on a sober but optimistic projection of growth and an undertaking by AEC to give adequate notice of cancellation plus protection of carrying charges of investment while the growth in demand of the sponsor group rose to relieve the AEC of any further responsibility.

5. The huge demand for fuel — 7.2 million tons of coal per year: This will be discussed in more detail later.

PROBLEMS STEMMING FROM THE CHARACTER OF THE PRINCIPAL RECIPIENT OF THE PRODUCT OF THE PROJECT

6. The need to minimize capacity and operating costs because of the always implied competition — governmentally owned supply: This in turn was reflected in setting up the special program in Problem 7.

7. Program to minimize cost of fuel: this is discussed in more detail later.

8. The need of a program to minimize the cost of capital. This too will be discussed in some detail later.

9. The handling of the tax component of capital costs to minimize the total capital cost.

10. The need for a long-term contract (to reduce the contingency factor) to reduce the cost of power and

These companies were asked to submit bids on the
[...] s of coal which they would have available. It later de-
[...]ped that these companies in each case submitted bids
[...]iderably higher in price than the bids which were
[...]lly accepted.

[...]roposals were received from a bare 21 coal companies,
[...] another 12 companies replied to say that, for one reason
[...]other, they could not bid at that time. Five bids were
[...]pted: two for the supply at Kyger Creek and three for
[...]upply at Clifty Creek. But the negotiations which were
[...]ed out following preliminary acceptance were in four
[...] five cases long, difficult, and at times almost frustrat-
[...]Particularly, difficulty was involved in developing fair
[...]tion clauses and especially ideas with regard to nega-
[...]scalation, that is, sharing in increased productivity.
[...]er OVEC knew where the coal would be produced,
[...]vitation to bid on the barging of coal from the mines to
[...]wer plants was prepared and sent to 20 barging com-
[...]. Contracts were awarded to three barge lines. Esca-
[...] was a formidable problem here also.

[...] complete working out of the OVEC coal supply
[...]ts inception before the first response to AEC's request
[...]ade to a time just prior to the first delivery of coal
[...]nted an interesting technoeconomic problem from
[...]standpoints. Although the indications were that we
[...] be flooded with offers of coal, we received serious
[...]es or expressions of interest only from less than 30%
[...]e addressed. Eventually, in spite of the putative
[...]f mines waiting for coal orders, more than half of the
[...]d to be obtained from brand-new mines tailored to
[...]jectives set and agreed upon in negotiations, and
[...]l another major mine had to be opened up to take
[...]another 20%.

the inability of the Commission to enter into any
agreement in excess of 25 years. This created a
double problem, the need to minimize the effect of
this short amortization period and also the need to
protect the residual interest of AEC in a fully amortized plant that was still productively operative.

11. The ineluctable need to work out provisions for
cancellation in view of the uncertainty with regard to
future need of enriched U-235 (see previous Item 4).

PROBLEMS STEMMING FROM THE NEED OF PROVIDING THE
UTMOST FLEXIBILITY

12. The cancellation provision and the need for multi-
party protection, fully rationalized, but designed not
to become an onerous burden (see Item 4).

13. The problem of reduced load factor and of replacing
load upon cancellation and the effect on ability to
meet minimum takes under long-term coal contracts.

14. The problem of transmitting large blocks of canceled
power to a diverse and extended group of 15 utility
sponsors without imposing an economic burden on
members located farthest from the two generating
plants.

15. The need to provide a mechanism for pickup of
load dropped or canceled by the Commission.

All of these problems were tackled on the basis of extended study, the development in some cases of carefully
evaluated projections of trends or experience and long
negotiation discussions which led to agreement among sponsor companies, between OVEC and sponsors, between
OVEC and suppliers such as the coal companies and the
barge lines, and between OVEC and AEC. But engineering

and economics and equities were all taken into fullest consideration. An early general agreement to look for and accept no windfalls materially helped in negotiations that led to contractual agreement which has stood the test of time remarkably well. A number of the most important problems are now discussed in more detail.

The Fuel Problem — Coal

It is fortunate that except for organization, the marshaling of a responsible sponsoring group, and financing, it was recognized from the very beginning — from the very early stages of the conception of OVEC — that the basic problem above all others was a fuel problem and that success or failure of the project would hinge in a very large measure on finding the proper sources of fuel and at proper prices for the power plants that would generate the power needed by the Commission at Portsmouth.

My visualization of the OVEC project in September–December 1951, some months prior to the first meeting with the AEC, definitely included the concept of a two-plant system to draw upon the southern Indiana–western Kentucky coal fields to supply a plant either in Indiana or Kentucky and on the eastern Ohio–western Pennsylvania–Kanawha Valley, West Virginia, coal fields to supply a second plant located in southeastern Ohio. Indeed, one of the strong bases for recommending to the AEC that the Ohio Valley be chosen as the general locus of a third diffusion plant by the special committee that was charged with the siting task was the general knowledge of the availability of reasonably priced coal in the areas mentioned. And this was further confirmed by Mr. Paul Weir, of Chicago, the coal consultant to the American Electric Power System whom the author brought into the problem shortly after January 23, 1952. But this needed more solid confirmation.

Hence, almost simultaneously wi[th]
the AEC toward the end of January
also held with various representatives
New York and Cleveland.

The meetings with the coal peopl[e]
become better acquainted with coal
ods, mining equipment, and the
prices and to see if the coal industr[y]
nishing the coal requirements of a p

It must be remembered that by
tainly by the standards of 1952, t
quired for OVEC's plants were
determined amounted to 3 millio[n]
Creek Plant and 4.2 million tons
Plant. It was not expected that co
would become available to any si
existing coal mines.

The coal industry at this tim[e]
from the flash of prosperity that
which had already begun to sl[ow]
given up the hope of a continui[ng]
did not yet get itself into focu[s]
utility market of the 1960s, an
conceive the great threat to its
was preparing to become. And
number, the opening of a new
coal mining management with
was the atmosphere into which
bids.

Invitations to bid were se[nt]
Weir and the AEP Service
analyzed the coal bids. Whil[e]
coal companies advised that t
not meet the coal specificatio[n]

In the case of one company that finally agreed to undertake the supply of 1.5 million tons of coal, the final agreement to do so came only after intensive and extensive hortatory pressures upon the chief executive. The solidity of the judgment to yield to that pressure was brought out by the statement of the same chief executive some ten years later that but for this contract he never would have survived the low period through which the coal industry passed for some ten years after this contract was executed.

What was true with regard to the coal was equally true with regard to the transportation facilities, and the three barging companies had to make an investment of over $20 million in barges and towing equipment in order to meet their contractual commitments to OVEC for coal transportation.

It is almost obvious that, under the conditions of indispensable extensive new mining operations and new transportation facilities, long-term contracts would become a must in order to provide a prudent basis for additional and large capital investment. This became involved by requirements to provide for cancellation by AEC when, even after the sponsors picked up the canceled capacity in full, they would still be unable to burn the previously contracted quantities of coal owing to the much poorer load factor of their system loads as contrasted with AEC's substantially 100% load factor. And this not only added to the burden of the negotiators, but offered a challenge also.

Finally, when all this investigation, appraisal, negotiation, and agreement with regard to coal and transportation had been completed, the coal situation stood as follows before and after the October 1952 escalation due to a UMW wage increase on October 1, 1952, and 14 years later at the end of 1966:

	Kyger Creek		Clifty Creek		Weighted System	
	$/ton	¢/MBtu	$/ton	¢/MBtu	$/ton	¢/MBtu
Cost before escalation of 10/1	4.265	18.98	3.83	17.40	3.97	18.06
Cost after escalation	4.705	20.47	4.02	18.24	4.15	18.91
Calendar year 1966					4.70	20.45
Total increase in 14 years						8.13%
Average increase per year — noncompound						0.58%

The answer to the question, how the cost increase was held to so low a level, lies mainly in three directions: (1) careful audit and analysis of books on escalation claims; (2) renegotiation at opportune times; and (3) negative escalation: sharing in savings due to increased productivity.

The OVEC Power System: Conversion Plant, Integrating Transmission System, Interconnections, Reliability, Performance

GENERAL ARRANGEMENT As has already been indicated from a discussion of the fuel problem, a decision had been made following the AEC's decision to locate the diffusion plant at Portsmouth to develop the generating capacity in two stations, both located on the Ohio River. The location of the two plants on the Ohio River made possible the taking advantage of experience gained in building major steam plants along the Ohio River or its principal tributaries for a period of over 30 years. Thus, the experience gained by many of the participating companies in building generating stations on the Ohio River and its tributaries (in particular those stations of the AEP System in which the size of units employed is practically the same as OVEC machines) was fully drawn upon in planning, designing, and constructing the two stations.

CONVERSION PLANT   The turbine generators selected were cross-compound with a gross capability of 217,250 kw at 1.5 inches of mercury back-pressure. They operate at 2000 psi, 1050°F at the throttle and reheat to 1050°F, the high-pressure unit at 3600 rpm and the low-pressure unit at 1800 rpm.

The thermodynamic cycle on which they operate is essentially the well-known regenerative cycle with seven stages of feed water heating to provide a final feed water temperature of 469°F and an over-all plant performance of close to 9200 Btu per kwh.

The search for an advanced and highly efficient plant was materially stimulated by the very high load factor at which the plants were planned to be operated and by the relatively low capital carrying charge owing to the low equity ratio and the almost complete elimination of the federal income tax component of cost. With an 18×10^9 kwh figure of annual generation and 20¢ per million Btu fuel, a saving in the two plants of one Btu per kwh could be prudently bought at a total capital cost of $50,000 — an extremely high figure, but fully justified under the prevailing conditions.

INTEGRATING TRANSMISSION   Transmission from the two generating stations to the diffusion plant runs approximately 140 and 30 miles. Except for certain 230-kv developments that were under way in the western part of Indiana, the highest existing transmission voltage level in this entire area was 138 kv—obviously inadequate to handle the transmission requirements of this over-all project economically. For the distance of transmission involved, comparative studies of both 230-kv and 330-kv transmission showed clear, even though small, advantages in cost in favor of 330-kv. The decision to go to the higher voltage was based on performance. The smaller number of circuits and rights of way required, together with the ability to interconnect at the same transmission voltage level with strategic generating and switching points on both ends of the project system, were telling advantages. The substantial capacity of even a single circuit was another. And we acted on the basic principle that, economics being approximately equal, sound judgment dictates choosing the highest practicable voltage.[6-8]

It was perhaps a fortunate circumstance, therefore, that the growing inadequacy of 138-kv transmission for the normal growth of load in much of this area had already been recognized some time before, and an extensive research and development program on higher voltage transmission had already been undertaken and largely brought to a conclusion prior to the inception of this new project. As a matter of fact, the 330-kv (350-kv maximum) transmission voltage had not only been decided upon, but the first section of 330-kv line some 60 miles long had already been completed and placed in operation, temporarily at 138-kv, during May 1952.

The carrying capability of some 500,000 kw per circuit, or 1 million kw per double-circuit line for line sections of

[6] Sporn, Philip, and A. C. Monteith, "Transmission of Electric Power at Extra High Voltages," Transactions AIEE, Vol. 66, pp. 1571–1582, 1947.

[7] East Central Area, Reliability Coordination Agreement, August 1, 1967.

[8] Sporn, Philip, H. P. St. Clair, and E. L. Peterson, "American Gas and Electric Company 330 Kv Extra-High Voltage Transmission System. Analysis of System Economics, Choice of Voltage, Basic Elements of System Design." Conférence Internationale des Grands Réseaux Electriques à Haute Tension, Paris, May 28–June 7, 1952.

50 to 75 miles in length, which 330-kv transmission supplied, made it possible to provide a very stiff and dependable transmission system consisting of two double-circuit
lines between each generating station and the diffusion
plant. The resulting transmission is shown in Figure 1: a
total of eight 330-kv circuits and four tower lines entering
the plant. Because of the distance of more than 140 miles
from the Clifty Creek station to the Portsmouth Area, it
was necessary to provide the intermediate Pierce switching
station. This reduced the length of line section that would
be switched out under fault conditions from approximately
145 to around 72 miles.

INTERCONNECTIONS   The strength and dependability of
this power supply system are assured not only by the two
generating stations and the adequate transmission between
these stations and the load, but are further and substantially
increased by the several interconnections with adjacent
power system networks. Both ends of the system are tied in
directly at 330-kv, the western connection at the Dearborn
switching station being tied directly to the 330-kv bus of
the Tanners Creek 500,000-kw generating station, and the
eastern connection, by means of a 12-mile double-circuit
330-kv line, being tied in directly to the 330-kv bus of the
Philip Sporn originally 600,000-kw, but at the present time
1,075,000-kw, generating station.

   In addition to the direct 330-kv interconnections, ties at
the lower voltage of 138-kv were established with the strategically located systems of four other sponsors.

RELIABILITY   The ability of the transmission system, as
finally decided upon in accordance with Figure 1, to carry
the required loads and maintain satisfactory voltage condi-

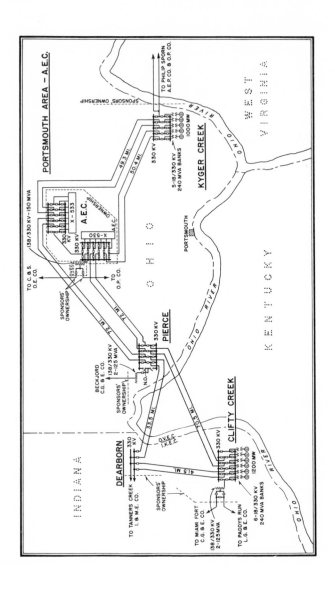

the inability of the Commission to enter into any agreement in excess of 25 years. This created a double problem, the need to minimize the effect of this short amortization period and also the need to protect the residual interest of AEC in a fully amortized plant that was still productively operative.

11. The ineluctable need to work out provisions for cancellation in view of the uncertainty with regard to future need of enriched U-235 (see previous Item 4).

PROBLEMS STEMMING FROM THE NEED OF PROVIDING THE UTMOST FLEXIBILITY

12. The cancellation provision and the need for multi-party protection, fully rationalized, but designed not to become an onerous burden (see Item 4).

13. The problem of reduced load factor and of replacing load upon cancellation and the effect on ability to meet minimum takes under long-term coal contracts.

14. The problem of transmitting large blocks of canceled power to a diverse and extended group of 15 utility sponsors without imposing an economic burden on members located farthest from the two generating plants.

15. The need to provide a mechanism for pickup of load dropped or canceled by the Commission.

All of these problems were tackled on the basis of extended study, the development in some cases of carefully evaluated projections of trends or experience and long negotiation discussions which led to agreement among sponsor companies, between OVEC and sponsors, between OVEC and suppliers such as the coal companies and the barge lines, and between OVEC and AEC. But engineering

and economics and equities were all taken into fullest consideration. An early general agreement to look for and accept
no windfalls materially helped in negotiations that led to
contractual agreement which has stood the test of time
remarkably well. A number of the most important problems
are now discussed in more detail.

The Fuel Problem — Coal

It is fortunate that except for organization, the marshaling of a responsible sponsoring group, and financing, it
was recognized from the very beginning — from the very
early stages of the conception of OVEC — that the basic
problem above all others was a fuel problem and that success
or failure of the project would hinge in a very large measure
on finding the proper sources of fuel and at proper prices
for the power plants that would generate the power needed
by the Commission at Portsmouth.

My visualization of the OVEC project in September–
December 1951, some months prior to the first meeting with
the AEC, definitely included the concept of a two-plant system to draw upon the southern Indiana–western Kentucky
coal fields to supply a plant either in Indiana or Kentucky
and on the eastern Ohio–western Pennsylvania–Kanawha
Valley, West Virginia, coal fields to supply a second plant
located in southeastern Ohio. Indeed, one of the strong
bases for recommending to the AEC that the Ohio Valley be
chosen as the general locus of a third diffusion plant by the
special committee that was charged with the siting task was
the general knowledge of the availability of reasonably
priced coal in the areas mentioned. And this was further
confirmed by Mr. Paul Weir, of Chicago, the coal consultant
to the American Electric Power System whom the author
brought into the problem shortly after January 23, 1952.
But this needed more solid confirmation.

Hence, almost simultaneously with the meetings with the AEC toward the end of January 1952, meetings were also held with various representatives of the coal industry in New York and Cleveland.

The meetings with the coal people were held in order to become better acquainted with coal reserves, mining methods, mining equipment, and the probable range of coal prices and to see if the coal industry was interested in furnishing the coal requirements of a project the size of OVEC.

It must be remembered that by any standards, and certainly by the standards of 1952, the quantities of coal required for OVEC's plants were very great and as finally determined amounted to 3 million tons per year at Kyger Creek Plant and 4.2 million tons per year at Clifty Creek Plant. It was not expected that coal in such great quantities would become available to any significant degree from any existing coal mines.

The coal industry at this time had not quite recovered from the flash of prosperity that followed the postwar boom which had already begun to slow down. It had not quite given up the hope of a continuing domestic heating market, did not yet get itself into focus to visualize the booming utility market of the 1960s, and definitely could not even conceive the great threat to its markets that atomic power was preparing to become. And except for a relatively small number, the opening of a new major mine filled the typical coal mining management with fear and apprehension. This was the atmosphere into which we launched our request for bids.

Invitations to bid were sent to 73 coal companies. Mr. Weir and the AEP Service Corporation summarized and analyzed the coal bids. While bidding was still open, a few coal companies advised that the coal from their mines would not meet the coal specifications outlined in the invitation to bid. These companies were asked to submit bids on the basis of coal which they would have available. It later developed that these companies in each case submitted bids considerably higher in price than the bids which were finally accepted.

Proposals were received from a bare 21 coal companies, and another 12 companies replied to say that, for one reason or another, they could not bid at that time. Five bids were accepted: two for the supply at Kyger Creek and three for the supply at Clifty Creek. But the negotiations which were carried out following preliminary acceptance were in four of the five cases long, difficult, and at times almost frustrating. Particularly, difficulty was involved in developing fair escalation clauses and especially ideas with regard to negative escalation, that is, sharing in increased productivity.

After OVEC knew where the coal would be produced, an invitation to bid on the barging of coal from the mines to the power plants was prepared and sent to 20 barging companies. Contracts were awarded to three barge lines. Escalation was a formidable problem here also.

The complete working out of the OVEC coal supply from its inception before the first response to AEC's request was made to a time just prior to the first delivery of coal represented an interesting technoeconomic problem from many standpoints. Although the indications were that we would be flooded with offers of coal, we received serious responses or expressions of interest only from less than 30% of those addressed. Eventually, in spite of the putative scores of mines waiting for coal orders, more than half of the coal had to be obtained from brand-new mines tailored to meet objectives set and agreed upon in negotiations, and later still another major mine had to be opened up to take care of another 20%.

In the case of one company that finally agreed to undertake the supply of 1.5 million tons of coal, the final agreement to do so came only after intensive and extensive hortatory pressures upon the chief executive. The solidity of the judgment to yield to that pressure was brought out by the statement of the same chief executive some ten years later that but for this contract he never would have survived the low period through which the coal industry passed for some ten years after this contract was executed.

What was true with regard to the coal was equally true with regard to the transportation facilities, and the three barging companies had to make an investment of over $20 million in barges and towing equipment in order to meet their contractual commitments to OVEC for coal transportation.

It is almost obvious that, under the conditions of indispensable extensive new mining operations and new transportation facilities, long-term contracts would become a must in order to provide a prudent basis for additional and large capital investment. This became involved by requirements to provide for cancellation by AEC when, even after the sponsors picked up the canceled capacity in full, they would still be unable to burn the previously contracted quantities of coal owing to the much poorer load factor of their system loads as contrasted with AEC's substantially 100% load factor. And this not only added to the burden of the negotiators, but offered a challenge also.

Finally, when all this investigation, appraisal, negotiation, and agreement with regard to coal and transportation had been completed, the coal situation stood as follows before and after the October 1952 escalation due to a UMW wage increase on October 1, 1952, and 14 years later at the end of 1966:

	Kyger Creek		Clifty Creek		Weighted System	
	$/ton	¢/MBtu	$/ton	¢/MBtu	$/ton	¢/MBtu
Cost before escalation of 10/1	4.265	18.98	3.83	17.40	3.97	18.06
Cost after escalation	4.705	20.47	4.02	18.24	4.15	18.91
Calendar year 1966					4.70	20.45
Total increase in 14 years						8.13%
Average increase per year — noncompound						0.58%

The answer to the question, how the cost increase was held to so low a level, lies mainly in three directions: (1) careful audit and analysis of books on escalation claims; (2) renegotiation at opportune times; and (3) negative escalation: sharing in savings due to increased productivity.

The OVEC Power System: Conversion Plant, Integrating Transmission System, Interconnections, Reliability, Performance

GENERAL ARRANGEMENT As has already been indicated from a discussion of the fuel problem, a decision had been made following the AEC's decision to locate the diffusion plant at Portsmouth to develop the generating capacity in two stations, both located on the Ohio River. The location of the two plants on the Ohio River made possible the taking advantage of experience gained in building major steam plants along the Ohio River or its principal tributaries for a period of over 30 years. Thus, the experience gained by many of the participating companies in building generating stations on the Ohio River and its tributaries (in particular those stations of the AEP System in which the size of units employed is practically the same as OVEC machines) was fully drawn upon in planning, designing, and constructing the two stations.

CONVERSION PLANT   The turbine generators selected were cross-compound with a gross capability of 217,250 kw at 1.5 inches of mercury back-pressure. They operate at 2000 psi, 1050°F at the throttle and reheat to 1050°F, the high-pressure unit at 3600 rpm and the low-pressure unit at 1800 rpm.

The thermodynamic cycle on which they operate is essentially the well-known regenerative cycle with seven stages of feed water heating to provide a final feed water temperature of 469°F and an over-all plant performance of close to 9200 Btu per kwh.

The search for an advanced and highly efficient plant was materially stimulated by the very high load factor at which the plants were planned to be operated and by the relatively low capital carrying charge owing to the low equity ratio and the almost complete elimination of the federal income tax component of cost. With an 18×10^9 kwh figure of annual generation and 20¢ per million Btu fuel, a saving in the two plants of one Btu per kwh could be prudently bought at a total capital cost of $50,000 — an extremely high figure, but fully justified under the prevailing conditions.

INTEGRATING TRANSMISSION   Transmission from the two generating stations to the diffusion plant runs approximately 140 and 30 miles. Except for certain 230-kv developments that were under way in the western part of Indiana, the highest existing transmission voltage level in this entire area was 138 kv—obviously inadequate to handle the transmission requirements of this over-all project economically. For the distance of transmission involved, comparative studies of both 230-kv and 330-kv transmission showed clear, even though small, advantages in cost in favor of 330-kv. The decision to go to the higher voltage was based on performance. The smaller number of circuits and rights of way required, together with the ability to interconnect at the same transmission voltage level with strategic generating and switching points on both ends of the project system, were telling advantages. The substantial capacity of even a single circuit was another. And we acted on the basic principle that, economics being approximately equal, sound judgment dictates choosing the highest practicable voltage.[6-8]

It was perhaps a fortunate circumstance, therefore, that the growing inadequacy of 138-kv transmission for the normal growth of load in much of this area had already been recognized some time before, and an extensive research and development program on higher voltage transmission had already been undertaken and largely brought to a conclusion prior to the inception of this new project. As a matter of fact, the 330-kv (350-kv maximum) transmission voltage had not only been decided upon, but the first section of 330-kv line some 60 miles long had already been completed and placed in operation, temporarily at 138-kv, during May 1952.

The carrying capability of some 500,000 kw per circuit, or 1 million kw per double-circuit line for line sections of

[6] Sporn, Philip, and A. C. Monteith, "Transmission of Electric Power at Extra High Voltages," *Transactions AIEE,* Vol. 66, pp. 1571–1582, 1947.

[7] East Central Area, Reliability Coordination Agreement, August 1, 1967.

[8] Sporn, Philip, H. P. St. Clair, and E. L. Peterson, "American Gas and Electric Company 330 Kv Extra-High Voltage Transmission System. Analysis of System Economics, Choice of Voltage, Basic Elements of System Design." Conférence Internationale des Grands Réseaux Electriques à Haute Tension, Paris, May 28–June 7, 1952.

50 to 75 miles in length, which 330-kv transmission supplied, made it possible to provide a very stiff and dependable transmission system consisting of two double-circuit lines between each generating station and the diffusion plant. The resulting transmission is shown in Figure 1: a total of eight 330-kv circuits and four tower lines entering the plant. Because of the distance of more than 140 miles from the Clifty Creek station to the Portsmouth Area, it was necessary to provide the intermediate Pierce switching station. This reduced the length of line section that would be switched out under fault conditions from approximately 145 to around 72 miles.

INTERCONNECTIONS The strength and dependability of this power supply system are assured not only by the two generating stations and the adequate transmission between these stations and the load, but are further and substantially increased by the several interconnections with adjacent power system networks. Both ends of the system are tied in directly at 330-kv, the western connection at the Dearborn switching station being tied directly to the 330-kv bus of the Tanners Creek 500,000-kw generating station, and the eastern connection, by means of a 12-mile double-circuit 330-kv line, being tied in directly to the 330-kv bus of the Philip Sporn originally 600,000-kw, but at the present time 1,075,000-kw, generating station.

In addition to the direct 330-kv interconnections, ties at the lower voltage of 138-kv were established with the strategically located systems of four other sponsors.

RELIABILITY The ability of the transmission system, as finally decided upon in accordance with Figure 1, to carry the required loads and maintain satisfactory voltage condi-

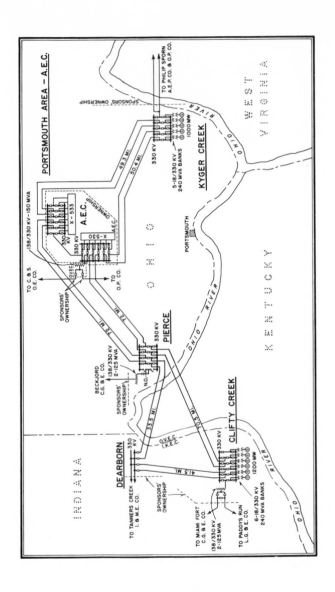

FIGURE 1

Diagrammatic view of OVEC-IKEC generation and transmission facilities. Essentially these consist of

(a) Two generating stations, Clifty Creek, near Madison, Ind., and Kyger Creek, near Gallipolis, Ohio.
(b) Two double-circuit 330-kv transmission lines from each station to the Portsmouth Area.
(c) Intermediate switching stations at Dearborn and Pierce.

The OVEC-IKEC system interconnects with the systems of the participating companies at 330 kv at Dearborn and Kyger Creek, and at 138 kv, at Clifty Creek, Pierce, and at the Portsmouth area.

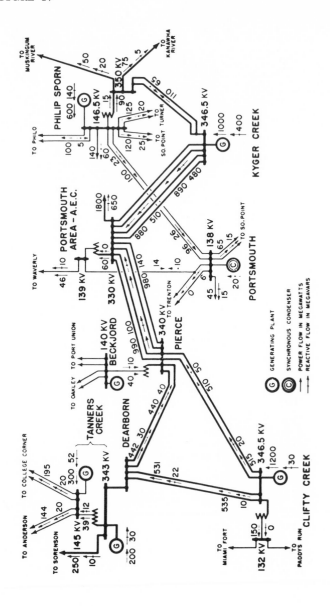

FIGURE 2

One of the scores of network analyzer studies made in planning the OVEC-IKEC system. This particular study shows power flow and voltage under normal conditions, with full generation of both power stations, and all lines in service.

Note:

(a) The flat level of 330-kv voltage at

    Dearborn   343 kv
    Clifty Creek   346.5 kv
    Pierce   340 kv
    Kyger Creek 346.5 kv
    Portsmouth Area 330 kv

(b) The balanced loading of the eight circuits delivering 1800 mw to Portsmouth.
(c) The light loading and direction of flow of the protective ties for OVEC.

tions under both normal and emergency conditions, was thoroughly tested by extensive network analyzer studies. It was necessary, of course, to include in network analyzer representations not only the OVEC system itself but also the actual or equivalent system networks to which the OVEC system was interconnected.

A typical study diagram showing power flow under normal line conditions and with all generating units in service is reproduced in Figure 2.

PERFORMANCE The success of the project's main objectives to strive for maximum economic yield consistent with and supportable by sound economics can be judged by the fact that over the past years 1957–1966 the thermal performance was at a thermal efficiency of 37% or better in five of the ten years and averaged 36.93% over the whole span. Over the entire period the plants have held the world record in high system efficiency, this on design projected and finalized in 1952.

ANTICIPATING AND DESIGNING TO AVOID POLLUTION The current preoccupation with air pollution has largely resulted from an awakening to the fact that the possible evil effects of pollution have been accumulating since the industrial revolution. From this has grown the almost panicky urge of governmental units to solve the air pollution problem by setting up rigid controls on chemical compositions of raw primary fuels.

But air pollution problems have been a concern for a long time to socially responsible organizations. They were not neglected by OVEC in 1952.

Not only were the two OVEC project plants going to be the two largest thermal electric plants in the United States at their completion, but due to the lower quality of coal (as much as 4% sulfur) that was dictated by the location of the plants it was expected that at least twice the amount of sulfur dioxide would be discharged as had been discharged by any previous plant. Further, their locations in predominantly rural areas ensured that any inadequacies in the disposal of the flue gases would be glaringly apparent. On top of that, not only would these new plants be larger than any of their predecessors but they would be built to their full capacities at one time. Where plant additions follow one another, there is the advantage of moving into the following stages more gradually and thus incorporating those changes which appear desirable on the basis of operating experience.

In order to deliver the power in accordance with the Atomic Energy Commission's schedule, all design decisions had to be made without the benefit of the experience of any portion of the plant being completed. Work on the pollution problem was, therefore, begun almost as soon as the AEC contract was executed. An extensive program of collecting meteorological information concerning the two sites was begun and completed. This was followed by extensive wind tunnel studies of the site at Madison, Indiana, subsequently named Clifty Creek.

The extensive program of wind tunnel tests led to a conclusion that at Clifty any stack height above 650 feet would be adequate for the disposal of, that is, sufficiently to clear and unfailingly to carry away, the stack effluent. This, however, was followed by a very thorough theoretical investigation by a group of air pollution and meteorological consultants on the diffusion of SO_2 to the upper atmosphere so as to provide a maximum SO_2 concentration at the ground level of 0.5 ppm, and on this basis a stack height of

690 feet was chosen. In order to get more expeditious approval of the Federal Aviation Agency this was later reduced to 683 feet. Even so, the Clifty stack became and for many years remained the tallest stack ever built for any purpose whatsoever. Its construction had, incidentally, an almost negligible adverse effect on the economics of the OVEC operation.

I have gone into this detailed discussion of the solution of the SO_2 problem at Clifty in 1952 because of the demonstration it provides that economics — in this case the ability to utilize low-cost heavy sulfur-bearing coal — can act as a great drive to technologically advanced solutions of many difficult problems, often in the absence of a substantial fund of science and any extensive operating experience. The solution adopted here not only solved our problem completely but today, 15 years later, has become one of the few milestones on the road presently available to a pragmatic solution of disposal of SO_2.[9,10]

The contractual arrangements that were developed that are specifically related to power supply are embodied in three principal agreements. The most important is the Power Agreement between OVEC and AEC. Another agreement is that made by OVEC with IKEC, a wholly owned subsidiary of OVEC, under which OVEC is obligated to purchase power produced by IKEC. And there was also executed an Inter-Company Power Agreement among OVEC and the 15 participating companies defining

[9] Sporn, Philip, "The Tall Stack for Air Pollution Control on Large Fossil-Fueled Power Plants," A Compilation of Papers including two, and an introduction, by Philip Sporn, May 1967. Printed by National Coal Policy Conference, Washington, D.C.

[10] Sporn, Philip, "Air Pollution Control Needs Taller Stacks," *Electrical World,* Vol. 168, No. 20, pp. 105–107, November 13, 1967.

respective rights and obligations. Only the Power Agreement between OVEC and AEC will, because of its many engineering-economic phases that were involved, be discussed in detail here. The other two almost explain themselves by title.

In determining the capacity required to provide adequate service to the Portsmouth Area, which was expected to operate at a load factor in excess of 95% (it actually was always well in excess of 99%) it was necessary to evaluate carefully the required generating capacity reserve. The operating range in load factors among existing utility systems extended from 50 to 72%; because of this every system is able to carry out a very considerable amount of maintenance during off-peak periods.

Because of the extreme importance of continuity of operation of the Portsmouth Area, and because of its very high load factor, analyses indicated that reserves considerably greater than those used for the normal utility system would be required. It was concluded that a reserve of 15% of the load should be built into the project generating stations and that the participating companies would provide a similar amount to be made up of the portion of the output of the project plants for which AEC does not pay a demand charge and from their own systems. The reserve furnished by the participating companies is known as supplemental power. This supplemental power is furnished at out-of-pocket cost and, in return for this service, the participating companies are allowed to use the reserve capacity that may exist from time to time in the project generating stations.

After taking into account the requirements at the Portsmouth Area, the electric losses incurred in delivering the power from the generating stations to the Portsmouth Area, and the reserve generating capacity to be included in the stations, it was concluded that the most economical solution was to install a total of eleven 200,000-kw units at the two generating stations. This resulted in the generating stations having a combined capability some 80,000 kw above the AEC requirements. The agreement provides that this portion of the project station capacity which is in excess of that required to meet AEC demands will be available to OVEC. All energy furnished to the Portsmouth Area generated at the project generating stations is called permanent power. The capacity in excess of AEC requirements which is available to OVEC is termed surplus power.

The contract between AEC and OVEC thus has, on the surface, the look of a typical cost-plus contract. In actuality it is about as far from being such an arrangement as the poles are apart. The principal reason is the flexibility clause which has made it possible for the Commission to bring its reservation from a one-time 1950 mw to a current 580 mw and scheduled to decline to 500 mw by December 1, 1968. Thus, from a matter of pure self-interest alone, OVEC could not possibly do anything less than the very optimum in design, construction, and operation of the OVEC facilities without imposing heavy burdens on its sponsors — currently, collectively three times as heavy on its sponsors as on AEC.

Under the Power Agreement, the rates payable by AEC are based on the distribution of fixed charges and operating costs allocated in the ratio of capacity required to meet the AEC requirements and the total capability of the plant. The rate for power is divided into two components: a demand charge and an energy charge. The demand charge represents fixed charges and operating expenses of OVEC on that portion of the plant and equipment needed to meet AEC's power requirements in the form of interest and amortization of the debt, insurance, property taxes, and a fixed return on the limited amount of equity capital. The energy charge represents the cost of fuel consumed in the generation of electric power and is limited to the energy generated for AEC's requirements. Under the terms of the contract, the demand and energy charges payable by AEC at the end of each calendar month are based on costs of the project estimated at the time the agreement was signed. These demand and energy charges were to be adjusted to reflect any changes from the estimates in the actual cost of new generating and transmission facilities and all operating expenses.

The contract is for a period of 25 years. AEC was, however, given the option of extending the contract after the initial 25-year period for three successive periods of five years each, under terms which will give full credit to AEC for the reduction or elimination of interest and amortization charges on borrowed capital during the initial term of the contract. The AEC would otherwise not be fully protected unless these options to extend were provided for, since the amortization charges are on the basis of 25 years, while the life of the facilities should be considerably longer. The contract is subject to cancellation or reduction for the convenience of AEC upon two years' notice, at any time after or prior to full-scale operation.

In the event of cancellation after full-scale operation, AEC may continue to receive power during the two-year notice period at the rates provided in the contract. However, it may also elect, at any time during the notice period, to serve a further notice to OVEC terminating all further power deliveries. In the latter event, AEC would, in lieu of the normal demand and energy charges, be liable during the unexpired portion of the notice period for a "modified" demand charge.

In addition to these payments during the notice period,

AEC is liable for a termination charge payable after the notice period. The termination charges are predicated on a fairly optimistic assumption that the participating companies' system load growth will be at a rate of about 6%, compounded annually. If load growth of the combined systems of the participating companies actually exceeds the 6% rate of growth, the contract provides for an appropriate reduction in the termination charges to be paid by AEC.

The cancellation provisions also recognize that OVEC entered into long-term coal agreements and arrangements for handling and shipping coal in order to insure the coal supply at a reasonable cost. They also recognize AEC's obligation to see OVEC whole in the event of contract cancellation or reduction. Because of differences in load factor, it may be necessary for OVEC to curtail the delivery of fuel, and to pay the costs and fixed charges for extending the delivery schedules under these contracts. While OVEC receives protection, it is obligated to exercise every reasonable effort to reduce these costs to a minimum by continuing long-term fuel supply contracts to the fullest amount consistent with the fuel requirements of the project generating stations after termination of the power contract by AEC and by arranging with the participating companies, to the extent practical, to utilize the fuel made available as a result of such termination.

The cancellation provisions are the result of extensive study, deliberations, and negotiations and exhaustive discussion in order to arrive at arrangements in which both parties would be fully and reasonably protected under all contingencies envisionable. They have been discussed at some length here to demonstrate the thesis that, provided rational economic objectives are the aim, they can be found and agreed upon if forthrightly sought as they were here.

It is significant that these cancellation arrangements were the subject of extensive hearings before the Eighty-third Congress and, in July 1953, the Atomic Act of 1946 was amended to permit the AEC to confirm the cancellation provisions.

Financing

The basic form and outline of the plan of financing was fairly well established by May 12, 1952. Several months before that, negotiations were commenced with representatives of several large insurance companies relating to the structure, terms, and provisions of the proposed debt securities. Essentially it was accepted that the project facilities could be built for a maximum amount of $400 million; working capital was needed to the extent of $20 million. In order to avoid difficult discussions in case of some unforeseen contingencies, it was also thought best to set up financing on the basis of a maximum cost of $440 million.

Since discussions with AEC brought forth the clear position that the Commission, in arriving at a decision as to the acceptability of the costs proposed, would prefer not to have to make any adjustments for taxes paid, it was decided to keep taxes paid down to the very minimum feasible; hence the decision to have $420 million debt and only $20 million of equity. But here we ran into an almost unmovable stumbling block. This was the problem presented by the possibilities of cancellation and the necessity under those conditions of reducing the debt ratio to something materially lower than 95%. This was rather neatly solved by introducing a new concept, $60 million of 14-year bank credit, obtained from a group of banks and pension trusts. The fact that this debt had to be amortized while main-

taining regular amortization payments in the bonded in-
debtedness was solved by maintaining the mortgage debt
level at 85% while amortizing the bank debt. An agreement by the equity owners to invest 50% of the 8% annual
return on their equity in notes subordinated to the bank
debt was of significant help in making possible the relatively long-term bank credit. And the other problems were
likewise resolved on the basis of proper recognition of the
preferred right of the mortgage lenders as against the un-
secured lenders, with their higher interest rate and much
earlier pay out.

Thus, in July 1953, after OVEC and the participating
companies had effected filing with, or secured approvals
from federal and state regulatory agencies in 17 separate
proceedings, 39 insurance companies and other institutional
investors committed themselves to purchase up to $360
million principal amount of first mortgage and collateral
trust bonds, bearing interest at the rate of 3.75% per annum,
and maturing in 1982. Also, 14 banking and similar institutions committed themselves to lend up to $60 million,
such loans bearing interest at the rate of 4% per annum
and maturing in 1967. Equity funds were to be made
available by the participating companies in an amount up
to $20 million so that an aggregate of $440 million became
available to supply funds for the construction of the facilities
and the required working capital. Later, in 1954, the interest
on the $60 million bank loan was renegotiated to a rate of
3⅝% and the $20 million of equity was reduced to $10
million or 2.5% and its place taken by a $10 million de-
mand loan. This, in essence, provided the capital for the
two plants' coal piles. The effect of these and of a number
of other minor steps was to reduce the price of energy to
AEC by $1,466,000 per year. These were but another

series of steps on the road to the set objective of delivered energy at 4 mills per kwh or less.[11,12]

It might be interesting to point out that this method of financing a project of this magnitude, while simple in basic concept (but because it involved the 15 participating companies), introduced many problems connected with regulatory agencies, involving not only the FPC and SEC but regulatory commissions in the seven states of Ohio, Indiana, Kentucky, Pennsylvania, Virginia, West Virginia, and Maryland. The fact that these separate proceedings could, and were, all in due course brought to satisfactory conclusion is, it is believed, a significant commentary on the workmanlike job in conceiving, negotiating, and bringing into effective agreement the many complex arrangements on the part of the managers of the project and on the workability of democratic procedures, even when brought to bear on a complex situation of this kind.

It is perhaps time to summarize:

A good question to ask in this particular situation is whether private enterprise, operating under all the rules and regulations to which private enterprise in electric power is subject in the United States, successfully fulfilled its basic and critical function of responding fully to the call of a vital national interest. Answering this question is not simple because the quality of the job cannot be measured by any precise or single standard. But there are, fortunately, many lesser measures, perhaps as many as half a dozen, and the question can be asked how the job stands up

[11] Waterman, Merwin H., "Ohio Valley Electric Corporation, A Case Study in Developing and Financing Private Power for a Public Purpose," Bureau of Business Research, University of Michigan, Ann Arbor, 1966.

[12] Ohio Valley Electric Corporation, Inter-Company Bond Agreement, July 10, 1963.

when measured by each of these. Out of an integration of these partial answers, I believe, one can reach a conclusion about the over-all effectiveness of private enterprise.

The first and the most natural measure that comes to mind is cost. Applying this test of performance, we find at the time the contract was signed in October 1952 the AEC was furnished an estimate of cost of 3.86 mills per kwh delivered. Some eight months later, in response to a question from the Congress, the AEC advised that body that on the basis of the current figures the energy cost would be about 4 mills per kwh. In general, almost from the beginning, it was known that, difficult as it might be to find general acceptability of the aim, the job cost over a 25-year period would have to be below 4 mills per kwh.

It is of course too soon to turn in the 25-year record, but in Table 1 there is shown a ten-year summary of per-

TABLE I

Ohio Valley Electric Corporation: A Ten-Year Comparative Summary of Performance

| | | Total Power Delivered by OVEC to AEC | |
Year	Kw Demand	Total Energy Delivered Billions of Kwh	Unit Cost-Basic Agreement Mills/kwh
1957	2,137,000	17.064	3.985
1958	1,960,000	17.040	4.029
1959	1,958,000	17.060	4.041
1960	1,957,000	17.122	4.018
1961	1,910,000	16.642	4.008
1962	1,809,000	15.769	3.985
1963	1,807,500	15.760	3.981
1964	1,806,750	14.241	3.947
1965	1,206,750	10.510	3.971
1966	1,203,000	7.923	3.864
Arithmetic mean			3.983

formance on cost during the period 1957–1966. During eight of these ten years a demand in excess of 1.8 million kw was supplied to the Commission, which in 1957 reached a peak of 2,137,000 kw. During this period the cost per kwh varied from a low of 3.864 mills to a high of 4.041, with an arithmetic mean for the ten years of 3.983 mills.

In the face of the many increases of the general level of prices in the United States throughout this period and the increase in local taxes, property and other, in Indiana, Ohio, and Kentucky, it is quite clear that this was a first-rate performance.[13]

One of the major reasons why the AEC decided to contract with OVEC for its power supply was OVEC's ability to obtain interim power from the facilities of its sponsoring companies while the special generating facilities were being constructed. Because the increased production of fissionable materials was at that time a matter of great urgency, the availability of this energy became an extremely important item.

How did OVEC perform on the delivery of interim power? Within two months after the execution of the contract between OVEC and the Commission, OVEC was able to deliver its first power to the Commission, starting with a modest 30 mw (all that the AEC could use) but reaching a figure in excess of 1 million kw in September 1955, even though the sponsoring companies were only committed to furnishing 465 mw. The outstanding performance in the delivery of a huge block of interim power made a significant contribution to the rapid build-up of the nation's stockpile of uranium-235. Here again, not only was the national defense aided significantly, but the job

[13] Ohio Valley Electric Corporation, Annual Report for the Year 1966.

was carried out to a degree beyond expectation and far beyond commitment.

What about service continuity — still another very important item in any evaluation of OVEC's performance from the Commission's standpoint? To provide an adequate amount of stand-by capacity the Commission had reserved, out of the facilities provided by OVEC, the capacity margin of 15% of its contractual demand, or a margin of 274 mw. Although this reserve was established in the original contract only after extensive studies had been carried out by the OVEC-AEC negotiating group, it was recognized that there would be many occasions when 15% would not be adequate to protect the continuity of supply. To provide this continuity the sponsoring companies agreed to deliver additional reserve out of their own systems when necessary. How did the OVEC supply system fare in this regard?

Since the placing of the full capacities of OVEC in commercial operation, there have been many occasions when grave and critical curtailment might have resulted at the gaseous diffusion plant had not surplus capacity been made available to it. The amounts of such emergency reserves that the sponsoring companies supplied to keep the AEC production schedule undisturbed ranged from as little as 132 mw in 1964 to over 820 mw in 1958. This record of completely backing up the facilities of the Commission without imposing additional fixed charges is still another outstanding example of the service which the interconnected system of the sponsoring companies was able to render to the Commission.

Perhaps the outstanding achievement in the record of OVEC's fulfillment of its contractual commitment to supply the Commission's energy requirements at Portsmouth has been the flexibility of the arrangement and its sharp responsiveness to the Commission's needs as they expanded beyond expected requirements — and as they later contracted.

Originally, the Commission felt that, while it could make a commitment for a period of 25 years for as large a block of power as 1800 mw, yet, because of the many uncertainties relating to defense planning, it would be necessary to include in the contract enough flexibility to permit cancellation of the entire commitment even before the facilities created for the sole purpose of supplying that demand had been completed. As it turned out, the AEC did not have to take advantage of this provision. Indeed, as matters developed, the reverse occurred; before the plant facilities were placed in commercial operation the Commission asked to be supplied with an additional block of capacity from OVEC facilities in the amount of 150 mw. Not only was this worked out satisfactorily, but included with it was an additional block of capacity which could be used to advantage to speed up the production of the stockpile of the necessary defense materials; so that the maximum capacity furnished to the Commission for more than a year ended up at 2100 mw, some 300 mw above the original demand.

This was the peak of demand and supply. However, as pressure on the Commission for production of enriched uranium material eased, the power demand was reduced first to 1950 mw and then to 1900 mw, until by January 1, 1962, it was brought back to the original contractual amount of 1800 mw. All this was worked out to the full satisfaction of the Commission's needs and with full accommodation to its requirements.

The final test of OVEC's flexibility occurred more than ten years after the plant went into commercial operation. As the demand on the Commission's material production facilities began to diminish, it became desirable for the Commission to reduce sharply its contractual power commitment. Although some indication of this likely development had been given OVEC, the definite notice did not come until the President's announcement in January 1964 of a 25% cut in fiscal 1965. This cut was deepened in the course of the subsequent year, and in the period 1964–1965 a number of involved discussions and rearrangements of contractual commitments took place. Following a notice received in February 1965, arrangements were worked out to the present schedule, under which the Commission was receiving 700 mw by August 1967, but which was reduced to 580 mw by September 1, 1967, with the expectation that it will be reduced to 500 mw by December 1, 1968. This will represent a decline in the Commission's demand of close to 75% as compared to the original 25-year contractual demand of 1800 mw. And all this was accomplished, as in other important matters, at OVEC's initiation, without the payment by the Commission of a single dollar of cancellation charge. Indeed, in the final arrangement, the sponsoring companies agreed that, if the Commission's requirements for enriched material should change owing to the development of a major peacetime demand for atomic fuel, the power released by AEC could be reinstated in accordance with agreed-upon and tested principles, as long as proper notice was given.

This has come into focus most recently and sooner than had been expected. The Commission has found itself confronted with the need to plan for increasing the diffusion plant production to take care of the need of the civilian requirements for fissionable material. Under that arrangement negotiations were completed early in August 1967 under which specific stepping up of capacity at Portsmouth to 700 mw has been agreed upon to become effective January 1, 1971, with a program of further increases that will call for the following schedule:

Period (Inclusive Dates)	Megawatts
1/1/71– 9/30/71	700
10/1/71– 9/30/72	1000
10/1/72– 9/30/73	1300
10/1/73– 9/30/74	1500
10/1/74–10/14/77	1700

The interesting development in this connection was an estimated cost of energy under such a renewal program submitted to the Commission for the period starting January 1, 1970. The estimated costs are given in Table 2.[14]

TABLE 2

Estimated Cost of Energy to AEC Upon Extension of Original Agreement

Period	Estimated Cost of Energy (Mills/kwh)
1/ 1/70–10/14/77	3.78–3.99
10/15/77– 3/12/81	3.99–4.10
3/13/81–10/15/87	3.15–3.30

Obviously the most interesting figure in that table is the one effective with March 13, 1981, that being the day after the completion of amortization of OVEC's funded debt. By that time it is estimated that the plant should have at least a decade of highly efficient, useful, and effective life left without any remaining burden of plant fixed charges. Consequently the figures indicate costs close to what is currently

[14] Sporn, Philip, "Nuclear Power Economics — 1962 through 1967," Report to the Joint Committee on Atomic Energy, Congress of the United States, February 1968.

143

projected for breeder reactors, whose expected introduction on a major basis into the United States energy economy is expected to come at about that time. This, too, I believe, is a most unusual achievement.

It is submitted that in this situation an extraordinarily unique result was obtained in a situation where assumptions, perceptions difficult to get hold of and even more difficult to evaluate had to be made, but yet, which having been grasped, did have a dominant say in the result: a successful demonstration of private enterprise, and more particularly the ability of the investor-owned power industry, to respond to the challenge of national need and to do so with a high sense of dedication to the public interest and a high level of competence. This result, an improvement over what was sought 15 years earlier, was achieved only by the tightest interlinkage of technology, engineering, and economics.

Index

Chromatographic Methods in Clinical Chemistry and Toxicology

Chromatographic Methods in Clinical Chemistry and Toxicology

Roger L. Bertholf, Ph.D., DABCC
*Department of Pathology, University of Florida
Health Science Center, Jacksonville, FL, USA*

Ruth E. Winecker, Ph.D., DABFT
*North Carolina Office of the Chief Medical Examiner,
Chapel Hill, NC, USA*

John Wiley & Sons, Ltd

Library of Congress Cataloging-in-Publication Data

Chromatographic methods in clinical chemistry and toxicology / [edited by] Roger Bertholf, Ruth Winecker.

p. ; cm.

Includes bibliographical references and index.
  ISBN-13 : 978-0-470-02309-9 (alk. paper)
  ISBN-10 : 0-470-02309-0 (alk. paper)
1. Chromatographic analysis. 2. Clinical chemistry. 3. Toxicological chemistry. I. Bertholf, Roger. II. Winecker, Ruth.
  [DNLM: 1. Clinical Chemistry Tests. 2. Chromatography–methods. 3. Toxicology–methods. QY 90 C557 20007]

RB40.7. C57 2007
616.07'56–dc22   2006024892

British Library Cataloguing in Publication Data

A catalogue record for this book is available from the British Library

ISBN-10 0 470 02309 0   ISBN-13 978-0-470-02309-9

Typeset in 10/12 pt. Times by Thomson Digital
Printed and bound in Great Britain by Antony Rowe, Chippenham, Wiltshire
This book is printed on acid-free paper responsibly manufactured from sustainable forestry
in which at least two trees are planted for each one used for paper production.

Contents

Preface

Among the vast array of methods available for biochemical analysis, chromatography occupies a venerable station. Few analytical methods have had such a vital impact on the development of clinical chemistry and toxicology. The roots of chromatography can be traced to the work of the Russian botanist Mikhail Semenovich Tswett, who separated plant pigments on a calcium carbonate column in 1903. Applications of this new technique to biochemical analysis were readily apparent. Chemical analysis of biological specimens is difficult because they contain a complex mixture of components, many of which react similarly with reagents selected to detect a specific compound. Analytical chemistry is not especially complicated when dealing with pure solutions, but the presence of many potentially interfering substances, as in biological matrices, creates a challenge. Chromatography answered that challenge by offering a method to separate components of a mixture, providing a pure specimen that could be measured by any of a number of analytical methods.

In this book, we have assembled a cross-section of contemporary applications of chromatographic methods used in clinical chemistry and toxicology. Chapter 1 focuses on quality assurance and quality control, emphasizing the importance of validating analytical methods that are used for clinical and forensic purposes. Merves and Goldberger cite the work of F. William Sunderman, to whom all clinical chemists and toxicologists owe a great debt of gratitude for his pioneering efforts to establish performance standards for clinical laboratories, including the first proficiency testing program ever devised. Sunderman was a fierce advocate of quality in laboratory medicine, and his dedication to that goal was pivotal in establishing national standards for laboratory practice in the USA, which were first codified in the Clinical Laboratory Improvement Act of 1967.

Just as any technology adapts to the applications for which it is suited, chromatography has evolved to meet the special needs of biochemical analysis. Packed column gas chromatography has been replaced by capillary column methods, which emerged in the late 1970s when fiber-optic technology provided a way to manufacture glass columns with micron-sized bores and lengths of 60 m and longer. Capillary gas chromatography revolutionized the analysis of drugs of abuse, with the ability to resolve congeneric compounds. Similarly, liquid chromatography has given way to the superior resolving ability of high-pressure (or -performance, depending on your etymological preference) liquid chromatography (HPLC). High-resolution derivatives of electrophoresis, a technique closely related to chromatography, have made the transition from research tools into commercial applications. Chromatography continues to evolve, and the versatility of this analytical method seems uniquely suited to clinical chemistry and toxicology applications.

Donald Catlin established the first Olympic drug testing laboratory for the 1982 Summer Games in Los Angeles, and he continues to operate the premier sports testing laboratory in the world. Catlin and co-workers Chang, Starcevic and Hatton, describe the application of liquid chromatography coupled with mass spectrometry for the detection of anabolic steroids in urine. Detection of exogenous steroids is challenging because structurally, the compounds are so closely related, and metabolic products of performance enhancing steroids that appear in the urine are often the same as naturally occurring products of steroid metabolism. Out of Catlins laboratory have come many innovations in the application of analytical chemistry to sports medicine, and as a consequence he has the distinction of making contributions both to clinical toxicology and to the forensic applications that ensure fairness in competitive sports.

Amitava Dasgupta applied HPLC to the detection and measurement of popular nutritional supplements. This is an important area of clinical investigation because the effects of these compounds on clinical laboratory tests are largely unknown. Dasgupta has explored this uncharted territory, providing an insight into the measurable effects dietary supplements have on laboratory tests. Le Bricon and colleagues used HPLC to measure L-dopa and L-tyrosine, as markers of malignant melanoma. Early detection of cancer is essential to the success of treatment, and the appearance of specific markers for this aggressive malignancy provides an important new tool in cancer therapy. Zhou and Johnston applied capillary zone electrophoresis, combined with liquid chromatography and mass spectrometry, to measure proteins in plasma. Proteomic analysis is a promising innovation in clinical biochemistry, and the work of Zhou and Johnston will make an important contribution to that field of study. Helander used HPLC to detect abnormally glycosolated transferrin, which is associated with alcohol abuse. Chan and Ho applied HPLC to the measurement of catecholamines, which are specific markers of neuroendocrine tumors.

This volume also contains several applications of toxicological interest. Larry Broussard describes the analysis of alcohols and inhalants by headspace gas chromatography. Wollersen and Musshoff developed liquid chromatographic methods for measuring organophosphorus pesticides, and Ropero-Miller used HPLC to detect and identify neurotoxins that are used as biochemical weapons. Marinetti developed an elegant GC-MS method for measuring γ-hydroxybutyrate and analogs, a regrettably popular group of abused drugs.

Analysis of heavy metals in biological specimens traditionally involves electrochemical or atomic absorption/emission techniques, but a few innovative methods involving chromatography have been pursued. Caruso and co-workers used a combination of liquid chromatography and an inductively coupled plasma detector to measure arsenic, mercury, selenium and platinum in a variety of biological matrices. Of particular interest is their ability to distinguish between the different compounds in which the metal occurs. Chromatography is ideally suited to their approach, in contrast to flame absorption and emission methods, which do not discriminate between various forms of the metal. Aggarwal, Fitzgerald and Herold pioneered a different approach to chromatographic measurement of heavy metals by generating volatile chelates that are separated by gas chromatography and detected by mass spectrometry.

The twelve applications described in this book illustrate the versatility of chromatographic methods and the range of applications to clinical chemistry and toxicology available with these powerful analytical techniques. Our intention was to provide an overview of useful methods, while emphasizing the contributions that chromatography continues to make in clinical laboratory medicine. We are indebted to our contributors for sharing their exemplary work in this field.

Roger L. Bertholf
Ruth E. Winecker

List of Contributors

Suresh K. Aggarwal, Bhabha Atomic Research Centre, Trombay, Mumbai, India

Larry Broussard, LSU Health Sciences Center, New Orleans, LA, USA

Joseph A. Caruso, University of Cincinnati, Cincinnati, OH, USA

Don H. Catlin, UCLA Olympic Analytical Laboratory, Los Angeles, CA, USA

Eric C. Y. Chan, National University of Singapore, Singapore

Yu-Chen Chang, UCLA Olympic Analytical Laboratory, Los Angeles, CA, USA

Amitava Dasgupta, University of Texas Health Sciences Center at Houston, Houston, TX, USA

Robert L. Fitzgerald, University of California, San Diego and VA San Diego Healthcare System, San Diego, CA, USA

Jean-Pierre Garnier, Hôpital Saint-Louis and Université Paris V, Paris, France

Bruce A. Goldberger, University of Florida, Gainesville, FL, USA

Caroline K. Hatton, UCLA Olympic Analytical Laboratory, Los Angeles, CA, USA

Anders Helander, Karolinska Institute and University Hospital, Stockholm, Sweden

David A. Herold, University of California, San Diego and VA San Diego Healthcare System, San Diego, CA, USA

Paul C. L. Ho, National University of Singapore, Singapore

Murray Johnston, University of Delaware, Newark, DE, USA

Thierry Le Bricon, Hôpital Saint-Louis, Paris, France

Sabine Letellier, Hôpital Saint-Louis and Université Paris V, Paris, France

Laureen Marinetti, Montgomery County Coroners Office and Miami Valley Regional Crime Laboratory, Dayton, OH, USA

Michele Merves, University of Florida, Gainesville, FL, USA

F. Musshoff, University of Bonn, Bonn, Germany

Jeri D. Ropero-Miller, Research Triangle Institute, Research Triangle Park, NC, USA

Borislav Starcevic, UCLA Olympic Analytical Laboratory, Los Angeles, CA, USA

Konstantin Stoitchkov, National Center of Oncology, Sofia, Bulgaria and EORTC DC, Brussels, Belgium

H. Wollersen, University of Bonn, Bonn, Germany

Katarzyna Wrobel, Universidad de Guanajauto, Guanajuato, Mexico

Kazimierz Wrobel, Universidad de Guanajauto, Guanajuato, Mexico

Feng Zhou, University of Delaware, Newark, DE, USA

1

Quality Assurance, Quality Control and Method Validation in Chromatographic Applications

Michele L. Merves and Bruce A. Goldberger

University of Florida, College of Medicine, Department of Pathology, Immunology and Laboratory Medicine, P.O. Box 100275, Gainesville, FL 32610-0275, USA

1.1 INTRODUCTION

Clinical chemistry and toxicology are also known as clinical biochemistry or clinical pathology, and they utilize a variety of analytical procedures to analyze body fluids to aid in medical prevention, diagnosis and treatment. Analytes commonly targeted include electrolytes, creatinine, albumin and other proteins, iron and prescription (therapeutic) and illicit drugs.

The focus of this chapter is on quality assurance and quality control issues facing clinical laboratories, with emphasis on chromatographic analysis. Chromatography is a versatile analytical technique with diverse applications including gas chromatography (GC), liquid chromatography (LC), GC or LC coupled with mass spectrometry (MS) and other qualitative and quantitative bioanalytical methods.

1.2 HISTORY

In his book, *Four Centuries of Clinical Chemistry*, Rosenfeld (1999) traces the evolution of laboratory medicine from medieval times to the present. Chemical testing in biological matrices began several centuries ago, with a very crude technique for detecting pregnancy. Egyptian women would urinate on wheat and other seeds, and then observe the seeds' growth patterns to reveal pregnancy, and it was supposed, the sex of the child. Subsequently,

Chromatographic Methods in Clinical Chemistry and Toxicology Edited by R. L. Bertholf and R. E. Winecker
© 2007 John Wiley & Sons, Ltd

Egyptians, Persians, East Indians, Chinese and other cultures observed changes in urine in order to assess disease states. Slowly, they developed a better understanding of normal and abnormal biological processes. The sixteenth and seventeenth centuries marked a major turning point in diagnostic medicine: it became clear that observation alone was not sufficient to make many diagnoses, paving the way for laboratory investigation to become an integral part of medical practice. As a result, research and clinical laboratory facilities were established, and their contributions led to fundamental technological advances in clinical laboratory medicine.

The early nineteenth century was an active time for clinical chemistry, when the role of biochemical processes in physiology began to emerge. Scientists and physicians started to realize the importance of biochemical imbalances in many diseases. However, interest in laboratory medicine languished in the middle of the nineteenth century, mostly due to the limited advancements in technology that could be applied to biochemical analyses. In its early stages, the laboratory did not provide much help to the clinician; therefore, few resources were invested in this new field. Eventually, technology would fuel a re-emergence of laboratory medicine and establish its essential role in medical practice.

The term 'clinical chemical laboratory' was first used by Johann Joseph Scherer in the mid-1800s. Scherer was appointed director of the first independent hospital laboratory, in Würzburg, Germany. Although the term did not become part of the general German vocabulary for another 100 years (German-speaking areas used the term 'pathological chemistry' before 'clinical chemistry' became popular), Scherer played a key role in the development of many assays important during his career.

Medicine has advanced exponentially with diagnostic testing, and has, in turn, prompted further development in clinical chemistry and toxicology. In the late 1800s, medical technologists were trained to perform laboratory tests, eventually leading to the creation of private medical laboratory services. Eventually, it became clear that for medical laboratories to succeed, they needed to adopt specifications that would ensure the reliability of the qualitative and quantitative results they report. Quality control and assurance benefit the laboratory, because they increase productivity (Westgard and Barry, 1986) and maximize the cost-to-benefit ratio of medical laboratory testing (Kallner et al., 1999).

The impact of analytical imprecision and bias on medical decisions is mostly unknown. A large margin of error may be acceptable in some circumstances, whereas other clinical scenarios demand more accurate and precise laboratory measurements. Often, clinicians interpret laboratory results within the larger context of the patient's history and physical examination, but the influence of imprecision in laboratory data on a physician's assessment of a patient is not well documented (Kaplan, 1999). Klee et al. (1999) evaluated the effect of analytical bias on medical decisions involving three common laboratory tests: serum cholesterol testing for risk assessment of cardiac disease, serum thyroid-stimulating hormone measurement for the detection of hypothyroidism and serum prostate-specific antigen testing for prostate cancer risk assessment. In each case, the authors concluded that tighter control of analytical variability resulted in a greater certainty in patient classification.

Ensuring quality results in clinical chemistry and toxicology requires optimization of every aspect of the laboratory operation, including maintenance, waste disposal, ana-lytical procedures, management, quality specifications, safety, cost management and overall strategies (Kallner et al., 1999). Kaplan (1999) suggested five tools that would be helpful to consider when writing guidelines for clinical laboratories: the effect of performance on specific clinical decisions, effect of performance on broad-based clinical decisions, survey of clinical opinions (i.e. when is an increase in error unacceptable), recommendations from professionals and societies and data from literature and proficiency testing.

Among clinical laboratories, there is a need for standardization to ensure that a result generated by one laboratory is comparable to the same measurement in another laboratory. As an example, therapeutic drug monitoring requires consistency of results over time, even when the measurements are being made in different laboratories (Groth, 1999). Several organizations are involved in establishing standards for clinical laboratory practice, including the National Academy of Clinical Biochemistry (NACB), the National Institute for Standards and Technology (NIST), the American Society of Clinical Pathologists (ASCP), the Clinical and Laboratory Standards Institute (CLSI), the International Federation of Clinical Chemistry and Laboratory Medicine (IFCC) and the International Organization for Standardization (ISO). These organizations are discussed in detail in a later section.

There has been significant progress in quality management over the last 50 years. For example, Levey and Jennings introduced statistical approaches to quality assurance in the 1950s. In the 1960s, Tonks, Barnett and Cotlove (among others) developed methods for controlling variability and established standards of quality (Westgard, 1999). In 1999, an international team of specialists with support from the International Union of Pure and Applied Chemistry (IUPAC) Clinical Chemistry Section, the International Federation of Clinical Chemistry and Laboratory Medicine (IFCC) and the World Health Organization (WHO) met for a 'consensus conference' to address laboratory quality goals (Kallner *et al.*, 1999). Other efforts towards improving laboratory quality are national, such as the set of guidelines for bioanalytical method validation in industry in the USA that was published by the Food and Drug Administration with help from the Biopharmaceutics Coordinating Committee in the Center for Drug Evaluation and Research in cooperation with the Center for Veterinary Medicine (FDA, 2001). There are also various guidelines written for different state and local jurisdictions.

Many challenges remain in establishing universal guidelines for laboratory quality. It should be noted that universal guidelines applicable to all analytical testing within clinical chemistry and toxicology cannot be set. Testing procedures have different goals that require unique specifications. For example, a semi-quantitative analytical protocol that involves a threshold concentration may only require the use of positive and negative controls, whereas quantitative assays have stricter specifications for linearity, accuracy and precision that require the use of multiple calibrators and controls.

1.3 DEFINITION OF QUALITY ASSURANCE AND QUALITY CONTROL

Quality assurance and quality control procedures are vital aspects of clinical chemistry and toxicology (Garfield *et al.* 2000; Tilstone, 2000; Ratliff, 2003). The terms quality assurance and quality control sometimes are used interchangeably, but the two terms have different meanings and should be used in different contexts. Quality assurance (QA) encompasses quality control, but also applies to virtually all aspects of laboratory operation. QA requires that all laboratory procedures are documented. Systematic internal and external review of these practices and documentation are essential to achieving optimal efficiency and accuracy (Rosenfeld, 1999). Key aspects of the QA process are laboratory accreditation, proficiency testing and staff education. Depending on the jurisdiction, there are many different types of guidelines under the umbrella of QA. For example, the laboratories or technologists may be required to be certified by the state or other agencies. In addition, continuing staff education is essential for maintaining and updating credentials, and proficiency testing may be required in order to maintain certification.

Quality control (QC) originated from the industrial goal of producing each batch of product with consistently high quality, detecting and correcting process errors that result in defective products. Therefore, QC is an essential component of a laboratory QA program, because it provides a mechanism for monitoring the quality of laboratory results. To ensure that a set of analytical results meets the desired specifications, control specimens are included in each step of the procedure. These controls should include both normal and abnormal concentrations of analyte (Goldberger *et al.*, 1997).

1.4 PROFESSIONAL ORGANIZATIONS

As mentioned in Section 1.2, various professional organizations are involved in establishing standards for clinical laboratory practices and key organizations include the NACB, NIST, ASCP, CLSI, IFCC and ISO.

The National Academy of Clinical Biochemistry (NACB) is the Academy of the American Association for Clinical Chemistry (AACC). Its mission is 'to elevate the science and practice of clinical laboratory medicine by promoting research, education and profes-sional development in clinical biochemistry and by serving as the leading scientific advisory group to AACC and other organizations' (http://www.nacb.org). It has six objectives which promote research, education, professional development, scientific advisory roles, recognition and infrastructure. Since the mid-1990s, the NACB has published several consensus-based laboratory medicine practice guidelines (LMPGs).

The National Institute for Standards and Technology (NIST) was established in 1901 as the National Bureau of Standards. The mission of NIST is 'to promote US innovation and industrial competitiveness by advancing measurement science, standards and technology in ways that enhance economic security and improve our quality of life' (http://www.nist.gov). The principal programs include NIST Laboratories, the Baldrige National Quality Program, the Manufacturing Extension Partnership and the Advanced Technology Program. As part of its commitment to the advancement of technology, NIST produces reference materials for bioanalysis. These standard reference materials are essential for validating the accuracy of analytical methods.

The American Society of Clinical Pathologists (ASCP) was founded in 1922 and its mission is to 'provide excellence in education, certification, and advocacy on behalf of patients, pathologists, and laboratory professionals' (http://www.ascp.org). The ASCP has developed many educational programs to advance the role of the laboratory in medical diagnosis (Rosenfeld, 1999).

In 1968, the National Committee for Clinical Laboratory Standards was established, and recently changed its name to the Clinical and Laboratory Standards Institute (CLSI) (http://www.nccls.org). The CLSI is a nonprofit organization that sponsors educational programs in order to promote national and international standards for clinical laboratories. This is a noteworthy task, especially since following these specifications is not mandatory and no enforcement agency exists to ensure each laboratory follows the CLSI guidelines. As part of its educational programs, the CLSI publishes consensus documents on laboratory procedures, reference methods and evaluation protocols. A major goal of CSLI is to encourage high quality in the laboratory in order to maximize patient care benefits (Rosenfeld, 1999).

The International Federation of Clinical Chemistry and Laboratory Medicine (IFCC) was established in the mid-1950s with goals to 'advance knowledge and promote the interests of biochemistry in its clinical (medical) aspects' (www.ifcc.org). IFCC was originally formed under IUPAC, but became independent in the late 1960s. However, it has maintained a close affiliation with IUPAC, and also WHO. Aside from its role in publishing guidelines and maintaining educational opportunities for its members, the IFCC plays a key role in developing national societies within countries without such resources.

The International Organization for Standardization (ISO) is a worldwide organization representing national standard societies from approximately 149 countries. ISO began in 1947 after a delegation of 25 countries met with the desire to 'facilitate the international coordination and unification of industrial standards' (http://www.iso.org). The ISO mission is to promote global standardization for exchange of services among nations. ISO publishes International Standards, which encourage intellectual, scientific, technological and economic exchange. For example, ISO 9000 is a set of four documents (9001–9004) that provide standards based on an international consensus for various aspects of quality assurance. In addition, ISO organizes Technical Advisory Groups to study quality control and assurance issues (Burtis and Ashwood, 2001).

1.5 INTERNAL QUALITY ASSURANCE AND CONTROL

Internal quality assurance and control procedures are designed to guarantee the reliability of the results generated by the laboratory; QA and QC ensure that the reported results contain the appropriate accuracy and precision through stringent monitoring of the analytical process, documentation of procedures and method validation.

1.5.1 Standard operating procedure manual

The standard operating procedure (SOP) manual contains the procedures validated by the laboratory; it is a complete set of instructions for pre-analytical, analytical and post-analytical methodology and also procedures for quality assurance/control, chain-of-custody and security. Each step in the handling of the specimen should be evaluated, optimized where possible and documented in the SOP. Important steps in the analytical process include collection, transport and accessioning of the specimen, sample preparation, isolation and detection of the analytes, production of the report and disposal of the specimen. This chapter focuses on the quality assurance and control issues for analytical method development and validation as well as statistical representation of the data.

1.5.2 Method development

Analytical methods are developed to meet specific needs. For example, pharmacodynamic and pharmacokinetic assessment of a new drug will require the ability to measure it in biological matrices. These methods should be developed under the 'Good Laboratory Practice' (GLP) standards issued by the US Food and Drug Administration (FDA), and provide accurate and precise results. Typically, constraints are placed on the new methods, such as pre-established administrative reporting limits (e.g. cut-off concentrations) or existing QA/QC requirements (Goldberger *et al.*, 1997). Therefore, the intent and analytical goals must be established for the new protocol before the analytical technique is optimized on the bench. The analytical goals (also known as analytical quality specifications, performance standards or performance goals) are required not only for method development and validation, but also to help justify the purchase of new instrumentation and assist manufacturers in the design, construction and marketing of new equipment and reagents. In addition, the analytical goals provide specifications for relevant external quality assurance programs (Fraser *et al.*, 1999).

1.5.3 Method validation

Once a new analytical method has been developed, it must be optimized and standardized to meet the purposes for which it was developed. Validation studies confirm that a new assay has met its required performance specifications, and must be done before the analytical procedure can become a routine laboratory protocol. There are multiple approaches to validating analytical methods. The most common is to compare results from the new method with results obtained using an established, or reference, assay. Alternatively, standard reference materials (e.g. standards certified by NIST) can be analyzed by the new method to demonstrate its accuracy. When available, authentic specimens should be tested and compared against a previously validated protocol.

Each analytical method will have a different purpose and therefore different analytical performance requirements. Some of the performance parameters are accuracy and precision, recovery, limits of detection (sensitivity) and quantitation, range of linearity, specificity, stability, carryover potential and ruggedness. Other methadological factors are selection of a reference standard, internal standard, derivatizing agent, ions for selected-ion monitoring in mass spectrometry, instrumentation and chromatographic performance (Goldberger *et al.*, 1997; Gerhards *et al.*, 1999; Wu *et al.*, 1999; Garfield *et al.*, 2000; Burtis and Ashwood, 2001; FDA, 2001; Jiménez *et al.*, 2002). These parameters can be assessed within the same batch (intra-assay, within-run) or between batches (inter-assay, between-run or day-to-day) using reference standards and quality control samples. Most standards and quality control materials have been verified and are available commercially, often in the form as lyophilized or liquid samples.

Accuracy and precision are the most important characteristics of an analytical method; they give the best indication of random and systematic error associated with the analytical measurement. Systematic error refers to the deviation of an analytical result from the 'true' value, and therefore affects the accuracy of a method. One the other hand, random errors influence the precision of a method (Kallner *et al.*, 1999). Ideally, accuracy and precision should be assessed at multiple concentrations within the linear range of the assay (low, medium and high concentration).

After a method has been validated, it can be used for routine laboratory analyses. There may be additional analytical considerations that are important in routine assays. Analysis of the specimens should be done in a timely manner, especially when the analyte is unstable.

Generally, specimens can be run singly unless duplicate measurements are required in order to improve precision and reliability. Quality controls should be included with each batch of specimens and the QC results monitored to detect trends in analytical performance that can potentially affect the accuracy of results.

1.5.4 Accuracy

Accuracy defines how closely the measured concentration agrees with the true (fortified) value. For most analytical applications, laboratories should achieve a level of accuracy within 5–10% of certified concentrations in reference materials. Accuracy is often expressed as a percentage difference from the true value:

$$\text{Accuracy}(\%) = [(\text{mean} - \text{true value})/\text{true value}] \times 100$$

Accuracy can also be expressed as the percentage of the mean to the true value:

$$\text{Accuracy}(\%) = (\text{mean/true value}) \times 100$$

Ideally, accuracy should be determined at multiple concentrations of the analyte. Since repeated measurements by any analytical method will produce a Gaussian distribution of results, Student's t-test provides a statistical way to assess the agreement of measured values with the target or 'true' value.

1.5.5 Precision

Precision describes the variability in a set of measurements, i.e. how closely repeated measurements on a single sample agree. Precision can be calculated for samples within an individual run (intra-assay precision) or across multiple runs (inter-assay precision). Precision is customarily expressed as the percentage coefficient of variation (%CV, or simply CV), which is the standard deviation expressed as a percentage of the mean:

$$\text{CV} = (\text{standard deviation/mean}) \times 100$$

Typically, laboratories should achieve a level of precision of <10%; however, in some instances, the CV may be >10%, especially at or near the lower limit of quantitation, when background noise is a larger portion of the total signal measured. The CV should be determined at multiple concentrations of the analyte.

1.5.6 Recovery

When a protocol requires some manipulation of the specimen prior to analysis, such as an extraction step, there may be opportunities for the analyte concentration to be diminished due to pre-analytical loss. Therefore, it may be important to calculate the recovery of the analyte. Analyte recovery can be calculated by comparing the detector response from the analyte after it has been extracted to the detector response from an unextracted sample. The unextracted sample represents 100% recovery because it contains the 'true' concentration of the analyte. It is not necessary to achieve 100% recovery, but the recovery should be reproducible and evaluated at three separate concentrations (FDA, 2001).

1.5.7 Lower limits of detection (sensitivity) and quantitation

The lower limit of detection (LLOD) and the lower limit of quantitation (LLOQ) are helpful parameters when validating an analytical method. The LLOD is the lowest concentration of the analyte that can be detected. Customarily, the LLOD corresponds to the concentration of analyte producing a signal at least three times greater than the analytical background noise. Establishing an LLOD requires knowing the signal-to-noise ratio, acceptable chromato-graphic separation, predicted retention time and, for mass spectrometric (MS) applications, acceptable m/z ion ratios.

The LLOQ is the lowest concentration of the analyte that is detected with a specified accuracy and precision. In practice, the LLOQ often corresponds to a signal 5–10 times greater than the analytical background noise. As an example, the LLOQ might correspond to the concentration meeting all of the LLOD criteria, be quantified within 10% of the target concentration and produce a CV of <10%. These parameters can be optimized by manipulating several factors, including the specimen volume, the detector threshold (gain), the type and condition of the chromatographic column, preconcentration of the analyte, the amount and type of internal standard, the extraction efficiency and the analytical method (Goldberger et al., 1997).

1.5.8 Range of linearity

Most, though not all, analytical methods produce results that vary linearly with analyte concentration, so the term calibration 'curve' is something of a misnomer. The range of linearity of an analytical method is determined by measuring standard solutions at multiple concentrations (calibrators), sometimes using an internal or external standard for quantitation. Calibrators should be prepared in a matrix similar to authentic specimens, to compensate for potential matrix effects. The practical linear range, which encompasses the expected concentrations in authentic specimens, should be assessed, i.e. challenged at its upper and lower limits, each time the assay is run. An acceptable linear range depends on various criteria, such as chromatographic resolution, correct retention time, appropriate m/z ion ratios and a correlation coefficient (r) or coefficient of determination (R^2) >0.99 (Goldberger et al., 1997).

There are various ways to derive a calibration curve. Multi-point calibration curves, for example, include a minimum of three different concentrations of the analyte. For semi-quantitative assays, a single-point calibration is common. The single point is usually the threshold concentration used to determine whether a specimen is positive or negative for the analyte of interest. Depending on the validation process and performance characteristics of the assay, a single-point calibration curve may also be used in quantitative applications over a limited range of linearity. A historical (pre-established) multi-point calibration curve may also be used, but only if the stability of the analytical method over time has been well established (Goldberger et al., 1997).

1.5.9 Specificity

Specificity, or selectivity, is defined as the capability of the methodology to identify the analyte of interest in the presence of matrix components and potentially interfering substances (FDA, 2001; Goldberger *et al.*, 1997). Analytical interferences include endogenous matrix components, decomposition products, metabolites and other structurally related compounds. FDA guidelines (FDA, 2001) suggest that six blank specimens from different sources be analyzed for matrix effects. Fewer samples may be required when using methodology with MS detection.

In contrast, there are instances in which decreased specificity is beneficial, such as in assays designed to detect an entire class of compounds. For example, antibody cross-reactivity is essential in immunoassays used to screen for multiple drugs or drug classes. Screening methods ordinarily are paired with confirmatory testing methods that use a different analytical methodology with increased specificity.

1.5.10 Stability

Some analytes are not stable under certain transport, storage and analytical conditions. Therefore, it is essential to evaluate the stability of an analyte under a variety of conditions. Specimens may be collected with preservatives (e.g. sodium fluoride) and stored refrigerated or frozen. During analysis, the addition of an internal standard and/or the use of quality control samples can be helpful in evaluating the stability of an analyte during the entire analytical procedure.

For assays requiring specimens to be frozen, the stability of the analyte in a freeze–thaw cycle should be validated. FDA guidelines (FDA, 2001) suggest that analytes should be tested in triplicate over three freeze–thaw cycles at two different concentrations.

1.5.11 Carryover

Carryover is defined as the 'contamination of a sample by a sample analyzed immediately prior to it' (Goldberger *et al.*, 1997). Therefore, the validation process should assess carryover potential in highly concentrated specimens and the acceptable limit of carryover into otherwise negative samples. Suggested ways to minimize carryover include the use of solvent washes after each injection, the use of solvent blanks between cases, dilution of specimens prior to extraction, dilution of extracts prior to injection, occasional reassessment of the minimum carryover limit and injection of specimen extracts in ascending concentration order (determined by initial screening). Carryover is most often due to contamination by the sampling syringe exposed to the previous specimen. Therefore, when carryover is suspected, the specimen in question should be re-extracted, since re-injection from the already contaminated vial will produce the same carryover error.

1.5.12 Ruggedness

Ruggedness refers to the ability of an analytical method to withstand minor procedural changes without affecting the overall performance of the assay. The ruggedness of an assay can be very important, particularly when multiple technicians or multiple laboratories perform the same procedure. Therefore, it is useful to determine the critical steps in the assay, such as pH, composition of solvent mixtures, derivatizing chemicals, temperature and incubation times (Goldberger *et al.*, 1997). In addition, specimen volume may be critical, so any adjustments to the volume of specimen should be validated. If specimen dilutions are required, the preferred diluent is blank (analyte-free) matrix (FDA, 2001).

1.5.13 Selection of a reference standard

Selection of a reference standard for most analytical methods is a straightforward task, involving acquisition of the relevant analyte in a state of known composition and purity. However, some analytes are not available in pure form, and structurally similar analogs (e.g. free base) may be used instead. Standard materials should be of a known purity, and the source, lot number, expiration date and certification of identity should be recorded (FDA, 2001).

1.5.14 Selection of an internal standard and standard addition

Analytes may be quantitated using an internal standard or by the method of standard addition. An internal standard is a structural analog (e.g. deuterium-labeled compound), a known amount of which is added to specimens prior to any preparatory steps such as extraction or hydrolysis. Ideally, the internal standard should be as chemically similar to the analyte of interest as possible, so that throughout the analytical procedure (e.g. extraction, chromatographic separation and ionization processes) and under the specified conditions (e.g. derivatization), its recovery will be similar to the analyte's. The concentration of analyte can then be determined from the ratio of the analyte response to the internal standard response.

The addition of an internal standard is advantageous because detection of the internal standard ensures that the extraction step successfully recovered analyte (if present). Any losses associated with preparatory steps are not critical, because this procedure relies on the ratio of concentrations, not absolute concentration.

Standard addition can be an alternative method for the quantitation of analytes. The general concept is to add known concentrations of the analyte to multiple aliquots of the unknown specimen. By extrapolation, the original concentration of the (unfortified) specimen can be determined. This technique is particularly helpful when a calibration curve cannot be generated using standards in a similar matrix.

1.5.15 Selection of derivatization agent

In the context of chromatographic applications, derivatization is the process of chemically altering a compound with the intent to improve the sensitivity with which it can be detected. Often, derivatization involves reacting polar groups (-NH, -OH and -SH) with perfluoryl or silyl derivatization reagents to increase the volatility of the compound. Some derivatives enhance the detector response, improving sensitivity. In other circumstances, it may be beneficial to increase the mass of a compound to distinguish its ions from background noise or similar structures (Goldberger *et al.*, 1997). There are some of disadvantages to derivatization, such as requiring an extra step (often manual) in the analytical method and the absence of many derivatized products in commercial mass spectral libraries. Therefore, laboratories may need to produce their own reference libraries when using unusual derivatives for MS applications.

1.5.16 Selection of ions for selected-ion monitoring or full-scan analysis

Assays that utilize MS require the selection of ions that uniquely represent the analyte and internal standard without any potential interferences (Goldberger *et al.*, 1997). It is helpful to identify ions for both quantitative and qualitative purposes. Ratios between the ions can be helpful to confirm the compound of interest. The base peak (ion with the highest abundance in the mass spectrum) is often preferred as the quantitative ion because it allows for the greatest sensitivity.

1.5.17 Chromatographic performance

Methods involving chromatographic separations prior to detection require optimization of the chromatographic conditions for best performance. A Gaussian peak shape (no tailing or fronting), resolution (baseline separation from nearby peaks) and reproducible retention time are necessary for identification of the analytes of interest (Goldberger *et al.*, 1997).

1.5.18 Statistical evaluation of quality control

Proper interpretation of quality control data requires presentation in a format that makes systematic bias readily apparent. Based on the work of Shewhart, Levey and Jennings, Westgard and co-workers proposed in 1981, an algorithm for monitoring quality control in clinical laboratories (Westgard *et al.*, 1981). 'Westgard's Rules' provide guidance for determining when control values are acceptable or, alternatively, when control measurements reflect changes in the performance of the analytical method (Westgard and Barry, 1986; Burtis and Ashwood, 2001).

In a Levey–Jennings control chart, the *x*-axis represents time or number of analytical runs and the *y*-axis represents the control results (Figure 1.1a). Control values are plotted on the Levey–Jennings chart, and should fall within a statistically predicted range, based on a Gaussian distribution. The acceptable range of values is typically indicated by lines on the *y*-axis representing the upper and lower limits for control results (commonly ±2 standard deviations or defined percentages from the mean control value). Westgard described a multi-rule procedure using the Levey–Jennings chart to detect deviations from Gaussian behavior that reflect analytical bias (Burtis and Ashwood, 2001).

Another way to present control data is the cumulative sum control chart. In this method, the *x*-axis is either time or number of control observations and the *y*-axis is the cumulative sum (cusum), or the difference between the measured control value and the predicted mean

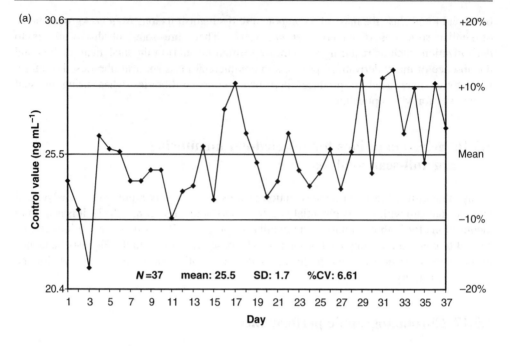

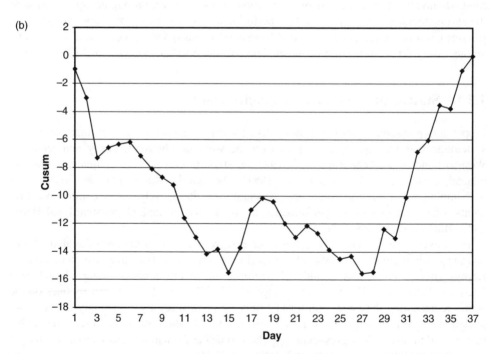

Figure 1.1 Examples of (a) Levey–Jennings and (b) cumulative sum (cusum) control charts using inter-assay quality control data from an alprazolam GC–MS assay. Since the cusum chart persents the cumulative sum of deviations from the mean, it is more sensitive to small biases that develop over time, whereas Levey–Jennings charts are most useful for detecting changes in the precision of the assay

concentration of the control material (Figure 1.1b). The cusum is calculated by adding this difference to the differences from previous control observations. Random errors should produce cusum values that cross the zero line (mean value) multiple times. However, if there are other sources of error, then the cusum will deviate from the mean value in one or the other direction. Other quality control charts, such as mean, standard deviation and range charts, are used in industry, but are not common in clinical chemistry and toxicology (Westgard and Barry, 1986).

1.6 EXTERNAL QUALITY ASSURANCE

External quality assurance (EQA) is fundamental to the standardization of clinical laboratory methods because it provides a means to compare results generated in one laboratory with those of peer laboratories subscribing to the same EQA program. EQA programs are especially beneficial since internal QA and QC procedures are limited in their ability to detect bias in analytical methods. Internal QA/QC can only detect errors that result in a deviation from the original method validation: inherent errors in the method may go unnoticed. Therefore, it is helpful to compare the results produced by a new method with those from other laboratories (Burtis and Ashwood, 2001). Monitoring the performance of laboratory procedures in a consistent manner keeps the laboratory accountable, and can reveal systematic errors that would otherwise be undetected. A prominent component of EQA is proficiency testing.

Proficiency testing in clinical laboratories began in the late 1940s as a response to lack of reproducibility among laboratories (Rosenfeld, 1999). In 1947, Belk and Sunderman published the results of a study in which they surveyed clinical laboratories in Pennsylvania for accuracy among common chemical testing procedures in hospital laboratories (Sunderman, 1992). Thereafter, other states conducted similar studies, and eventually monthly samples were mailed to various laboratories worldwide and the statistical summaries were returned to these laboratories, comparing their results with other laboratories performing the same analyses. The Sunderman Proficiency Testing Service diligently continued this program for 36 years until 1985, when the American Society of Clinical Pathologists took over the service. Since then, many other organizations, such as College of American Pathologists, the American Association of Bioanalysts and American Association for Clinical Chemistry have provided various forms of proficiency testing. Combined, these external testing procedures have helped to standardize testing among laboratories and enhance the overall quality of laboratory practices.

REFERENCES

Burtis, C. A., Ashwood, E. R. (Eds) (2001). *Tietz Fundamentals of Clinical Chemistry*, 5th edn. Philadelphia, PA: W. B. Saunders.

FDA (2001). *Guidance for Industry: Bioanalytical Method Validation*. US Department of Health and Human Services, Food and Drug Administration, Center for Drug Evaluation and Research (CDER), Center for Veterinary Medicine (CVM). http://www.fda.gov/cder/guidance/4252fnl.pdf.

Fraser, C. G., Kallner, A., Kenny, D., Hyltoft Petersen, P. (1999). Introduction: strategies to set global quality specifications in laboratory medicine. *Scand J Clin Lab Invest*, **59**, 477–478.

Garfield, F. M., Klesta, E., Hirsch, J. (2000). *Quality Assurance Principles for Analytical Laboratories*, 3rd edn. Gaithersburg, MD: AOAC International.

Gerhards, P., Bons, U., Sawazki, J., Szigan, J., Wertmann, A. (1999). *GC/MS in Clinical Chemistry*. New York: Wiley-VCH.

Goldberger, B. A., Huestis, M. A., Wilkins, D. G. (1997). Commonly practiced quality control and quality assurance procedures for gas chromatography/mass spectrometry analysis in forensic urine drug-testing laboratories. *Forensic Sci Rev*, **9**, 59–80.

Groth, T. (1999). Series analyses and quality specifications required for monitoring over time. *Scand J Clin Lab Invest*, **59**, 501–508.

Jiménez, C., Ventura, R., Segura, J. (2002). Validation of qualitative chromatographic methods: strategy in antidoping control laboratories. *J Chromatogr B*, **767**, 341–351.

Kallner, A., McQueen, M., Heuck, C. (1999). The Stockholm Consensus Conference on quality specifications in laboratory medicine, 25–26 April 1999. *Scand J Clin Lab Invest*, **59**, 475–585.

Kaplan, L. A. (1999). Determination and application of desirable analytical performance goals: the ISO/TC 212 approach. *Scand J Clin Lab Invest*, **59**, 479–482.

Klee, G. G., Schryver, P. G., Kisabeth, R. M. (1999). Analytic bias specifications based on the analysis of effects on performance of medical guidelines. *Scand J Clin Lab Invest*, **59**, 509–512.

Ratliff, T. A. (2003). *Laboratory Quality Assurance System: a Manual of Quality Procedures and Forms*, 3rd edn. Hoboken, NJ, John Wiley & Sons, Inc.

Rosenfeld, L. (1999). *Four Centuries of Clinical Chemistry*. Amsterdam: Gordon and Breach.

Sunderman, F. W. Sr. (1992). The history of proficiency testing/quality control. *Clin Chem*, **38**, 1205–1209.

Tilstone, W. J. (2000). Quality assurance. In *Encyclopedia of Forensic Sciences*, J. A. Siegel, P. J. Saukko, G. C. Knupfer, G. C. (Eds). San Diego, CA: Academic Press, pp. 1307–1314.

Westgard, J. O. (1999). The need for a system of quality standards for modern quality management. *Scand J Clin Lab Invest*, **59**, 483–486.

Westgard, J. O., Barry, P. L. (1986). *Cost-effective Quality Control: Managing the Quality and Productivity of Analytical Processes*. Washington, DC: AACC Press.

Westgard, J. O., Barry, P. L., Hunt, M. R., Groth, T. (1981). A multi-rule Shewhart chart for quality control in clinical chemistry. *Clin Chem*, **27**, 493–501.

Wu, A. H. B., Hill, D. W., Crouch, D., Hodnett, C. N., McCurdy, H. H. (1999). Minimal standards for the performance and interpretation of toxicology tests in legal proceedings. *J Forensic Sci*, **44**, 516–522.

2

Liquid Chromatographic-Mass Spectrometric Measurement of Anabolic Steroids

Don H. Catlin, Yu-Chen Chang, Borislav Starcevic and Caroline K. Hatton

UCLA Olympic Analytical Laboratory, 2122 Granville Avenue, Los Angeles, CA 90025, USA

2.1 INTRODUCTION

The growing availability of liquid chromatography–tandem mass spectrometry (LC-MS-MS) instruments to life science researchers has sparked an exponential flurry of efforts to discover new possibilities for the detection and quantitation of anabolic steroids in a wide variety of samples. Of all the directions and applications that have been explored to date, the area of greatest potential impact on science focuses on natural androgens in human samples, as opposed to synthetic anabolic steroids or animal samples.

The quantitation of natural androgens has been attempted in a wide variety of human samples, from urine and serum to prostate biopsy samples and testicular fluid. Analytical studies have also been conducted on standards and various *in vitro* media. Because the incidence of one particular androgen-dependent disease, prostate cancer, is on the rise, many studies revolve around its diagnosis and monitoring. In the wake of research connected to prostate cancer and the related benign prostatic hyperplasia, we can hope to gain insights into more basic aspects of human physiology and pathology. After all, androgens are sex steroids associated with a number of physiological developments and responses such as the development of reproductive systems, sexual differentiation and maintenance of secondary sexual characteristics. The balance of male and female hormones can elicit biological responses with a profound impact on health and behavior.

Testosterone (T), the male hormone *par excellence*, is primarily produced in the testis. It is metabolized to the more potent dihydrotestosterone (DHT). Physiological T concentrations reflect the balance between biosynthesis and metabolism. Therefore, T is a prime

Chromatographic Methods in Clinical Chemistry and Toxicology   Edited by R. L. Bertholf and R. E. Winecker

target of quantitative assays because T measurements make it possible to determine its pharmacokinetics, production rate and clearance. In turn, these can be used in clinical research studies to study androgen metabolism after physiological or pharmacological interventions or to fine tune replacement doses.

T is most commonly measured in serum, and it is one of the hormones most commonly measured in serum. T circulates in plasma non-specifically bound to albumin, specifically bound to sex hormone binding globulin and unbound (free). Clinicians use serum T measurements to diagnose and monitor various disorders such as hypogonadism (androgen deficiency), testicular dysfunction, hirsutism, virilization, alopecia, prostate disease, adrenal hyperplasia and aging, in addition to anomalies caused by exposure to chemicals or hormone disruptors.

Most of these measurements are still routinely performed with immunoassays because of their simplicity, rapidity and relatively low cost, and despite occasional concerns about their reliability and validity. Serum total T has also been quantitated by gas chromatography–mass spectrometry (GC-MS), liquid chromatography–mass spectrometry (LC-MS) and LC-MS-MS. MS-based methods are often used in research studies or to confirm immunoassay results. The advantage of chromatography coupled with MS is the high specificity not available with immunoassays because they are susceptible to cross-reactions. An immunoassay might measure a host of structurally related compounds in addition to the target analyte.

Recent clearance studies use stable isotope dilution methods. MS methods make it possible to measure deuterated and endogenous species simultaneously and specifically. LC-MS methods offer advantages over GC-MS methods, such as streamlined sample preparation (no derivatization necessary), high recovery and sensitivity and superior specificity.

This chapter will focus primarily on natural anabolic steroids in human samples because of the potential impact of improved analysis and quantitation by LC-MS-MS on our understanding of reproduction and disease. However, LC-MS methods for synthetic steroids or animal samples have also been published, therefore they will be reviewed first, but only briefly.

2.2 LC-MS ANALYSIS OF SYNTHETIC STEROIDS OR ANIMAL SAMPLES

In veterinary medicine, boldenone, a synthetic anabolic steroid (Figure 2.1), is commercially available, hence it is a concern in the horseracing industry. Pu *et al.* (2004) used ion-trap LC-MS analysis to detect boldenone conjugates (sulfate and glucuronide) and their 17-epimers in horse urine after intramuscular administration of boldenone undecylenate. Soon afterwards, Ho *et al.* (2004) reported the occurrence of endogenous boldenone sulfate in the urine of uncastrated male horses, and quantitated it by quadrupole time-of-flight (Q-TOF) LC-MS-MS.

Also known as endogenous in male horses and prohibited as a doping agent is 19-nortestosterone (nandrolone; Figure 2.1). Kim *et al.* (2000a) validated an LC-MS-MS method for the detection and quantitation of three different commercially available esters, but 19-nortestosterone esters are rapidly hydrolyzed in horse plasma, which limits the usefulness of this method in racehorse doping control.

Another synthetic anabolic steroid available as a veterinary product is stanozolol (Figure 2.1), whose metabolites were investigated by McKinney *et al.* (2004) by ion-trap LC-MS in horse urine after intramuscular injection.

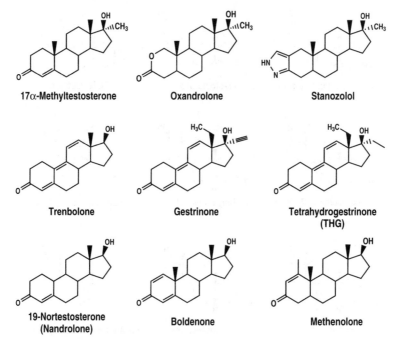

17α-Methyltestosterone   Oxandrolone   Stanozolol

Trenbolone   Gestrinone   Tetrahydrogestrinone (THG)

19-Nortestosterone (Nandrolone)   Boldenone   Methenolone

Figure 2.1 Examples of synthetic androgen analytes

Yu *et al.* (2005) developed an LC-MS-MS screen for deconjugated anabolic steroids in horse urine and characterized the method using horse urine samples spiked with 15 prohibited anabolic steroids. In an excretion study in two horses, methenolone acetate was administered by mouth, methenolone (Figure 2.1) was detected in urine and the 17-epimethenolone metabolite was detected for a longer time.

Racehorse anti-doping work is not the only area that requires screening for prohibited anabolic steroids in animals. The drugs act as growth promoters in cattle and this use is prohibited by European regulations, hence analytical methods have been developed to detect residues in meat for enforcement purposes. Joos and Van Ryckeghem (1999) described a procedure for the LC-MS-MS analysis of 36 anabolic steroids found in animal kidney fat matrices. The method's limits of detection are low enough to meet regulations. In Japan, where maximum residue limits were established in 1995, Horie and Nakazawa (2000) developed a method for the determination of trenbolone (Figure 2.1) and zeranol (an anabolic agent but not a steroid) in bovine muscle and liver tissues by LC-MS with selected-ion monitoring (SIM).

Draisci *et al.* (2000) used LC-MS-MS to quantitate T (Figure 2.2), 19-nortestosterone and their 17-epimer metabolites in bovine serum and urine, and subsequently stanozolol and its major metabolite (Draisci *et al.*, 2001) and boldenone (Draisci *et al.*, 2003) in bovine urine. Van de Wiele *et al.* (2000) also worked on stanozolol in cattle urine and feces, with or without derivatization before LC-MS-MS analysis. For trenbolone, Buiarelli *et al.* (2003) developed and characterized an LC-MS-MS method for bovine urine and serum. Van Poucke and co-workers extended the list of LC-MS-MS target analytes to four anabolic steroids (Van Poucke and Van Peteghem, 2002), then 21 anabolic steroid residues in bovine urine (Van Poucke *et al.*, 2005).

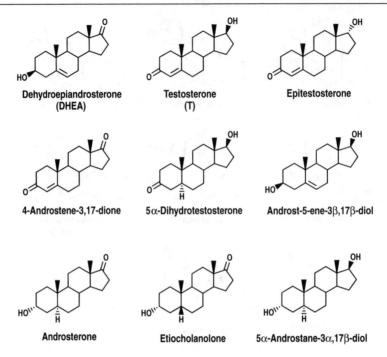

Figure 2.2 Examples of natural androgen analytes in human samples. Glucuronide or sulfate conjugates of these steroids are formed with the C-3 or C-17 hydroxyl group

Poelmans *et al.* (2002) reviewed analytical approaches to the detection of stanozolol and its metabolites.

Nielen *et al.* (2001) focused on a related facet of the illegal use of growth promoters in cattle, namely the detection of anabolic steroids in illegal cocktails. They presented a Q-TOF LC-MS-MS method with the aim of measuring accurate mass and calculating elemental composition for identification purposes.

Turning to the abuse of anabolic steroids by human competitors, Shackleton *et al.* (1997) conducted LC-MS analyses of pharmaceutical T esters in human plasma because of the potential applications to doping control, especially if sensitivity could be improved.

Leinonen *et al.* (2002) compared LC-MS-MS with electrospray ionization (ESI), atmospheric pressure chemical ionization (APCI) and atmospheric pressure photoion-ization (APPI) for unconjugated (free) anabolic steroids in human urinary extracts. The selected analytes were synthetic steroids or their metabolites, known to be misused in sports and known to be excreted in urine unconjugated, namely oxandrolone (Figure 2.1), the 3'-hydroxy metabolite of stanozolol and the 6β-hydroxy metabolite of 4-chlorodehydro-methyltestosterone.

Kim *et al.* (2000b) used GC-MS, LC-MS and LC-MS-MS to study the metabolites of gestrinone (Figure 2.1), a contraceptive and potential sports doping agent, in human urine.

Anti-doping scientists had suspected for years that underground chemists supply performance-enhancing substances to athletes compelled to cheat, including 'designer' drugs made and used only to beat the test. This was proven when the anabolic steroid tetrahydrogestrinone (THG; Figure 2.1), a new chemical entity, was identified by Catlin *et al.* (2004) in a used syringe anonymously turned in to the US Anti-Doping Agency. LC-MS and LC-MS-MS played key roles in the identification and urine sample screening procedures. Thevis *et al.* (2005) reasoned that chemical modifications of steroids usually alter their molecular weights, used to detect and identify them in doping control samples, but might not alter their nuclei or some of their characteristic fragments or product ions. When suspicious product ions have been detected, precursor ion scan experiments can help identify unknown steroids.

Another reality that anti-doping scientists must face is that some athletes accused of using banned anabolic steroids offer explanations such as the inadvertent consumption of various alleged sources of the substances. To assess the theory that uncastrated boar meat consumption can result in an adverse finding in a doping control urine test, Le Bizec *et al.* (2000) tested volunteers and detected 19-norandrosterone and 19-noretiocholanolone for about 24 h after intake. The same team identified and quantitated the boar meat 19-norsteroids responsible for this finding (De Wasch *et al.*, 2001).

To help provide anti-doping scientists with reference standards for LC-MS work, Kuuranne *et al.* (2002) carried out the enzyme-assisted syntheses of the glucuronides of methyltestosterone and nandrolone (Figure 2.1), and developed a new LC-MS method to control the synthetic product purity.

Watching the world become awash in steroid sex hormones, abused in animals and humans, Lopez de Alda *et al.* (2003) felt the need to review LC-MS and LC-MS-MS methods for the determination of these and other analytes in the aquatic environment.

In the only publication on the LC-MS analysis of synthetic steroids or animal samples that is unrelated to any steroid misuse, Magnusson and Sandstrom (2004) reported on a quantitative analysis of eight T metabolites in rat intestine mucosa.

2.3 LC-MS ANALYSIS OF NATURAL ANDROGENS IN HUMAN SAMPLES

Table 2.1 summarizes publications describing the measurement of natural anabolic steroids as standards or in human samples. Listed are analytes, samples, sample preparation, LC conditions, MS-MS conditions, whether the assay is quantitative or qualitative and references.

In anticipation of potential studies involving endogenous steroids, including neurosteroids, or exogenous steroids, or metabolites, Ma and Kim (1997) reported results obtained by LC-MS analysis by APCI or ESI of 29 steroid standards. Twelve of those had an androstane nucleus. They were either 3-one-4-enes or unconjugated ketones or they had only hydroxyl groups. The use of different LC mobile phases for each ionization mode helped to optimize the sensitivity.

Under the heading of natural androgens in human samples falls one kind of drug abuse – sports doping or the use of athletic performance- enhancing substances – which is prohibited by major sports organizations. If athletes take pharmaceutical T, the level of T in their urine will increase, but the level of epitestosterone, a natural isomer with no known function, will not, therefore the ratio of the two, of T/E, will increase. In the 1980s, the International Olympic Committee defined a T/E ratio of >6 as an adverse finding that anti-doping laboratories ought to report, and in 2005 the cut-off was lowered to 4. To help study and

Table 2.1 Summary of LC-MS analyses relevant to natural androgens in human samples

Analyte	Sample	Sample preparation[a]	LC injection volume, stationary phase	MS-MS	Quantitative	Comments	Ref.
29 steroids (12 androstane-type), 3-one,4-enes, unconjugated ketones, hydroxyls only	Standards	None	C_{18}	LC-MS APCI and ESI full-scan and SIM	NA		Ma and Kim (1997)
T glucuronide, T sulfate, epitestosterone glucuronide, epitestosterone sulfate	Standards	None	C_{18}	LC-MS and LC-MS-MS triple quadrupole ESI	NA	Explored potential to detect T misuse in sports doping	Bowers and Sanaullah (1996)
8 glucuronide standards: T, epitestosterone, nandrolone, androsterone, 5α-estran-3α-ol-17-one, 5β-estran-3α-ol-17-one, 17α-methyl-5α-androstane-3α,17β-diol, 17α-methyl-5β-androstane-3α, 17β-diol	Standards	None	No LC	Triple quadrupole ESI and APCI negative and positive ion mode	NA	Potential application to sports doping control analysis	Kuuranne et al. (2000)
T, dihydrotestosterone, 4-androstene-3,17-dione	100 μL cell culture medium	No IS, no sample preparation except automated: column-switching C_4-alkyldiol-silica restricted access SPE column	100 μL, C_{18}	Triple quadrupole ESI positive ion mode	Y	Comparison with RIA	Chang et al. (2003)
T, hydrocortisone, SR 27417	1 mL spiked human plasma	IS d_3-T C_8 disc		Ion-trap APCI positive ion mode	Y		Tiller et al. (1997)

Compound	Sample	Sample preparation	Column	MS		Application	Reference
Dehydroepiandrosterone and sulfate, androstenedione, T	Standards and 760 µL spiked serum	IS deuterated analogs, precipitation by adding 1140 µL IS solution, analyze supernatant	1700 µL, C_{18}	Triple quadrupole APPI positive ion mode	Y	Comparison with 7 RIAs	Guo et al. (2004)
Androstenediol 3-sulfate, dehydroepiandrosterone sulfate	100 µL human serum	Deuterated analogs, protein precipitation, SPE, 30 µL acetonitrile–ammonium bicarbonate reconstitution	C_{18}	LC-MS ion trap ESI negative ion mode	Y	Comparison of healthy subjects with prostate cancer or benign prostate hyperplasia patients	Mitamura et al. (2003)
Dehydroepiandrosterone sulfate	5 µL human plasma	IS deuterated analog, add aqueous ethanol, supernatant through C_{18}, reconstitute in 100 µL liquid phase	Capillary LC	Triple quadrupole ESI negative ion mode	Y		Liu et al. (2003)
T	2 mL human serum	IS d_3-T C_8 disc		Triple quadrupole APCI positive ion mode	Y	Comparison with 6 IAs	Wang et al. (2004a)
T, d_3-T	2 mL human serum	IS 19-nortestosterone, sodium acetate buffer, 5-mL ether extractions 100 µL ethanol reconstitution	15 µL, C_{18}	Triple quadrupole APCI positive ion mode	Y	Assay applied to 38 samples from normal volunteers	Starcevic et al. (2003)
						Influence of ethnicity and age in normal men	Wang et al. (2004b)
T	50 µL human serum	IS d_2-T, 100 µL zinc sulfate–methanol precipitation	40 µL, C_{18}	Triple quadrupole APCI positive ion mode	Y	Sample preparation and GC-MS vs LC-MS-MS comparison, T measurement in women	Cawood et al. (2005)

(continued)

Table 2.1 (*Continued*)

Analyte	Sample	Sample preparation[a]	LC injection volume, stationary phase	MS-MS	Quantitative	Comments	Ref.
T glucuronide, dihydrotestosterone glucuronide	50 μL human urine	IS d_3-T, filtration	C$_{18}$	Ion-trap ESI negative ion mode	Y	Comparison of benign prostate hyperplasia and normal men	Choi et al. (2003)
T glucuronide, T sulfate, epitestosterone glucuronide, epitestosterone sulfate	2 mL human urine	IS d_3-T glucuronide, d_3-epitestosterone sulfate, SPE, aqueous methanol elution, 20 μL water reconstitution	5 μL	Triple quadrupole ESI positive ion mode	Y	To detect T misuse in sports doping?	Bean and Henion (1997)
T glucuronide, T sulfate, epitestosterone glucuronide, epitestosterone sulfate	Standards and 3 mL human urine	IS deuterated analogs, SPE C$_{18}$ cartridge, aqueous methanol elution, 100 μL mobile phase reconstitution	10 μL, C$_{18}$	Triple quadrupole ESI positive ion mode	Y	To detect T misuse in sports doping?	Borts and Bowers (2000)
T, epitestosterone, dehydroepiandrosterone, androsterone, etiocholanolone, their glucuronides and their sulfates	2 mL human urine	IS d_3-epitestosterone glucuronide, d_3-T sulfate, d_3-T, SPE, methanol elution, 50 μL aqueous methanol reconstitution	5 μL	Triple quadrupole ESI positive and negative ion mode	Ratios	To profile urinary steroids for doping control	Buiarelli et al. (2004)

Analytes	Sample	Sample preparation	Column	MS method	Derivatization	Application	Reference
Dehydroepiandrosterone metabolites: androst-5-ene-3β,17β-diol, androst-4-ene-3,17-dione, T, 5α-dihydrotestosterone, androsterone, 7α-hydroxydehydroepiandrosterone	Human prostate homogenate incubation mixture	IS d_4-DHEA, hexane–ethyl acetate extraction, C_{18}, aqueous acetonitrile	C_{18}	LC-MS ion-trap APCI full-scan mode	No	Comparison with GC-MS results	Mitamura et al. (2002)
Dihydrotestosterone	Human serum and prostate biopsy tissue	Alkaline extraction, SPE, N-methylpyridinium polar derivatization		LC-MS ESI positive ion mode	Y	Cancer patients before and after androgen deprivation therapy	Nishiyama et al. (2004)
T, dihydrotestosterone	10 mg human prostate tissue	IS d_3-T, aqueous methanol, Oasis HLB cartridge, N-methylpyridinium polar derivatization	C_{18}	Triple quadrupole ESI positive ion mode	Y	Comparison of sujects with benign prostate hyperplasia and prostate cancer	Higashi et al. (2005)
T, dihydrotestosterone, estradiol, 5α-androstane-3α,17β-diol	20 μL human testicular fluid	IS dimethylbenzoyl-phenylurea, one-step ether extraction	C_{18}	Triple quadrupole ESI positive ion mode	Y	Comparison with RIA	Zhao et al. (2004)

[a]IS, internal standard.

understand such cases, and to help resolve basic questions about steroid excretion, Bowers and Sanaullah (1996) worked out a direct method for the quantitation of four analyte standards, namely the conjugates (glucuronides and sulfates) of T and its naturally occurring but inactive epimer, epitestosterone (Figure 2.2), by LC-MS and LC-MS-MS. The following year, Bean and Henion (1997) reported the quantitation of the same four analytes in human urine. Using [16,16,17-^{2}H$_3$]testosterone glucuronide and [16,16, 17-^{2}H$_3$]epitestosterone sulfate as internal standards, they extracted 2 mL of urine by solid-phase extraction and monitored m/z 465–289, 465–271 and 465–253 for T glucuronide and epitestosterone glucuronide, and m/z 468–292 and 372–273 for the trideuterated internal standard analogs. The positive ion mode was substantially more sensitive than the negative ion mode for T glucuronide, and almost as sensitive for epitestosterone sulfate. Switching mode would have been impractical with the instrument used by the authors because the analytes eluted too close to each other. Calibration curves were prepared in urine for T glucuronide and epitestosterone glucuronide from 50 to 1000 nmol L^{-1} and for T sulfate and epitestosterone sulfate from 10 to 200 nmol L^{-1}. LODs for all four target analytes and for the two internal standards were in the low nanomolar range. Subsequently, Borts and Bowers (2000) tried to overcome ion suppression by using deuterated analogs of all four analytes. They extracted 3 mL of urine and extended the calibration curves to cover 0–1500 nmol L^{-1}. Comparisons with a GC-MS method and with published population data showed agreement. These methods targeting the urinary glucuronides and sulfates of T and epitestosterone were published between 1996 and 2000, but their potential applications in human sports doping analysis, to control the abuse of T, have not become a reality. Perhaps this is in part because the method fails to close a loophole in the detection of T abuse. Athletes determined to cheat might take both T and epitestosterone, to benefit from T while keeping T/E below the cut-off. Quantitating urinary conjugates might not have helped to distinguish such users from innocent athletes, because like GC-MS, LC-MS-MS is unable to distinguish natural from pharmaceutical T. By the late 1990s, this loophole was closed by the introduction of isotope ratio mass spectrometry, which is able to detect a measurable difference in carbon-13 between natural and pharmaceutical T.

Buiarelli *et al.* (2004) extended the above analytical approach to many more related steroids when they published a method for the direct analysis of 15 urinary anabolic steroids in a single run, namely T, epitestosterone, dehydroepiandrosterone (DHEA), androsterone, etiocholanolone, their sulfates and their glucuronides (Figure 2.2). They extracted 2 mL of human urine by solid-phase extraction with methanol elution and reconstituted the residue in aqueous methanol in the presence of deuterated internal standards (d_3-epitestosterone glucuronide, [16,16,17-^{2}H$_3$]testosterone sulfate and [16,16,17-^{2}H$_3$]testosterone), then monitored, for example, m/z 289–97 and 109 for T and epitestosterone, m/z 367–97 for their sulfates, and m/z 463–113 and 287 for their glucuronides. The method does not achieve quantitation, but it allows the estimation of ratios, which makes it possible to monitor the urinary steroid profile, which is useful for monitoring the abuse of anabolic steroids.

In a study also designed to pave the way for LC-MS-MS of steroid glucuronides, Kuuranne *et al.* (2000) compared ESI and APCI, in positive and negative ion modes, MS-MS of eight steroids synthesized by an enzyme-assisted approach. They determined the best ionization method, ESI in the positive ion mode, and the MS data allowed them to distinguish isomers such as T glucuronide and epitestosterone glucuronide or 5α- and 5β-nandrolone glucuronide.

Turning to cell culture media, Chang *et al.* (2003) worked out a specific and sensitive method for the quantitation of anabolic hormone residues without any sample preparation except for two automated steps. The tandem MS transitions used were m/z 289.2 to 97.1 to monitor 4-androstene-3,17-dione (Figure 2.2) and T and m/z 291.3 to 255.3 for DHT. Calibration curves were prepared in the range 0.05–100 ng mL^{-1} for T ($\sim$0.17–347 nmol L^{-1}) and 4-androstene-3,17-dione ($\sim$0.17–350 nmol L^{-1}) and 1.0–100 ng mL^{-1} ($\sim$3.4–345 nmol L^{-1}) for DHT. Recoveries ranged from 86.6 to 91% depending on the analyte and concentration. Matrix effects were assessed by varying the proportions of cell culture medium and loading buffer used to prepare the steroid standards. The method was validated. Comparison with radioimmunoassay (RIA) indicated that LC-MS-MS provided better accuracy, specificity and sensitivity and a wider dynamic range while sparing workers the handling of and exposure to radiochemicals. The method has been applied in a study of the effects of chemical exposure on male sex hormone homeostasis in cultured human cell lines. In that the method targets free steroids, if it were applied to urine samples it would require a deconjugation step.

Plasma and serum are biological samples commonly collected from subjects in clinical studies for basic research. Tiller *et al.* (1997) spiked plasma extracts with T, hydrocortisone, a corticosteroid (not an anabolic steroid), and SR 27417, a platelet-activating factor, to prepare a calibration curve with nine concentrations for analysis by LC–ion trap MS. For T, the best ions for quantitation were deemed to be m/z 97, 109 and 253. The authors found the full-scan mode helpful in identifying the best ions for quantitation and matrix contaminants.

Guo *et al.* (2004) took advantage of the improved sensitivity achieved by using APPI to develop a method for profiling nine steroid hormones, including three androgens, and applied it to a spiked plasma pool. The androgen analytes were dehydroepiandrosterone (DHEA; Figure 2.2), DHEA sulfate, androstenedione and T. The other steroids were estrogen, progestins and corticoids. Sample preparation consisted of protein precipitation by adding, to 760 μL of plasma, a solution containing all deuterated analog internal standards. The transitions monitored were m/z 271 to 213, 287 to 97 and 289 to 97 for DHEA and DHEA sulfate, androstenedione and T, respectively. The report includes a comparison with seven RIAs and assessments of precision, recovery and accuracy.

Liu *et al.* (2003) focused their method paper on the instrumental set-up for on-line capillary LC coupled with MS-MS via a low flow-rate interface, hoping that miniaturization would improve sensitivity for many biomedical applications. The system design and performance were tested on DHEA sulfate and pregnenolone sulfate quantitation from 5 μL of plasma from one male volunteer. Sample preparation was simple – addition of aqueous methanol and of the deuterated analog internal standards, protein precipitation, supernatant filtration through a C_{18} bed, evaporation and reconstitution in 100 μL of mobile phase, of which 20 μL were injected for LC-MS-MS analysis. The DHEA sulfate concentration found was in agreement with literature values based on GC-MS and LC-MS studies.

Working on actual serum samples, Wang *et al.* (2004a) measured morning serum T concentrations in samples from 62 eugonadal and 60 hypogonadal men (25 baseline samples, 35 after transdermal T replacement therapy) in order to compare a reference method by LC-MS-MS with six commonly used immunoassays, four or them automated and two manual. In this study, the analyte was T, therefore it was possible to use d_3-T as internal standard for LC-MS-MS analysis. The method was validated using protocols specified by the

US Food and Drug Administration and considered the reference method in accordance with the literature and in view of the widely accepted superior specificity. The calibration curve for T was linear up to 2000 ng dL^{-1} ($\sim$69 nmol L^{-1}) and the limit of quantitation (LOQ) was 20 ng dL^{-1} ($\sim$0.69 nmol L^{-1}). Recovery ranged from 71.4 to 77.0% depending on the concentration.

Because the serum T values were spread over a wide range ($<$50–1500 ng dL^{-1} or $\sim$2–52 nmol L^{-1}) and the number of samples was large ($>$100), the method comparison should not be affected by individual variability and it should hold true in studies by other workers.

The immunoassays were biased towards either lower or higher values, or they over-estimated or underestimated serum T levels. The results confirmed previously published opinions that for the immunoassays, adult male reference ranges need to be established in each individual laboratory, as opposed to being provided by the automated instrument manufacturer. At low concentrations, the immunoassays were biased and suffered from a lack of precision and accuracy. Therefore, it seemed that they cannot be used to measure T levels accurately in serum from females or pre-pubertal males, unless those level are abnormally high.

The same team developed an LC-MS-MS method to quantitate T and trideuterated T ([16,16,17-^{2}H$_3$]testosterone or d_3-T) in human serum, for clinical studies involving administration of d_3-T (Starcevic et al., 2003). To a serum aliquot (2 mL), 19-nortestosterone internal standard, sodium acetate buffer (pH 5.5) and diethyl ether (5 mL) were added, the mixture was shaken and centrifuged and the ether recovered and evaporated to dryness. The residue was reconstituted in methanol (100 μL) and injected (15 μL) into an LC–triple quadrupole MS system operated in the positive ion mode. After chromatography on a C$_{18}$ column with gradient elution, T was monitored with the transition m/z 289 to 97, the calibration curve was linear in the range 0.5–20 ng mL^{-1} ($\sim$1.7–69 nmol L^{-1}), the LOQ was 0.5 ng mL^{-1} ($\sim$1.7 nmol L^{-1}) and the recovery was 91.5%. d_3-T was monitored with the transition m/z 292 to 97, the calibration curve was linear in the range 0.05–2 ng mL^{-1} ($\sim$0.17–7 nmol L^{-1}), the LOQ was 0.05 ng mL^{-1} ($\sim$0.17 nmol L^{-1}) and the recovery was 96.4%. Accuracy and precision were determined. For the 19-nortestosterone internal standard, the transition was m/z 275 to 109. The serum T concentrations in 38 healthy subjects ranged from 2.5 to 14.0 ng mL^{-1} ($\sim$8.7–48.6 nmol L^{-1}) in baseline samples and samples collected during an infusion of d_3-T. Good sensitivity was achieved for d_3-T. The ideal internal standard would have been trideuterated T, but it was the analyte, hence 19-nortestosterone was used, and the authors hoped that ion suppression or instrument instability would affect it to the same extent as T and d_3-T. The transitions to monitor were selected so as to avoid interfering peaks. Immunoassays cannot be used to quantitate T and d_3-T separately because they cannot distinguish between the two, therefore LC-MS-MS provides a unique advantage.

Wang et al. (2004b) then used the Starcevic et al. method to determine T metabolic clearance and production rates in normal men by stable isotope dilution and LC-MS-MS, to assess the influence of ethnicity and age. Subjects underwent a constant infusion of d_3-T, serum d_3-T concentrations were measured by LC-MS-MS and serum total T ($= T + d_3$-T) was measured by RIA. There were no ethnic differences. Values were lower in older than younger men. This is the first report of the use of LC-MS-MS to quantitate labeled and unlabeled analogs of an anabolic androgenic steroid in a clearance study. The amount of

d_3-T infused was small and the LC-MS-MS method has sufficient sensitivity to allow the measurement of production rate and metabolic clearance rate with a small enough d_3-T dose to avoid perturbing the endogenous production of T.

Cawood et al. (2005) improved Wang et al.'s method by reducing the plasma or serum sample volume 40-fold from 2 mL to 50 μL. The estimated T functional LOQ of 0.3 nmol L^{-1} is not far from the d_3-T LOQ of 0.05 ng mL^{-1} (~0.17 nmol L^{-1}) in the Starcevic et al. method, which targeted low d_3-T levels but had no need to measure particularly low T levels (T LOQ 0.5 ng mL,$^{-1}$ ~1.7 nmol L^{-1}). In Cawood et al.'s study, to prepare the samples, d_2-T internal standard and precipitating reagent containing zinc sulfate and methanol were added (100 μL) and the supernatant (40 μL) was analyzed. After chromatography on a C_{18} column with isocratic elution, T was monitored with the transition from m/z 289.1 to 96.7. The calibration curve was linear from 0.25 to 100 nmol L^{-1}. The method is simple, quick, robust, accurate and validated. Not only does it make it possible to measure low T levels (<8.0 nmol L^{-1} serum, low enough for most clinical purposes) in women, but also it is suitable for routine high-throughput use.

In view of the major role played by DHT in the development of benign prostatic hyperplasia, Choi et al. (2003) set out to quantitate simultaneously T glucuronide and the glucuronide of its active metabolite, DHT, in human urine, in hope of assessing the activity of 5α-reductase, the enzyme that converts T to DHT. Their method used [16,16,17-^{2}H$_3$]testosterone-17β-glucuronide as internal standard. A calibration curve was prepared using urine stripped of steroids, then spiked with 0.2–400 μg L^{-1} (~0.4–860 nmol L^{-1}) for T glucuronide and 3–200 μg L^{-1} (~6–430 nmol L^{-1}) for DHT glucuronide. Recoveries were 94 and 97% for T glucuronide and DHT glucuronide, respectively. Limits of quantitation were 1 μg L^{-1} (~2 nmol L^{-1}) for T glucuronide and 3 μg L^{-1} (~6 nmol L^{-1}) for DHT glucuronide. The assay was applied to 24-h urine samples from 27 patients, 19 with benign prostatic hyperplasia and eight healthy. For T glucuronide, there was no significant difference between patients and controls. In contrast, the levels of DHT glucuronide were significantly lower in the patients. This seems contrary to published results where patients with benign prostatic hyperplasia had higher plasma and tissue concentrations of DHT, unless the lower urinary concentration indicates lower glucuronidation and higher free DHT in plasma and tissue. The ability to measure urinary T glucuronide and DHT glucuronide may help in assessing a patient's androgen status in any condition where abnormal 5α-reductase activity is suspected.

Mitamura et al. (2003) quantitated androstenediol 3-sulfate and DHEA sulfate in human serum, in the presence of the deuterated analogs as internal standards, by LC–ion-trap MS in the negative ion mode. Calibration curves were linear in the range 10–400 ng mL(~26–1020 nmol L^{-1}) for the diol sulfate and 0.05–8 μg mL^{-1} (~0.1–21 nmol L^{-1}) for DHEA sulfate. The assay was applied to samples from 14 healthy men, 19 prostate cancer patients and seven patients with benign prostatic hyperplasia. The concentration ranges overlapped but were lower in the cancer patients. This showed that the method was practical. It needs to be applied to larger populations despite the challenge of finding healthy controls among older men.

In an earlier study, Mitamura et al. (2002) used LC-MS in the full-scan mode and GC-MS to identify six dehydroepiandrosterone metabolites formed by incubation with a human prostate homogenate of androst-5-ene-3β,17β-diol (major metabolite), androst-4-ene-3,17-dione, T, 5α-DHT (identified by GC-MS but not LC-MS due to insufficient sensitivity), androsterone (Figure 2.2) and 7α-hydroxydehydroepiandrosterone.

Nishiyama *et al.* (2004) set out to quantitate DHT in human serum and prostate biopsy tissue using LC-MS with charged derivatization to the N-methylpyridinium derivative. Prostate tissue DHT was dissolved in alkaline solution, extracted by SPE and derivatized before LC-MS analysis [ESI, positive ion mode, selected reaction monitoring (SRM)]. Pretreatment values in patients with ($N = 69$) or without ($N = 34$) prostate cancer were, for serum DHT, 423.9 pg mL^{-1} ($\sim$1.46 nmol L^{-1}) with SD 243.2 pg mL^{-1} ($\sim$0.84 nmol L^{-1}) or 462.5 pg mL^{-1} ($\sim$1.60 nmol L^{-1}) with SD 274.6 pg mL^{-1} ($\sim$0.95 nmol L^{-1}), respectively, and for prostate DHT the values were 5.61 ng g^{-1} ($\sim$0.0193 nmol g^{-1}) with SD 1.96 ng g^{-1} ($\sim$0.0067 nmol g^{-1}) or 5.19 ng g^{-1} ($\sim$0.0179 nmol g^{-1}) with SD 2.50 ng g^{-1} ($\sim$0.0086 nmol g^{-1}), respectively. In 30 cancer patients who received androgen deprivation therapy with castration and flutamide, serum DHT fell from 503.4 pg mL^{-1} ($\sim$1.74 nmol L^{-1}) with SD 315.9 pg mL^{-1} ($\sim$1.09 nmol L^{-1}) to 38 pg mL^{-1} ($\sim$0.13 nmol L^{-1}) with SD 31.2 pg mL^{-1} ($\sim$0.11 nmol L^{-1}), and prostate DHT from 5.44 ng g^{-1} ($\sim$0.0188 nmol g^{-1}) with SD 2.84 ng g^{-1} ($\sim$0.0098 nmol g^{-1}) to 1.35 ng g^{-1} ($\sim$0.0047 nmol g^{-1}) with SD 1.32 ng g^{-1} ($\sim$0.0046 nmol g^{-1}). These results showed that not all of the prostate DHT was gone after treatment. This was the first report on changes in prostate DHT upon androgen deprivation therapy.

Higashi *et al.* (2005) quantitated DHT and T in prostate tissue using LC-MS-MS with the same charged derivatization. Although the samples in this study were obtained after prostatectomy, the 10-mg sample size is consistent with the amount of tissue expected from a needle biopsy. After adding the internal standard, 19,19,19-[^{2}H$_3$]T, the tissue was homogenized, including in the extraction solvent, aqueous methanol, in which it was heated. After purification through an Oasis HLB cartridge and derivatization, LC-MS-MS analysis with ESI in the positive ion mode was applied. Recoveries were quantitative. Limits of detection were 1.0 ng g^{-1} ($\sim$0.003 nmol g^{-1}) of tissue for both analytes. The method was used to quantitate T and DHT in the prostates of patients with benign prostatic hyperplasia ($N = 7$) and prostate cancer ($N = 3$). The range for DHT was 5.18 $\pm$ 1.32 ng g^{-1} ($\sim$0.0179 $\pm$ $\sim$0.0046 nmol g^{-1}) in the first seven patients' tissues. That it was not detected in the three cancer patients is consistent with the expected effects of neoadjuvant hormone therapy. T was below the LOQ of 1.0 ng mL^{-1} ($\sim$3 nmol L^{-1}) in all samples, an observation consistent with the literature and with the knowledge that T is rapidly converted to DHT in the prostate. DHT is indeed the more important analyte of the two since its elevated levels are believed to be a major factor in the pathogenesis of prostate cancer.

Zhao *et al.* (2004) determined the steroid composition of human testicular fluid using LC-MS-MS. After a one-step diethyl ether extraction, they monitored T (m/z 289.0–108.8), DHT (m/z 291.0–255.0) and 5α-androstane-3α,17β-diol (m/z 275.2–257.1) (Figure 2.2), and also estradiol, which is an estrogen, not an anabolic steroid. Calibration curves were linear in the range 0.1–50 ng mL^{-1} ($\sim$0.3–170 nmol L^{-1}) for T, 0.02–1 ng mL^{-1} ($\sim$0.07–3 nmol L^{-1}) for DHT and 2–10 ng mL^{-1} ($\sim$7–34 nmol L^{-1}) for 5α-androstane-3α,17β-diol. Androgen recoveries ranged from 60 to 86% depending on the compound. In samples obtained by percutaneous testicular aspiration from 10 male volunteers, T concentrations were 572 $\pm$ 102 ng mL^{-1} ($\sim$1990 $\pm$ 350 nmol L^{-1}), a range similar to that previously measured by RIA by the same authors.

For the other two androgens, the concentrations were below the limit of detection by RIA but measurable by LC-MS-MS and therefore reported for the first time. This method should be of help in studying the relationship between intratesticular T and spermatogenesis in men.

2.4 CONCLUSION

The earliest clinical studies concerned with androgens measured them mostly by RIA, therefore they suffered from potential cross-reactions. Chromatographic methods coupled with mass spectrometry, namely GC-MS and LC-MS, brought much needed specificity, and the ability to distinguish deuterated, non-radioactive androgens from unlabeled analogs. As a result, stable isotope dilution studies made it possible to give to volunteers small enough doses of labeled T to avoid perturbing endogenous production and yet assess production and clearance rates. The advent of LC-MS-MS brought such higher throughputs than GC-MS and even superior specificity that it is now considered the gold standard. LC-MS-MS measurement of anabolic steroids is a sharper research tool that opens new windows into the understanding of issues of global importance such as human reproduction, contraception, prostate cancer and aging.

REFERENCES

Bean, K.A. and Henion, J.D. (1997). Direct determination of anabolic steroid conjugates in human urine by combined high-performance liquid chromatography and tandem mass spectrometry. *J. Chromatogr. B*, **690**, 65–75.

Borts, D.J. and Bowers, L.D. (2000) Direct measurement of urinary testosterone and epitestosterone conjugates using high-performance liquid chromatography/tandem mass spectrometry. *J. Mass Spectrom.*, **35**, 50–61.

Bowers, L.D. and Sanaullah (1996). Direct measurement of steroid sulfate and glucuronide conjugates with high-performance liquid chromatography-mass spectrometry. *J. Chromatogr. B*, **687**, 61–68.

Buiarelli, F., Cartoni, G.P., Coccioli, F., De Rossi A. and Neri, B. (2003). Determination of trenbolone and its metabolite in bovine fluids by liquid chromatography–tandem mass spectrometry. *J. Chromatogr. B*, **784**, 1–15.

Buiarelli, F., Coccioli, F., Merolle, M., Neri, B. and Terracciano, A. (2004). Development of a liquid chromatography–tandem mass spectrometry method for the identification of natural androgen steroids and their conjugates in urine samples. *Anal. Chim. Acta*, **526**, 113–120.

Catlin, D.H., Sekera, M.H., Ahrens, B., Starcevic, B., Chang, Y.-C. and Hatton, C.K. (2004). Tetrahydrogestrinone: discovery, synthesis, and detection in urine. *Rapid Commun. Mass Spectrom.*, **18**, 1245–1249.

Cawood, M.L., Field, H.P., Ford, C.G., Gillingwater, S., Kicman, A., Cowan, D. and Barth, J.H. (2005). Testosterone measurement by isotope-dilution liquid chromatography–tandem mass spectrometry: validation of a method for routine clinical practice. *Clin. Chem.*, **51**, 1472–1479.

Chang, Y.-C., Li, C.-M., Li, L.-A., Jong, S.-B., Liao, P.C. and Chang, L.W. (2003). Quantitative measurement of male steroid hormones using automated on-line solid phase extraction liquid chromatography–tandem mass spectrometry and comparison with radioimmunoassay. *Analyst*, **128**, 363–368.

Choi, M.H., Kim, J.N. and Chung, B.C. (2003). Rapid HPLC–electrospray tandem mass spectrometric assay for urinary testosterone and dihydrotestosterone glucuronides from patients with benign prostate hyperplasia. *Clin. Chem.*, **49**, 322–325.

De Wasch, K., Le Bizec, B., De Brabander, H., Andre, F. and Impens, S. (2001). Consequence of boar edible tissue consumption on urinary profiles of nandrolone metabolites. II. Identification and quantification of 19-norsteroids responsible for 19-norandrosterone and 19-noretiocholanolone excretion in human urine. *Rapid Commun. Mass Spectrom.*, **15**, 1442–1447.

Draisci, R., Palleschi, L., Ferreti, E., Lucentini, L. and Caammarata, P. (2000). Quantitation of anabolic hormones and their metabolites in bovine serum and urine by liquid chromatography–tandem mass spectrometry. *J. Chromatogr. A*, **870**, 511–522.

Draisci, R., Palleschi, L., Marchiafava, C., Ferreti, E. and delli Quadri, F. (2001). Confirmatory analysis of residues of stanozolol and its major metabolite in bovine urine by liquid chromatography–tandem mass spectrometry. *J. Chromatogr. A*, **926**, 69–77.

Draisci, R., Palleschi, L., Ferreti, E., Lucentini, L. and delli Quadri, F. (2003). Confirmatory analysis of 17β-boldenone and androsta-1,4-diene-3,17-dione in bovine urine by liquid chromatography–tandem mass spectrometry. *J. Chromatogr. B*, **789**, 219–226.

Guo, T., Chan, M. and Soldin, S.J. (2004). Steroid profiles using liquid chromatography–tandem mass spectrometry with atmospheric pressure photoionization source. *Arch. Pathol. Lab. Med.*, **128**, 469–475.

Higashi, T., Yamauchi, A., Shimada, K., Koh, E., Mizokami, A. and Namiki, M. (2005). Determination of prostatic androgens in 10 mg of tissue using liquid chromatography–tandem mass spectrometry with charged derivatization. *Anal. Bioanal. Chem.*, **382**, 1035–1043.

Ho, E.N., Yiu, K.C., Tang, F.P., Dehennin, L., Plou, P., Bonnaire, Y. and Wan, T.S. (2004). Detection of endogenous boldenone in the entire male horses. *J. Chromatogr. B*, **808**, 287–294.

Hori, M. and Nakazawa, H. (2000). Determination of trenbolone and zeranol in bovine muscle and liver by liquid chromatography–electrospray mass spectrometry. *J. Chromatogr. A*, **882**, 53–62.

Joos, P.E. and Van Ryckeghem, M. (1999). Liquid chromatography–tandem mass spectrometry of some anabolic steroids. *Anal. Chem.*, **71**, 4701–4710.

Kim, J.Y., Choi, M.H., Kim, S.J. and Chung, B.C. (2000a). Measurement of 19-nortestosterone and its esters in equine plasma by high-performance liquid chromatography with tandem mass spectrometry. *Rapid Commun. Mass Spectrom.*, **14**, 1835–1840.

Kim, Y., Lee, Y., Kim, M., Yim, Y.-H. and Lee, W. (2000b). Determination of the metabolites of gestrinone in human urine by high performance liquid chromatography, liquid chromatography/mass spectrometry and gas chromatography/mass spectrometry. *Rapid Commun. Mass Spectrom.*, **14**, 1717–1726.

Kuuranne, T., Vahermo, M., Leinonen, A. and Kostiainen, R. (2000). Electrospray and atmospheric pressure chemical ionization tandem mass spectrometric behavior of eight anabolic steroid glucuronides. *J. Am. Soc. Mass Spectrom.*, **11**, 722–730.

Kuuranne, T., Aitio, O., Vahermo, M., Elovaara, E. and Kostiainen, R. (2002). Enzyme-assisted synthesis and structure characterization of glucuronide conjugates of methyltestosterone (17 alpha-methylandrost-4-en-17 beta-ol-3-one) and nandrolone (est-4-en-17 beta-ol-3-one) metabolites. *Bioconjug. Chem.*, **13**, 194–199.

Le Bizec, B., Gaudin, I., Monteau, F., Andre, F., Impens, S., De Wasch, K. and De Brabander, H. (2000). Consequence of boar edible tissue consumption on urinary profiles of nandrolone metabolites. I. Mass spectrometric detection and quantification of 19-norandrosterone and 19-noretiocholanolone in human urine. *Rapid Commun. Mass Spectrom.*, **14**, 1058–1065.

Leinonen, A., Kuuranne, T. and Kostiainen, R. (2002). Liquid chromatography/mass spectrometry in anabolic steroid analysis – optimization and comparison of three ionization techniques: electrospray ionization, atmospheric pressure chemical ionization and atmospheric pressure photoionization. *J. Mass Spectrom.*, **37**, 693–698.

Liu, S., Griffiths, W.J. and Sjövall, J. (2003). Capillary liquid chromatography/electrospray mass spectrometry for analysis of steroid sulfates in biological samples. *Anal. Chem.*, **75**, 791–797.

Lopez de Alda, M.J., Diaz-Cruz, S., Petrovic, M. and Barcelo, D. (2003). Liquid chromatography–(tandem) mass spectrometry of selected emerging pollutants (steroid sex hormones, drugs and alkylphenolic surfactants) in the aquatic environment. *J. Chromatogr. A*, **1000**, 503–526.

Ma, Y. and Kim, H.J. (1997). Determination of steroids by liquid chromatography/mass spectrometry. *J. Am. Soc. Mass Spectrom.* **8**, 1010–1020.

Magnusson, M.O. and Sandstrom, R. (2004). Quantitative analysis of eight testosterone metabolites using column switching and liquid chromatography/tandem mass spectrometry. *Rapid Commun. Mass Spectrom.*, **18**, 1089–1094.

McKinney, A.R., Suann, C.J., Dunstan, A.J., Mulley, S.L., Ridley, D.D. and Stenhouse, A.M. (2004). Detection of stanozolol and its metabolites in equine urine by liquid chromatography–electrospray ionization ion trap mass spectrometry. *J. Chromatogr. B*, **811**, 75–83.

Mitamura, K., Nakagawa, T., Shimada, K., Namiki, M., Koh, E., Mizokami, A. and Honma, S. (2002). Identification of dehydroepiandrosterone metabolites formed from human prostate homogenate using liquid chromatography–mass spectrometry and gas chromatography–mass spectrometry. *J. Chromatogr. A*, **961**, 97–105.

Mitamura, K., Nagaoka, Y., Shimada, K., Honma, S., Namiki, M., Koh, E. and Mizokami, A. (2003). Simultaneous determination of androstenediol 3-sulfate and dehydroepiandrosterone sulfate in human serum using isotope diluted liquid chromatography–electrospray ionization–mass spectrometry. *J. Chromatogr. B*, **796**, 121–130.

Nielen, M.W., Vissers, J.P., Fuchs, R.E., van Velde, J.W. and Lommen, A. (2001). Screening for anabolic steroids and related compounds in illegal cocktails by liquid chromatography/time-of-flight mass spectrometry and liquid chromatography/quadrupole time-of-flight tandem mass spectrometry with accurate mass measurement. *Rapid Commun. Mass Spectrom.*, **15**, 1577–1585.

Nishiyama, T., Hashimoto, Y. and Takahashi, K. (2004). The influence of androgen deprivation therapy on dihydrotestosterone levels in the prostatic tissue of patients with prostate cancer. *Clin. Cancer Res.*, **10**, 7121–7126.

Poelmans, S., De Wasch, K., De Brabander, H.F., Van De Wiele, M., Courtheyn, D., van Ginkel, L.A., Sterk, S.S., Delahaut, Ph., Dubois, M., Schilt, R., Nielen, M., Vercammen, J., Impens, S., Stephany, R., Hamoir, T., Pottie, G., Van Poucke, C. and Van Peteghem, C. (2002). Analytical possibilities for the detection of stanozolol and its metabolites. *Anal Chim. Acta*, **473**, 39–47.

Pu, F., McKinney, A.R., Stenhouse, A.M., Suann, C.J. and McLeod, M.D. (2004). Direct detection of boldenone sulfate and glucuronide conjugates in horse urine by ion trap liquid chromatography–mass spectrometry. *J. Chromatogr. B*, **813**, 241–246.

Shackleton, C.H., Chuang, H., Kim, J., de la Torre, X. and Segura, J. (1997). Electrospray mass spectrometry of testosterone esters: potential for use in doping control. *Steroids*, **62**, 523–529.

Starcevic, B., DiStefano, E., Wang, C. and Catlin, D.H. (2003). Liquid chromatography–tandem mass spectrometry assay for human serum testosterone and trideuterated testosterone. *J. Chromatogr. B*, **792**, 197–204.

Thevis, M., Geyer, H., Mareck, U. and Schänzer, W. (2005). Screening for unknown synthetic steroids in human urine by liquid chromatography–tandem mass spectrometry. *J. Mass Spectrom.*, **40**, 955–962.

Tiller, P.R., Cunniff, J., Land, A.P., Schwartz, J., Jardine, I., Wakefield, M., Lopez, L., Newton, J.F., Burton, R.D., Folk, B.M., Buhrman, D.L., Price, P., Wu, D. (1997). Drug quantitation on a benchtop liquid chromatography–tandem mass spectrometry system. *J. Chromatogr. A* **771**, 119–125.

Wiele Van de, M., De Wasch, K., Vercammen, J., Courtheyn, D., De Brabander, H. and Impens, S. (2000). Determination of 16β-hydroxystanozolol in urine and faeces by liquid chromatography-multiple mass spectrometry. *J. Chromatogr. A*, **904**, 203–209.

Van Poucke, C. and Van Peteghem, C. (2002). Development and validation of a multi-analyte method for the detection of anabolic steroids in bovine urine with liquid chromatography–tandem mass spectrometry. *J. Chromatogr. B*, **772**, 211–217.

Van Poucke, C., Van de Velde, M. and Van Peteghem, C. (2005). Combination of liquid chromatography/tandem mass spectrometry and gas chromatography/mass spectrometry for the detection of 21 anabolic steroid residues in bovine urine. *J. Mass Spectrom.*, **40**, 731–738.

Wang, C., Catlin, D.H., Demers, L.M., Starcevic, B. and Swerdloff, R.S. (2004a). Measurement of total serum testosterone in adult men: comparison of current laboratory methods *versus* liquid chromatography–tandem mass spectrometry. *J. Clin. Endocrinol. Metab.*, **89**, 534–543.

Wang, C., Catlin, D.H., Starcevic, B., Leung, A., DiStefano, E., Lucas, G., Hull, L. and Swerdloff, R.S. (2004b). Testosterone metabolic clearance and production rates determined by stable isotope dilution/tandem mass spectrometry in normal men: influence of ethnicity and age. *J. Clin. Endocrinol. Metab.*, **89**, 2936–2941.

Yu, N.H., Ho, E.N.M., Leung, D.K.K. and Wan, T.S.M. (2005). Screening of anabolic steroids in horse urine by liquid chromatography–tandem mass spectrometry. *J. Pharm. Biomed. Anal.*, **37**, 1031–1038.

Zhao, M., Baker, S.D., Yan, X., Zhao, Y., Wright, W.W., Zirkin, B.R. and Jarow, J.P. (2004). Simultaneous determination of steroid composition of human testicular fluid using liquid chromatography–tandem mass spectrometry. *Steroids*, **69**, 721–726.

3

High-performance Liquid Chromatography in the Analysis of Active Ingredients in Herbal Nutritional Supplements

Amitava Dasgupta

Department of Pathology and Laboratory Medicine, University of Texas Health Sciences Center at Houston, Houston, TX 77030, USA

3.1 INTRODUCTION

Herbal dietary supplements (herbal remedies) are readily available in the USA from herbal stores without prescriptions. Chinese medicines are an important component of herbal remedies available today. Ayurvedic medicines are widely used in India and some preparations are available in the USA. Unlike Western medicines, herbal supplements are crude plant extracts and both active ingredients and undesired components are present side by side. Moreover, the amounts of active ingredients may vary widely among different batches of the same products or between different manufacturers.

Herbal remedies can be toxic and have side-effects, or may show significant drug–drug interactions with Western medicines. Ginseng, St John's wort, ma huang, kava, ginkgo biloba, Danshen, chan su, feverfew, garlic, ginger, saw palmetto, comfrey, pokeweed, hawthorne, dong quai and cat's claw are used by the general population in the USA. Common herbal remedies and their intended uses are summarized in Table 3.1. Gulla *et al.*[1] published a survey of 369 patient–escort pairs and reported that 174 patients used herbs. Most common was ginseng (20%), followed by echinacea (19%), ginkgo biloba (15%) and St John's wort (14%).[1] Klepser and Klepser reported their opinions regarding safe and unsafe herbal products. Many herbs that have been classified as unsafe include comfrey, life root, borage, calamus, chaparral, licorice and ma huang. Relatively safe herbs are feverfew,

Chromatographic Methods in Clinical Chemistry and Toxicology Edited by R. L. Bertholf and R. E. Winecker
© 2007 John Wiley & Sons, Ltd

Table 3.1 Intended use of some herbal medicines

Herbal medicine	Intended use
Ginseng	Tonic capable of invigorating users physically and, mentally and also used for stress relief
Danshen	Stimulation of heart
Chan Su (toxic)	Heart tonic
St John's wort	Treatment of mood disorders, particularly depression
Ginkgo biloba	Promoted mainly to sharpen mental focus in otherwise healthy adults and also in people with dementia
	Improvement of blood flow in brain and peripheral circulation
Valerian	Treatment of insomnia
Echinacea	Immune stimulant that helps increase resistance to cold, influenza and other infections, wound healing
Feverfew	Relief from migraine headache and arthritis
Garlic	Lowering cholesterol and blood pressure; preventing heart attacks and stroke
Aloe	Healing wounds, burns, skin ulcers. Also used as a laxative
Senna	Laxative
Kava[a]	Promoted for relief of anxiety and stress; sedative
Pokeweed	Antiviral and antineoplastic. Eating uncooked berry or root may cause serious poisoning
Comfrey[a]	Repairing of bone
Chaparral[a]	General cleansing
Ephedra[a]	Herbal weight loss products

[a]These herbal products may cause serious toxicity and there are also some reported cases of death due to use of such herbal supplements.

garlic, ginkgo, Asian ginseng, saw palmetto, St John's wort and valerian.[2] Common adverse reactions due to use of herbal remedies are summarized in Table 3.2.

The US Food and Drug Administration (FDA) mandates that only medicines have to be proven to be safe before release to market. Herbal products are classified as 'dietary supplements' and are marketed pursuant to the Dietary Supplement Health and Education Act of 1994. Herbal products are regulated differently in other countries. In the UK, for

Table 3.2 Toxicity of commonly used herbs

Toxicity	Herbal product
Allergic reaction	Aloe, chamomile, echinacea, garlic, cat's claw
Cardiovascular	Ephedra, danshen, chan su[a], oleander
Carcinogenic	Aloe, chaparral, comfrey, senna
Dermatological	Garlic, kava, St John's wort
Hepatic	Chaparral, comfrey[a], valerian
Hematological	Feverfew, garlic, ginkgo
Neurological	Ephedra, ginkgo, kava, St John's wort, valerian
Renal	Cat's claw, chaparral, ephedra, licorice

[a]Banned by the FDA.

example, any product not granted a license as a medical product by the Medicines Control Agency is treated as a food and cannot carry any health claim or medical advice on the label. Similarly, herbal products are sold as dietary supplement in The Netherlands. In Germany, documents called the German Commission E monographs are prepared by an interdisciplinary committee using historical information, chemical, pharmacological, clinical and toxicological studies, case reports, epidemiological data and manufacturer's unpublished data. If a herb has an approved monograph, it can be marketed.

Active components of herbal supplements can be measured using various analytical techniques, including high-performance liquid chromatography (HPLC) with photodiode-array detection, HPLC combined with mass spectrometry (MS), gas chromatography (GC), thin-layer chromatography (TLC) and immunoassays. Understanding concentrations of active ingredients maybe useful in evaluating herbal remedy-induced toxicity or the magnitude of interactions between herbal supplements and Western drugs.

3.2 St JOHN'S WORT

St John's wort is a very popular herbal antidepressant. The most widely available St John's wort preparation in the USA comes from the dried alcoholic extract of hypercian, a perennial aromatic shrub with bright yellow flowers that bloom from June to September. The flowers are believed to be most abundant and brightest around 24 June, the day traditionally believed to be the birthday of John the Baptist. Therefore, the name St John's wort became popular. The German Commission E monograph indicates that St John's wort can be used in supportive treatment of anxiety and depression. Many chemicals have been isolated from St John's wort, including hyperforin, adhyperforin, hypericin, pseudohypericin, protohypericin, protopseudoquercetin, isoquercitrin, rutin, amentoflavone, flavonoids and xanthones. However, hypericin, hyperforin and 1,3,5,7-tetrahydroxyxanthone are unique to St John's wort.[3]

3.2.1 Drug interactions with St John's wort

Interactions between St John's wort and digoxin are clinically significant. Johne *et al.* reported that 10 days' use of St John's wort could result in a 33% decrease in trough serum digoxin concentrations and a 26% increase in peak concentrations.[4] Durr *et al.* also confirmed the lower digoxin concentrations in healthy volunteers who concurrently took St John's wort.[5] Because St John's wort increases the metabolism of drugs by inducing liver enzymes, co-ingestion of St John's wort with warfarin, cyclosporin, oral contraceptives, protease inhibitors and other drugs has led to reported interactions and reduced therapeutic efficacy. Barone *et al.* reported two cases where renal transplant recipients started self-medication with St John's wort. Both patients experienced sub-therapeutic concentrations of cyclosporin and one patient developed acute graft rejection due to low cyclosporin level. In both patients, termination of the use of St John's wort returned their cyclosporin concentrations to therapeutic levels.[6] Another report describes a kidney transplant patient with a steady-state cyclosporin trough level between 100 and 130 ng mL^{-1} who suddenly had a sub-therapeutic cyclosporin level of 70 ng mL^{-1}, despite an increase in daily dose. The patient was using a tea containing St John's wort. After stopping St John's wort, his

Table 3.3   Common Drug–Herb Interactions

Herbal product	Interacting drug	Comments
Ginseng	Warfarin	Ginseng may decrease effectiveness of warfarin
	Phenelzine	Toxic symptoms: headache, insomnia, irritability
St John's wort	Paxil	Lethargy, incoherent, nausea
	Digoxin	Decreased AUC, peak and trough concentration of digoxin, may reduce effectiveness of digoxin
	Cyclosporin/FK 506	Lower cyclosporin/FK 506 concentrations due to increased clearance may cause transplant rejection
	Theophylline	Lower concentration thus decreases the efficacy of theophylline
	Indinavir	Lower concentration may cause treatment failure in patients with HIV
	Oral contraceptives	Lower concentration/failed birth control
Ginkgo biloba	Aspirin	Bleeding because ginkgo can inhibit clotting factors
	Warfarin	Hemorrhage
	Thiazide	Hypertension
Kava	Alprazolam	Additive effects with CNS depressants, alcohol
Garlic	Warfarin	Increases effectiveness of warfarin, bleeding
Ginger	Warfarin	Increases effectiveness of warfarin, bleeding
Feverfew	Warfarin	Increases effectiveness of warfarin, bleeding
Dong quai	Warfarin	Dong quai contains coumarin
		Dong quai increases INR[a] for warfarin/bleeding
Danshen	Warfarin	Increase effectiveness of warfarin due to reduced elimination
Borage oil	Phenobarbital	May lower seizure threshold
Evening primrose oil	Phenobarbital	May lower seizure threshold
Licorice	Spirolactone	May offset the effect of spirolactone

[a]International normalized ratio for prothrombin time.

cyclosporin level increased to 170 ng mL^{-1} after 5 days.[7] A significant reduction in the area under the plasma concentration curve (AUC) for tacrolimus was also observed in 10 stable renal transplant patients receiving St John's wort. Interestingly, no interaction was observed with mycophenolic acid.[8] St John's wort also reduced the AUC for the HIV-1 protease inhibitor indinavir by a mean of 57% and decreased the extrapolated trough by 81%. A reduction in indinavir exposure of this magnitude could lead to treatment failure.[9] A reduced plasma level of methadone was observed in the presence of St John's wort, resulting in reappearance of withdrawal symptoms.[10] Major interactions between St John's wort and Western medicines, and common drug–herb interactions, are summarized in Table 3.3.

3.2.2   Measurement of active ingredients of St John's wort using HPLC

Commercially available St John's wort is not prepared following rigorous pharmaceutical standards and wide variations in concentrations of active ingredients have been reported. One report involving the determination of hypericin and hyperforin content among eight brands of commercially available St John's wort preparations indicated that the hypericin content varied from 0.03 to 0.29% and the hyperforin content from 0.01 to 1.89%. One product (Nature's balance) had the lowest concentration of hypericin (0.03%). Although some products had hypericin and hyperforin contents comparable to the reported concentrations, other products did not match the claim of hypericin concentration on the package insert.[11] Li and Fitzloff described a rapid reversed-phase HPLC method for the determination of the major constituents of St John's wort, rutin, hyperoside, isoquercitrin, quercitrin, quercetin, pseudohypericin, hypericin and hyperforin. The specimens were extracted into methanol by two sonication steps (30 min each) at low temperature. The extraction efficiency for the major active ingredients of St John's wort was around 99%. HPLC analysis was carried out using a reversed-phase C_{18} column and a mobile phase gradient of water–acetonitrile–methanol–trifluoroacetic acid. The run time was 60 min and the analytes were measured with a photodiode-array detector.[12] Ruckert et al. described an HPLC method with electrochemical detection for the determination of hyperforin in St John's wort preparations. They used an isocratic mobile phase consisting of 10% ammonium acetate buffer (0.5 M, pH 3.7)–methanol–acetonitrile (10:40:50 v/v/v) and a flow-rate of 0.8 mL min^{-1}. Hyperforin was detected ampherometrically with a glassy carbon electrode at a potential of $+1.1$ V versus a silver/silver chloride/3 M potassium chloride reference electrode. The limit of detection was 0.05 ng of hyperforin on-column.[13] Mauri and Pietta used HPLC coupled simultaneously with a diode-array detector and an electrospray mass spectrometer for the analysis of hypercium extract. Hypericin, psudohypericin, hyperforin and adhyperforin were separated and identified based on their UV and mass spectra.[14]

Several studies indicate that hypericin and hyperforin are unstable and may degrade on exposure to light, heat or air.[15] Ang et al. studied the effects of pH and light exposure on the active ingredients of St John's wort, including pseudohypericin, hypericin, hyperforin and adhyperforin, in aqueous buffer solution and non-alcoholic, non-carbonated fruit-flavored beverages. The components were separated using a Luna C_8 column (250×2 mm i.d., particle size 5 µm) and a mobile phase delivered at a rate of 0.2 mL min^{-1} consisting of a 30-min linear gradient from 50 to 80% acetonitrile with a constant 3 mM ammonium formate, which was maintained for an additional 30 min. The authors used mass spectrometric detection in the negative electrospray mode and found that components of St John's wort were unstable under acidic aqueous conditions. Major degradation products of hyperforin in acidic aqueous solutions were identified as furohyperforin, furohyperforin hydroperoxide and furohyperforin isomer a.[16] Active ingredients of St John's wort are more stable when stored in the dark; under exposure to light, hypericin, hyperforin and adhyperforin all decomposed rapidly. Liu et al. reported the influence of light and solution pH on the stability of phloroglucinols (hyperforin and adhyperforin) and naphthodianthrones (hypericin, pseudohypericin, protohypericin and protopseudohypericin) extracted with methanol from St John's wort powder. They used liquid chromatography combined with mass spectrometry for analysis and found that when exposed to light, both hyperforin and adhyperforin in the extract rapidly degraded regardless of pH and complete degradation was observed within 12 h. In contrast, when protected from light, minimal degradation was observed even after 36 h. Under light and at neutral pH, phloroglucinols and naphthodianthrones showed different stability behaviors. The marked increase in oxidation when hyperforin is exposed to light even at neutral pH may indicate a susceptibility to light-induced free radical reactions where one of the end products is hyperforin hydroperoxide. However, hypericin (closed-ring structure) is not susceptible to light-induced formation of free radicals.[17]

3.2.3 Analysis of St John's wort extract with other analytical techniques

Although HPLC, in addition to HPLC combined with mass spectrometry (MS), is the most common method for the analysis of active components of St John's wort, Seger *et al.* used both HPLC-MS and GC-MS for the analysis of a supercritical fluid extract of St John's wort. Supercritical fluid extraction of plant material with carbon dioxide yields extracts enriched with lipophilic components. In addition to the dominating phloroglucinols hyperforin (36.5 ± 1.1%) and adhyperforin (4.6 ± 0.1%), the extracts mainly contained alkanes (predominately nonacosane), fatty acids and wax esters. The non-polar components tended to accumulate in a waxy phase resting at the top of the hyperforin-enriched phase. Highly polar compounds (naphthodianthrones) were not found. For the GC-MS analysis, the authors used electron ionization MS analysis (scan range: 40–640 amu). Ten oxygenated hyperforin derivatives were identified.[18]

Bilia *et al.* reported the efficiency of two-dimensional homonuclear ^{1}H–^{1}H correlated spectroscopy and two-dimensional reverse heteronuclear shift correlation spectroscopy in evaluating the composition of phloroglucinols, flavonols and naphthodianthrones in a dried extract of St John's wort. They successfully assigned carbon resonances for these three classes of compounds and also identified shikimic and chlorogenic acids, sucrose, lipid, polyphenols and traces of solvent (methanol) during the extraction process. This rapid technique is an alternative to HPLC, TLC or capillary GC for the analysis of St John's wort preparations.[19] Another report utilized near-infrared reflectance spectroscopy (NIRS) for the determination of two major constituents of St John's wort, including hyperforin.[20] HPLC was used as the reference method.

3.2.4 Measurement of hypericin and hyperforin in human plasma using HPLC

Several reports indicate that the magnitude of interactions between a Western drug and St John's wort depends on the concentrations of the active ingredients of St John's wort in plasma. Mai *et al.* reported that a St John's wort product containing a low amount of hyperforin did not affect the cyclosporin pharmacokinetics in renal transplant patients, but the group of patients who received St John's wort containing higher amounts of hyperforin showed significantly lower concentrations of cyclosporin and needed a 65% increase in their cyclosporin dose in order to maintain therapeutic levels.[21] Hyperforin induces cytochrome P450 mixed function oxidase, the liver enzyme responsible for the metabolism of many drugs.[22] Mannel also concluded that hyperforin is probably responsible for the induction of liver enzyme (CYP3A4) via activation of nuclear/pregnane and the xenobiotic receptor.[23]

Hyperforin concentration in human plasma can be measured using HPLC with UV detection at 287 nm. A Luna C_{18} column (150 × 4.6 mm i.d., particle size 3 μm, Phenomenex) was used and the mobile phase was prepared by adding a methanol–acetonitrile organic phase (3:2 v/v) to water so that the final composition of the organic phase was 92% and the aqueous phase 8%. Finally, 2 mL of formic acid and 2 mL of triethylamine were added to 1000 mL of mobile phase and the pH of the final mobile phase was 3.2. Hyperforin-containing or spiked plasma was mixed with acetonitrile and finally hyperforin was extracted using a solid-phase extraction column. Benzo[*k*]fluoranthene was used as the internal standard. The limit of detection was 4 ng mL^{-1} and the limit of quantitation was 10 ng mL^{-1}.[24] Bauer *et al.* described HPLC combined with UV detection for the determination of hyperforin and HPLC combined with fluorimetric detection for the determination of hypericin and psudohypericin in human plasma. They used liquid–liquid extraction. The limit of quantitation was 10 ng mL^{-1} for hyperforin and 0.25 ng mL^{-1} for both hypericin and pseudohypericin.[25]

HPLC-MS can also be utilized for analysis of active components of St John's wort in human plasma. Pirker *et al.* used liquid–liquid extraction and HPLC-MS-MS for the simultaneous determination of hypericin and hyperforin in human plasma and serum. For sample preparation, 1 mL of plasma containing hypericin and hyperforin was mixed with 0.4 mL of DMSO and 0.15 mL of acetonitrile, followed by mixing for 30 s and extraction into 1 mL of ethyl acetate–hexane (70:30 v/v). After removing the ethyl acetate–hexane layer, the residue was extracted again and both organic phases were combined and concentrated under nitrogen for further analysis. The recovery of hyperforin (89.9–100.1%) was much higher than that of hypericin (32.2–35.6%). The assay was linear for hypericin concentrations between 8.4 and 28.7 ng mL^{-1} and for hyperforin concentrations from 21.6 to 242.6 ng mL^{-1}.[26] Riedel *et al.* also described an HPLC method combined with MS-MS for the determination of hypericin and hyperforin concentrations in human plasma. They used liquid–liquid extraction with ethyl acetate and hexane and used a reversed-phase (RP-18) column for analysis. The limit of quantitaion was 0.05 ng mL^{-1} for hypericin and 0.035 ng mL^{-1} for hyperforin. The hypericin assay was linear between 0.05 and 10 ng mL^{-1} and the hyperforin assay between 0.035 and 100 ng mL^{-1}.[27]

3.3 HERBAL SUPPLEMENTS WITH DIGOXIN-LIKE IMMUNOREACTIVITY

Several herbal supplements, including several Chinese medicines, have digoxin-like immunoreactivity because some components have structural similarity to digoxin. Digoxin immunoassays available commercially may use either a monoclonal antibody specific to digoxin or polyclonal antibodies. In general, assays that employ polyclonal antibodies against digoxin, such as the fluorescence polarization immunoassay (rabbit polyclonal antibody) and microparticle enzyme immunoassay (both marketed by Abbott Laboratories), are subject to more interference by Chinese medicines than other digoxin immunoassays based on monoclonal antibodies. Several Chinese medicines such as chan su, lu-shen-wan, danshen and Asian and Siberian ginseng interfere with various digoxin immunoassays.[28–35]

The Chinese medicine chan su is prepared from the dried white secretion of the auricular and skin glands of Chinese toads (*Bufo melanostictus* Schneider and *Bufo bufo gargarzinas* Gantor). Chan su is also a major component of traditional Chinese medicines lu-shen-wan and kyushin.[36] These medicines are used for the treatment of tonsillitis, sore throat, furuncle, palpitations, etc., because of their anesthetic and antibiotic action. Chan su, given in small doses, also stimulates myocardial contraction, has an anti-inflammatory effect and is analgesic. The cardiotonic effect of chan su is due to its major bufadienolides such as bufalin, cinobufagin and resibufogenin.[37] At high dosages, chan su causes cardiac arrhythmia, breathlessness, convulsion and coma. Death of a Chinese woman after ingestion of Chinese herbal tea containing chan su has been reported.[28] Structural similarity between bufadienolides and digoxin accounts for the digoxin-like immunoreactivity of chan su. Fushimi and Amino reported a serum digoxin concentration of 0.51 nmol L^{-1} (0.4 ng mL^{-1}) in a healthy volunteer after ingestion of kyushin tablets containing chan su as the major component.[29] Panesar reported an apparent digoxin concentration of $1124 \, \text{pmol L}^{-1}$ ($0.88 \, \text{ng mL}^{-1}$) in healthy volunteers who ingested lu-shen-wan pills.[30]

Ingestion of chan su and related drugs prepared from toad venom cause digoxin-like immunoreactivity in serum. Moreover, it caused positive interference (falsely elevated serum digoxin levels) in serum digoxin measurement by fluorescence polarization immunoassay (FPIA) (Abbott Laboratories) and negative interference (falsely lowered serum digoxin values) by the microparticle enzyme immunoassay (MEIA) (Abbott Laboratories). Other digoxin immunoassays, such as EMIT 2000, Synchron LX system (Beckman), Tina-quant (Roche Diagnostics) and turbidimetric (Bayer Diagnostics), are also affected but the magnitudes of interference are less significant than in the FPIA assay. The chemiluminescent assay (CLIA) marketed by the Bayer Diagnostics is free from such interferences. The interfering components in chan su are very strongly bound to serum proteins and are absent in the protein-free ultrafiltrates. On the other hand, digoxin is only 25% bound to serum protein and is present in the ultrafiltrate. Therefore, monitoring free digoxin concentration eliminated this interference of chan su in serum digoxin measurement. Another way to eliminate this interference is to use a specific CLIA (Bayer Diagnostics).[31]

Danshen is a Chinese medicine prepared from the root of the Chinese medicinal plant *Salvia miltiorrhiza*. This herb has been used in China for many centuries for treating various cardiovascular diseases, including angina pectoris. Danshen caused modest interference with polyclonal-based digoxin immunoassays such as MEIA and FPIA. A digoxin CLIA (Bayer), EMIT 2000 digoxin assay and Roche and Beckmann digoxin assays are also free from interference from Danshen.[32]

McRae reported a case where ingestion of Siberian ginseng was associated with an elevated digoxin level in a 74-year-old man. In this patient, the serum digoxin level had been maintained between 0.9 and $2.2 \, \text{ng mL}^{-1}$ over a period of 10 years. After ingestion of Siberian ginseng, his serum digoxin level increased to $5.2 \, \text{ng mL}^{-1}$, although the patient did not experience any sign of digoxin toxicity.[33] Another report indicated that Siberian ginseng produces only modest interference in the FPIA and MEIA digoxin assays. Asian ginseng also showed modest positive (FPIA) and modest negative (MEIA) interference. Again, the EMIT 2000, Bayer (both turbidimetric and CLIA), Roche (Tina-quant) and Beckman (Synchron LX system) digoxin assays were free from interference from both Asian and Siberian ginseng.[34,35]

3.3.1 Use of HPLC for the determination of chan su, danshen and ginsengs

Major bufadienolide components of chan su and related Chinese medicines can be analyzed by TLC and HPLC. Hong *et al.* described an HPLC protocol for the determination of bufalin, cinobufagin and resibufogenin, the major active ingredients of chan su and liu-shen-wan. They extracted bufadienolides from Chinese medicines using chloroform and ultrasonication and then applied HPLC with an RP-18 column and methanol–water (74:26 v/v) as mobile phase. They reported differences in composition of bufadienolides in 11 different pills that they analyzed and concluded that the variability of active constituents in products available to the public in Hong Kong may represent a hazard to public health.[38] Another report described the use of TLC using silica gel for the analysis of chan su. The TLC plates were developed using chloroform–methanol–water (75:20:5 v/v/v). The authors also developed an HPLC technique using a linear water–methanol gradient and an RP-18 column. The compounds eluted were detected at 220 and 300 nm.[39] Wang et al. described an HPLC method for the simultaneous determination of four bufadienolides in human liver. They used solid-phase extraction, a reversed-phase column and photodiode-array detection. The detection limits of the method were 0.4 ng for cinobufotalin and bufalin and 0.5 ng for cinobufagin and resibufogenin.[40]

An extract of Radix Salvia miltiorrhiza is used in the Chinese herbal product danshen and related medicines used as heart tonics. Shi et al. successfully separated components of danshen (tanshinones including cryptotanshinone, tanshinone I and tanshinone IIA) using a C_{18} column (150 × 4.6 mm i.d., particle size 5 μm). The mobile phase was methanol–tetrahydrofuran–water–glacial acetic acid (20:35:44:1 v/v/v/v), employing isocratic elution at a flow-rate of 1.0 mL min^{-1}. The analytes were detected by measuring their UV absorption at 254 nm. The method was successfully applied to the analysis of five kinds of Chinese herbal medicines containing danshen.[41] Hu et al. used LC-MS for the quantitative determination of dihydrotanshinone I, cryptotanshinone, tanshinone I and tanshinone IIA. These components were separated using a reversed-phase C_{18} column and quantification was based on the $[M + H]^+$ fragments produced under collision activation conditions and the selected reaction monitoring mode.[42] Zhang et al. studied the metabolism of phenolic acids from Salvia miltiorrhiza roots in rats using HPLC-UV and HPLC-MS and identified danshenu, caffeic acid, ferulic acid, isoferulic acid and methylated ferulic acid as metabolites.[43]

The analysis of the active components of ginseng using HPLC combined with UV or MS detection has been extensively studied. Harkey et al. analyzed 25 commercial ginseng products available in the USA for the presence of marker compounds using HPLC-MS-MS. They concluded that although all products were labeled correctly and marker compounds were found in all preparations, there were wide variations in the concentrations of marker products, suggesting poor quality control and standardization in manufacturing such products.[44] Bonfill et al. described a reversed-phase HPLC assay for the simultaneous quantitative determination of several ginsenosides, Rb(1), Rb(2), Rc, Rg(1), Re and Rf, in ginseng products. Chromatographic separation can be achieved in less than 20 min using a diol column and UV detection at 203 nm.[45] Another method employed HPLC combined with negative ion electrospray MS for determination of three ginsenosides [Rb(1), Rc, and Re] in six different samples of ginseng, including a liquid extract, capsules, tea bags and instant tea. The authors found at least one ginsenoside in four of the six products studied.[46] Zhu et al. reported a comparative study on the triterpene saponins of 47 samples of ginseng using HPLC. They selected 11 ginsenosides as markers and found 10-fold variations in ginsenoside concentrations between different products.[47]

Sun et al. studied ginsenoside concentrations in rat plasma using LC-MS. After solid-phase extraction and HPLC separation, the chloride adduct anions of Rg(1), Rh(1) and aglycone protopanaxatriol (PPT) were analyzed in the selected-ion monitoring mode. The detection limit was 20 pg for Rg(1), 100 pg for Rh(1) and 10 pg for PPT. Chromatographic separation was achieved in less than 8 min.[48]

3.4 HERBAL REMEDIES AND ABNORMAL LIVER FUNCTION TESTS

Consumption of kava has been associated with increased concentrations of γ-glutamyl-transferase (GGT), suggesting potential hepatotoxicity. Escher and Desmeules described a case in which severe hepatitis was associated with kava use. A 50-year-old man took three or four kava capsules daily for 2 months and liver function tests showed 60–70-fold increases in AST and ALT. Serology was negative for hepatitis A, B and C, CMV and HIV. The patient eventually received a liver transplant.[49] Humberston et al. also reported a case of acute hepatitis induced by kava-kava.[50] Other cases of hepatotoxicity due to the use of kava have been documented.[51] In January 2003, kava extracts were banned in the entire European Union and Canada. The FDA strongly cautioned against using kava. Eleven cases of serious hepatic failure and four deaths have been reported in association with kava use. There are also 23 reports indirectly linking kava-kava with hepatotoxicity.[52]

Chaparral can be found in health food stores as capsules and tablets and is used as an antioxidant and anti-cancer herbal product. Leaves, stems and bark in bulk are also available for brewing tea. However, this product can cause severe hepatotoxicity. Several reports of chaparral-associated hepatitis have been reported. A 45-year-old woman who took 160 mg of chaparral per day for 10 weeks presented with jaundice, anorexia, fatigue, nausea and vomiting. Liver enzymes and other liver function tests showed abnormally high values (ALT 1611 $U L^{-1}$, AST 957 $U L^{-1}$, alkaline phosphatase 265 $U L^{-1}$, GGT 993 $U L^{-1}$ and bilirubin 11.6 mg dL^{-1}). Hepatitis, CMV and EBV were ruled out. A liver biopsy showed acute inflammation with neutrophil and lymphoplasmocytic infiltration, hepatic disarray and necrosis. The diagnosis was drug-induced cholestatic hepatitis, which in this case was due to the use of chaparral.[53] Gordon et al. reported a case where a 60-year-old woman took chaparral for 10 months and developed severe hepatitis for which no other cause was found. On admission her bilirubin was 12.4 mg dL^{-1}, ALT 341 $U L^{-1}$, AST 1191 $U L^{-1}$ and alkaline phosphatase 186 $U L^{-1}$. Her prothrombin time was 15.9 s, but all tests for viral causes were negative. Eventually she received a liver transplant.[54]

Comfrey is a perennial herb used for the prevention of kidney stones; nourishing and repairing bone and muscle and for the treatment of injuries such as burns and bruises. In Australia, comfrey is classified as a poison and its sales have been restricted in several regions. Many different commercial forms of comfrey are marketed, including oral and external products. Commercial comfrey is usually derived from the leaves or roots of Symphytum officinale (common comfrey). However, some products are also derived from Russian comfrey. Russian comfrey contains a very toxic pyrrolizidine alkaloid, echimidine, which is not found in common comfrey. However, common comfrey contains other hepatotoxic alkaloids, namely 7-acetylintermedine, 7-acetyllycopsamine and symphytine. The metabolites of these alkaloids are very toxic to the liver.[55] Ridker et al. documented hepatic venocclusive disease associated with consumption of comfrey root.[56] Long-term studies in animals have also confirmed the carcinogenicity of comfrey in animal models.[57]

Germander has been used as a remedy for weight loss and general tonic. Germander tea made from the aerial parts of the plant has been used for many centuries. Twenty-six cases of germander-induced liver toxicity have been reported in Europe. A 55-year-old woman taking 1600 mg per day of germander became jaundiced after 6 months. Her bilirubin was 13.9 mg dL^{-1}, AST 1180 $U L^{-1}$, ALT 1500 $U L^{-1}$ and ALP 164 $U L^{-1}$. Serological tests for hepatitis viruses were negative. A liver biopsy suggested drug-induced hepatitis. Germander therapy was discontinued and the hepatitis resolved in 2 months.[58] Bosisio et al. described an HPLC method for detection of teucrin A, the active component of germander, in beverages.[59]

3.4.1 Use of GC-MS and HPLC for the measurement of active components

Duffield *et al.* used GC combined with chemical ionization MS for the identification of human urinary metabolites of kava lactones following ingestion of kava prepared by the traditional method of aqueous extraction from the plant *Piper methysticum*. All seven major and several minor kava lactones were identified in human urine. Metabolic transformations include reduction of the 3,4-double bond and/or demethylation. Ring-opening products of kava lactones were also detected in human urine.[60] Gaub *et al.* used HPLC combined with coordination ionspray MS, where charged complexes are formed through addition of central complexing ions such as sodium, silver and cobalt, for the analysis of kava extracts.[61]

Phenolic components of chaparral, including lignans and flavonoids, can be analyzed using HPLC combined either with UV or MS detection.[62] Mroczek *et al.* utilized cation-exchange solid-phase extraction and ion-pair HPLC for the simultaneous determination of *N*-oxides and free base pyrrolizidine alkaloids of comfrey. The recoveries were 80% for retrorsine *N*-oxide, 90% for retrorsine and 100% for senkirkine as assessed by both TLC and HPLC.[63] Schaneberg *et al.* developed a reversed-phase HPLC method with evaporative light scattering detection for the simultaneous determination of hepatotoxic pyrrolizidine alkaloids including riddelline, riddelline *N*-oxide, senecionine, senecionine *N*-oxide, seneciphylline, retrorsine, integerrimine and lasisocarpine, and also heliotrine.[64]

Teucrin A is the major ingredient of germander responsible for its toxicity and it is also considered a marker for germander use. Teucrin A accounts for approximately 70% of the *neo*-clerodane diterpenoids found in the extract of germander (*Teucrium chamaedrys*). Avula *et al.* reported an HPLC method for the simultaneous analysis of nine *neo*-clerodane diterpenoids from germander. They used a reversed-phase Phenomenex Luna C_{18} (2) column (150 × 4.6 mm i.d., particle size 5 μm) and an acetonitrile–water gradient at a flow-rate of 1 mL min^{-1} for the separation. The limit of detection was 0.24–0.9 μg mL^{-1} using photodiode-array detection.[65]

3.5 GINKGO BILOBA

Ginkgo biloba is prepared from dried leaves of the ginkgo tree (Ginkgo biloba) by organic extraction (acetone–water). After the solvent has been removed, the extract is dried and standardized to contain a constant faction of flavonoids (usually 24%) and terpenes (usually 6%). Most commercial dosage forms contains 40 mg of this extract. Ginkgo biloba is sold in the USA as a dietary supplement intended to improve blood flow in brain and peripheral circulation. It is used mainly to sharpen mental focus and to improve diabetes-related circulatory disorders. The German Commission E approved the use of ginkgo for memory deficit, disturbances in concentration, depression, dizziness, vertigo and headache. Ginkgo leaf contains kaempferol-3-rhamnoglucoside, ginkgetin, isoginketin and bilobetin. Several flavone glycosides have also been isolated from ginkgo (ginkgolide A and B). Other substances isolated from ginkgo include shikimic acid, D-glucarica acid and anacardic acid. Several chemicals found in ginkgo extracts, especially ginkgolide B, are potent antagonists to coagulation factors, and also have antioxidant effects.

A common reported adverse effect of ginkgo is bleeding. Spontaneous intracerebral hemorrhage occurred in a 72-year-old woman who had been taking 50 mg of ginkgo three times per day for 6 months.[66] Fessenden *et al.* reported a ginkgo-associated case of postoperative bleeding after laparoscopic cholecystectomy.[67] Concurrent use of ginkgo and non-steroidal anti-inflammatory drugs (NSAIDs) or anticoagulants should be avoided because ginkgolide B is a potent inhibitor of platelet activating factors. Hauser *et al.* reported a case of bleeding complications after liver transplant in a 59-year-old patient who was using of ginkgo biloba. Seven days after a second liver transplant, subpherenic hematoma appeared in this patient. Three weeks later, an episode of vitreous hemorrhage was documented. No further bleeding occurred after the patient stopped taking ginkgo biloba.[68]

3.5.1 Analysis of components of ginkgo biloba by HPLC

Ginkgolic acids in ginkgo biloba extract can be analyzed by HPLC after liquid–liquid extraction of an aqueous commercially available extract of ginkgo biloba with ethyl acetate or aliphatic hydrocarbons such as hexane. Analysis can be carried out using HPLC combined with UV detection with a photodiode array (200–550 nm) or by HPLC-MS.[69] Tang *et al.* combined reversed-phase HPLC (C_{18} column) and a mobile phase composed of methanol and water (33:67 v/v) for the analysis of ginkgolides and bilobalide in ginkgo extract. Samples were extracted with ethyl acetate and purified by passage through an aluminum oxide column. They used evaporative light scattering detection of the compounds eluting from the column.[70] On-line dialysis is an alternative to the conventional extraction technique for isolating compounds of interest from a complex matrix. Chiu *et al.* developed a method for measuring ginkgolide A and B and bilobalide from ginkgo biloba extract using a self-assembled microdialysis device coupled to an HPLC instrument. They dialysis efficiencies for ginkgolide A and B and bilobalide were between 97.8 and 100.7%. They used a Zorbax SB-C_{18} column (150 × 4.6 mm i.d., particle size 5 µm) and the detection wavelength was set at 219 nm. The mobile phase was methanol–acetonitrile–0.01 M phosphate buffer (pH 5.0) (30:5:65 v/v/v).[71] HPLC combined with photodiode-array detection can also be used for the quantitative determination of five selected flavonol compounds (rutin, quercitrin, quercetin, kaempferol and isorhamnetin) which can be used as markers for quality control of ginkgo biloba extracts. Separation of these compounds was achieved using a Phenomenex Luna C_{18} (2) column (250 × 2.0 mm i.d., particle size 5 µm). The temperature of the column was maintained at 45 °C and the mobile phase was acetonitrile–formic acid (0.3%) with a one-step linear gradient and a flow-rate of 0.4 ml min^{-1}. The limits of detections were 2.76, 0.77, 1.11, 1.55 and 1.03 µg ml^{-1} for rutin, quercitrin, quercetin, kaempferol and isorhamnetin, respectively.[72] Dubber and Kanfer applied HPLC-MS-MS for the accurate determination of two flavonolglycosides, rytin and quercitrin, together with quercetin, kaempferol and isorhamnetin in several ginkgo biloba oral formulations. A one-step gradient of acetonitrile–formic acid (0.3%) at a flow-rate of 0.5 mL min^{-1} was used and the column temperature was maintained at 45 °C. Baseline separations of these compounds were achieved using a run time of 20 min.[73]

Wang *et al.* studied the disposition of quercetin and kaempferol in 10 adult volunteers following oral administration of ginkgo biloba extract. Quercetin and kaempferol were determined in human urine using reversed-phase HPLC. Quercetin and kaempferol were excreted from urine mainly as glucuronides.[74] An HPLC protocol using the ion-pairing technique has been reported for the rapid analysis of 4-*O*-methylpyridoxine in human serum. This compound is present in ginkgo seeds and, when consumed in large quantities, can cause vomiting and convulsions. The authors used fluorescence detection (excitation wavelength 290 nm, emission wavelength 400 nm) and achieved a detection limit of 5 pg. The analysis time was 30 min.[75] This method can be applied to the detection of 4-*O*-methylpyridoxine in human serum.

3.6 ECHINACEA

Echinacea, a genus of flowering plants including nine species that grow in the USA, is a member of the daisy family. Three species are found in common herbal preparations: *Echinacea angustifolia*, *Echinacea pallida* and *Enchinacea purpurea*. Native-American Indians considered the plant a blood purifier. In the USA, echinacea is used in oral dosage as an immune stimulant that helps increase resistance to cold and influenza. Fresh herb, freeze-dried herb and the alcoholic extract of the herb are all commercially available. The ability of oral echinacea preparations to prolong the time of onset of an upper respiratory infection, was compared with placebo in 302 volunteers from an industrial plant and several military institutions. Although subjects felt better after taking echinacea, there was no statistically significant difference with regard to the time of onset of upper respiratory tract infection.[76] Another study also examined the efficacy of echinacea in preventing colds and upper respiratory tract infection. The authors found no difference between the placebo group and the group that took echinacea.[77]

The Australian Adverse Drug Reaction Advisory Committee received 11 reports of adverse reactions associated with echinacea use between July 1996 and September 1997. There were three reports of hepatitis, three of asthma, one of rash, myalgia, and nausea, one of utricaria and one of anaphylaxis. There are other published reports of echinacea associated with contact dermatitis and anaphylaxis.[78]

3.6.1 Analysis of active components of echinacea by HPLC

A simple TLC method was reported for the identification of a marker compound in echinacea that demonstrated a blue fluorescence at an excitation wavelength of 366 nm after staining with a spray reagent containing ethanol, trifluoroacetic acid and zinc.[79] Chicoric acid is the main phenolic compound present in *Echinacea purpurea* roots and another component is caftaric acid. Echinacoside and cynarin are found in *Echinacea padilla* and *Echinacea angustifolia*. Perry *et al.* reported an HPLC protocol for analysis of these components.[80] Molgaard *et al.* reported a reversed-phase HPLC protocol for the analysis of caffeic acid derivatives in echinacea; naringenin was used as an internal standard.[81] Pellati *et al.* used an 80% methanol in water solution and magnetic stirring for extraction of caffeic acid derivatives (caftaric acid, chlorogenic acid, caffeic acid, cynarin, echinacoside and cichoric acid) from echinacea roots. The extracts were analyzed on a HPLC column with gradient elution and photodiode-array detection.[82] Luo *et al.* used HPLC coupled with UV photodiode-array detection and electrospray ionization MS for the simultaneous analysis of caffeic acid derivatives and alkamides in the roots and extracts of *Echinacea puppurea*.[83]

3.7 VALERIAN

Valerian is a perennial herb that grows in North America, Europe and western Asia. The crude valerian root, or rhizome, is dried and used as is or as an extract. Valerian is available as a capsule, oral solution or tea, and is used as a sleeping pill. The recommended dose is one or two tablets before bedtime, depending on the amount of valerenic acid in the preparation. The German Commission E has approved valerian as a sleep-promoting and calmative agent. The chemical components of valerian are valeranone, valerenic acid, isoeugenyl isovalarerate, valepotriate, isovaltrate and didrovaltrate. At least 37 valepotriates and seven alkaloids have been isolated from valerian. It is believed that valepotriates are responsible for the sedative activity. Valerenic acid can exert pentobarbital-like central nervous system depressant activities. However, valerian's mechanism of action has not been fully elucidated.

Leathwood *et al.* conducted a double-blind crossover study with 128 volunteers and concluded that, compared with placebo, valerian significantly improved subjective sleep quality in habitually poor or irregular sleepers.[84] However, placebo effects were marked in some studies, and in some cases beneficial effects of valerian were not seen until after 2–4 weeks of therapy. The adverse effects of valerian include gastrointestinal upset, allergies, restless sleep, headache and mydriasis. Valerian overdoses have the major effect of central nervous system depression.[85]

3.7.1 Analysis of components of valerian by HPLC

Shohet *et al.* analyzed 31 commercial valerian preparations available in Australia, including tea, tablets, capsules and liquids, by HPLC for valepotriates, valerenic acid and valerenic acid derivatives. The concentrations of valerenic acid and its derivatives ranged from 0.01 to 6.32 μg g^{-1} of the product; powdered capsules on average contained the highest concentrations of valerenic acid and liquid preparations had the lowest concentrations.[86] Torrado studied the *in vitro* release of valerenic and hydroxyvalerenic acids from valerian tablets. The valerenica acid and hydroxyvalerenic acid concentrations were measured by HPLC using a C$_{18}$ Kromasil column (200 × 4.6 mm i.d., particle size 5 μm) and a mobile phase of methanol–aqueous 0.5% (v/v) orthophosphoric acid (75:25 v/v). The flow-rate was 1 mL min^{-1}. The uncoated tablets had the fastest release profile whereas the coated tablets showed very different release patterns, depending on the type of formulation.[87]

3.8 FEVERFEW

Feverfew (*Tanacetum parthenium*) is a short perennial shrub that grows along fields and roadsides. In the 1970s, the use of feverfew as an alternative to traditional medicines for relief from arthritis and migraine headache gained popularity. Feverfew is available as the fresh leaf, dried powdered leaf, in capsules and tablets, as a fluid extract and in oral drops. During a migraine episode, serotonin is released from platelets, so serotonin antagonists (for example, methysergide) are used to treat migraine symptoms. An *in vitro* study using a bovine platelet bioassay has shown that parthenolide, and also other sesquiterpene lactones found in feverfew, inhibit serotonin release by platelets.[88] Feverfew may inhibit serotonin release from platelets in the same manner as methysergide, an ergot alkaloid.[89] Feverfew may also irreversibly inhibit prostaglandin synthesis by interfering with cyclooxygenase and phospholipase A_2. Parthenolide and epoxyartemorin, found in feverfew, have been shown to inhibit irreversibly thromboxane B_2 and leukotrine B_4.[90] Johnson evaluated the efficacy of feverfew in 17 patients with at least a 2-year history of migraines. The patients took 50 mg of feverfew or placebo. Patients in the placebo group had migraines more often than patients receiving feverfew.[91] Murphy *et al.* conducted a randomized double-blind study of feverfew for the prophylaxis of migraine in 60 patients who had had migraines for at least 2 years. All migraine-related medicines were stopped before the study. The patients received one feverfew capsule (70–114 mg) or placebo four times daily for 4 months. The number of migraine attacks was significantly lowered by feverfew.[92]

Adverse effects associated with feverfew use include dizziness, heartburn, indigestion, bloating and ulceration of oral mucosa. Some 18% of 300 feverfew users reported adverse effects, with mouth ulceration reported by 11.3% of subjects. Discontinuation of feverfew may produce muscle and joint stiffness, rebound of migraine symptoms and anxiety and disrupt sleeping patterns. Feverfew should be avoided in pregnancy because it is purportedly associated with spontaneous abortion in cattle and uterine contraction in term human pregnancy. Because of the potential of feverfew to inhibit cyclooxygenase, it interacts with anticoagulants and increases the antiplatelet effect of aspirin.

3.8.1 Analysis of parthenolide by HPLC

Zhou *et al.* described a sensitive method for the quantification of parthenolide in feverfew using HPLC. The compound was extracted into acetonitrile–water (90:10 v/v) and separated on a Comosil C_{18} HPLC column (150 × 4.6 mm i.d., particle size 5 μm). The mobile phase was acetonitrile–water (55:45 v/v) at a flow-rate of 1.5 mL min^{-1}, and parthenolide was detected by its UV absorption at 210 nm. The analysis time was only 6 min and the detection limit was 0.10 ng of parthenolide on-column. The spiked recovery of parthenolide was 99.3%.[93] Nelson *et al.* studied variations in the parthenolide content of feverfew products available commercially using a similar HPLC method, and observed wide variations in parthenolide content in a single dose (0.02–3.0 mg) of feverfew among the various preparations.[94] Curry *et al.* measured plasma concentrations of parthenolide by solid-phase extraction and mass spectrometric detection. Although the limit of detection by this method was 0.5 ng mL^{-1}, the authors did not observe any detectable concentration of parthenolide even after a daily oral dose containing 4 mg of parthenolide.[95]

3.9 GARLIC

Garlic has been promoted as a dietary constituent that lowers cholesterol and blood pressure, thereby reducing the risk of heart attacks and stroke. Garlic contains various sulfur-containing compounds derived from allicin. Allicin is formed from alliin by the action of allinase. Allinase is released when garlic is chopped. Allicin then produces diallyl sulfide, diallyl disulfide, diallyl trisulfide and other related sulfur compounds. Cooking destroys allinase. Allicin produces the characteristic odor of garlic. There have been several studies evaluating the medical efficacy of garlic. Silagy and Neil performed a meta-analysis of eight prospective randomized studies. All of the studies used dried garlic powder

(600–900 mg daily, equivalent to 1.8–2.7 g of fresh garlic). The overall mean systolic blood pressure reduction was 7.7 mmHg (range 50–17.2) with garlic, and the overall mean reduction in diastolic pressure was 5.0 mgHg (range 3.4–9.6). None of the studies assessed compliance and some studies did not mention the position (e.g. supine) when blood pressure was measured.[96] The reduction in blood pressure was variable and small, and more clinical trials are needed in order to determine whether garlic is an effective treatment for mild hypertension. Jabbari *et al.* studied the difference between swallowing and chewing garlic on levels of serum lipids, lipid peroxidation, creatinine and cyclosporine in 50 renal transplant recipients, and concluded that swallowed garlic had no effect on the lipid profile in serum but chewed garlic reduced lipids in serum, a lipid peroxidation marker (malondialdehyde), in addition to blood pressure.[97]

Hypersensitivity to garlic has been reported. Topically applied garlic can cause garlic burn and an allergic dermatitis. Garlic can increase the effectiveness of warfarin, causing bleeding. A person taking warfarin should avoid garlic. The use of garlic should be stopped 7–10 days before surgery because it can prolong bleeding time. Postoperative bleeding has been reported with garlic alone.[98,99]

3.9.1 Measurement of components of garlic by HPLC

Arnault *et al.* described an ion-pair chromatographic method for the simultaneous analysis of allin, deoxyallin, allicin and dipeptide precursors in garlic products. They developed a rapid HPLC protocol using a mobile phase containing heptanesulfonate as an ion-pairing reagent and photodiode-array UV detection, and also electrospray ionization ion-trap MS detection.[100] Rosen *et al.* determined allicin, *S*-allylcysteine and volatile metabolites of garlic in breath, plasma and simulated gastric fluid using headspace sampling and GC-MS the for analysis of volatiles from breath and HPLC-MS for the determination of *S*-allylcysteine in plasma. For the determination of concentrations of volatiles in breath, a short-path thermal desorption device was used to volatilize the analytes. The absorption trap was spiked with toluene-d_8 and naphthalene-d_8 as internal standards. The desoprtion time was 3 min and the desoprtion temperature was 150 °C. The GC column used was a DB-1 methylsilicone capillary column (0.25 μm film thickness, 60 m × 0.32 mm i.d.), and the mass spectrometer was operated in electron ionization mode (70 eV). The mass scanning range was 35–450 amu. The HPLC system for the analysis of allicin used a Supelco 25 cm × 4.6 mm i.d. C_{18} column with photodiode-array detection (195 nm) and an isocratic mobile phase composition of acetonitrile–water (30:70 v/v). The major volatile component found in breath was allyl methyl sulfide. After consumption of raw garlic, limonene and *p*-cymene were also detected in breath.[101]

3.10 EPHEDRA (MA HUANG) AND RELATED DRUGS

Ma huang (ephedra) is commonly found in herbal weight loss products that are often referred to herbal fen-phen. Some weight loss clinics and herbal outlets promote 'Herbal fen-phen' as an alternative to fenfluramine, the prescription drug that has been withdrawn from the market due to toxicity. Herbal fen-phen products sometimes contain St John's wort and are sold as 'herbal prozac'. Ephedra-containing products are also marketed as decongestants, bronchodilators and stimulants. Other promoted uses include bodybuilding and enhancement of athletic performance. 'Herbal ecstasy' is also an ephedrine-containing product, which can induce a euphoric state. Ephedra is a small perennial shrub with thin steams. The plant rarely grows over 30 cm tall. Some of the better known species include *Ephedra sinica* and *Ephedra equisentina* (collectively called ma huang from China. The German Commission E report recommended against the use of ephedra in patients with high blood pressure, glaucoma or thyrotoxicosis.

(−)-Ephedrine is the predominant alkaloid of ephedra plants. Other phenylalanine-derived alkaloids found in ephedra plants are (+)-pseudoephedrine, (−)-norephedrine, (+)-norpseudoephedrine, (+)-*N*-methylephedrine and phenylpropanolamines. Ephedrine is a potent central nervous system (CNS) stimulant. Because ephedra is both an α- and β-adrenergic agonist, ingestion of quantities over 50 mg lead to a rise in blood pressure, heart rate and cardiac output.

Haller and Benowitz evaluated 140 reports of ephedra-related toxicity and concluded that 31% of the cases were definitely related to ephedra toxicity and a further 31% were possibly related; 47% of reports of ephedra toxicity involved cardiovascular problems and 18% involved problems with the CNS. Hypertension was the single most frequent adverse reaction, followed by palpitation, tachycardia, stroke and seizure. Ten events resulted in death and 13 events caused permanent disability. The authors concluded that use of dietary supplements that contains ephedra may pose a health risk.[102]

3.10.1 Analysis of active components of ephedra-containing products

(+)-Pseudoephedrine and (−)-ephedrine in ephedra-containing herbs can be analyzed using ion-pair reversed-phase HPLC with sodium dodecyl sulfate after solid-phase extraction from herbal preparations. The mobile phase was water–acetonitrile–phosphoric acid (650:350:1 v/v/v) containing 0.5% sodium dodecyl sulfate, at a flow-rate of 1 mL min^{-1}. The column temperature was maintained at 50 °C and detection was achieved using a photodiode array at 210 nm. *N*-Benzyldiethylamine was used as the internal standard.[103] A simple GC-MS method for the determination of ephedrine alkaloids and 2,3,5,6-tetramethylpyrazine in ephedra has also been reported. The sample was extracted with diethyl ether and analyzed by GC without derivatization. A capillary column (30 m × 250 μm i.d.) coated with 5% phenylmethylsilicone was used, and the detection limits were 0.4, 0.7 and 0.02 ng for ephedrine, pseudoephedrine and 2,3,5,6-tetramethylpyrazine, respectively.[104] Cottiglia *et al.* isolated and characterized two novel phenolic glycosides, 4-hydroxy-3-(3-methyl-2-butenyl)phenyl β-D-glucopyranoside and *O*-coumaric acid β-D-allopyranoside, from *Ephedra nebrodensis* using NMR and MS.[105]

Ephedrine and pseudoephedrine have been measured in guinea pig plasma using HPLC with fluorescence detection, following precolumn derivatization with 5-dimethylamino-napthalene-1-sulfonyl chloride in acetonitrile. The mobile phase was 0.6% phosphate buffer (pH 6.5)–methanol (3:8 v/v).[106] Jacob *et al.* developed an LC–atmospheric pressure chemical ionization MS-MS method for the quantitaion of various alkaloids found in ephedra-containing dietary supplements and also in plasma and urine from subjects using these supplements. Using this method, the concentrations of ephedrine, pseudoephedrine, norephedrine, norpseudoephedrine, methylephedrine, methylpseudoephedrine and caffeine were determined in low nanogram quantities in plasma and urine. The analytical cycle time for this method was 12 min.[107]

3.11 CONCLUSIONS

Despite many reported toxic effects and drug interactions resulting from the use of alternative medicines, they remain popular and most users consider them safe. Labeling of herbal products may not accurately reflect their content, and adverse events or interactions attributed to a specific herb may be due to misidentification of the plant or contamination of the plant with pharmaceuticals or heavy metals. The addition of pharmaceuticals to Chinese herbal products is a serious problem. Of 2069 samples of traditional Chinese medicines collected from eight hospitals in Taiwan, 23.7% contained pharmaceuticals, including caffeine, acetaminophen, indomethecin, hydrochlorothiazide and prednisolone.[108,109] A fatal case of hepatic failure due to contamination of a herbal supplement with nitrosofenfluramine has been reported. Analysis of the herbal supplement also revealed the presence of fenfluramine.[110] Cole and Fetrow reported the presence of colchicine in gingko biloba and echinacea preparations. They also reported the case of a 23-year-old woman who had a serum digoxin concentration of $3.66 \, \text{ng mL}^{-1}$ after taking a herbal product that was found to contain digitalis lantana as an unlabeled constituent.[111] Heavy metal contamination is a major problem with Asian medicines. Ko reported that 24 of 254 Asian patent medicines collected from herbal stores in California contained significant quantities of lead, 36 contained arsenic and 35 contained mercury.[112] Contamination with lead and other heavy metals is common in Indian Ayurvedic medicines.[113] Analysis of active ingredients in serum and other biological fluids is an important step towards understanding the toxicity of herbal products, the magnitude of interactions between herbal supplements and pharmaceuticals and the detection of the presence of contaminants in these preparations.

REFERENCES

1. Gulla J, Singer AJ, Gaspari R. Herbal use in ED patients. *Acad Emerg Med* 2001; **8**: 450.
2. Klepser TB, Klepser ME. Unsafe and potentially safe herbal therapies. *Am J Health Syst Pharm* 1999; **56**: 125–141.
3. Raffa R. Screen of receptor and uptake site activity of hypericin components of St John's wort reveal σ receptor binding. *Life Sci* 1998; **62**: 265–270.
4. Johne A, Brockmoller J, Bauer S, *et al.* Pharmacokinetic interaction of digoxin with an herbal extract from St John's wort (*Hypericum perforatum*). *Clin Pharmacol Ther* 1999; **66**: 338–345.
5. Durr D, Stieger B, Kullak-Ublick GA, *et al.* St. John's wort induces intestinal P-glycoprotein/MDR1 and intestinal and hepatic CYP3A4. *Clin Pharmacol Ther* 2000; **68**: 598–604.
6. Barone GW, Gurley BJ, Ketel BL, Abul-Ezz SR. Herbal supplements; a potential for drug interactions in transplant recipients. *Transplantation* 2001; **71**: 239–241.
7. Alscher DM, Klotz U. Drug interactions of herbal tea containing St. Kohn's wort with cyclosporine. *Transpl Int* 2003; **16**: 543–544.
8. Mai I, Stormer E, Bauer S, *et al.* Impact of St. John's wort treatment on the pharmacokinetics of tacrolimus and mycophenolic acid in renal transplant patients. *Nephrol Dial Transplant* 2003; **18**: 819–822.
9. Piscitelli SC, Burstein AH, Chaitt D, Alfaro RM, Fallon J. Indinavir concentrations and St. John's wort. *Lancet* 2000; **355**: 547–548.
10. Eich-Holchli D, Oppliger R, Golay KP, Baumann P, Eap CB. Methadone maintenance treatment and St. John's wort: a case study. *Pharmacopsychiatry* 2003; **36**: 35–37.

11. de Los Reyes GC, Koda RT. Determination of hyperforin and hypericin content in eight brands of St. John's wort. *Am J Health Syst Pharm* 2002; **59**: 545–547.

12. Li W, Fitzloff JF. High performance liquid chromatographic analysis of St. John's wort with photodiode array detection. *J Chromatogr B* 2001; **765**: 99–105.

13. Ruckert U, Eggenreich K, Wintersteiger R, Wurglics M, Likussar W, Michelitsch A. Development of a high performance liquid chromatographic method with electrochemical detection for the determination of hyperforin. *J Chromatogr A* 2004; **1041**: 181–185.

14. Mauri P, Pietta P. High performance liquid chromatography/electrospray mass spectrometry of *Hypercium perforatum* extract. *Rapid Commun Mass Spectrom* 2000; **14**: 95–99.

15. Bilia AR, Bergonzi MC, Morgenni F, MazziG, Vinciceri FF. Evaluation of chemical stability of St. John's wort commercial extract and some preparations. *Int J Pharm* 2001; **123**: 199–208.

16. Ang CY, Hu L, Heinze TM, Cui Y, *et al.* Instability of St. John's wort (*Hypericum perforatum* L) and degradation of hyperforin in aqueous solutions and functional beverages. *J Agric Food Chem* 2004; **52**: 6156–6164.

17. Liu F, Pan C, Drumm P, Ang CY. Liquid chromatography–mass spectrometry studies of St. John's wort methanolic extraction: active constituents and their transformation. *J Pharm Biomed Anal* 2005; **37**: 303–312.

18. Seger C, Rompp H, Sturm S, Haslinger E, Schmidt PC, Hadacek F. Characterization of supercritical fluid extracts of St. John's wort (*Hypercium perforatum* L) by HPLC-MS and GC-MS. *Eur J Pharm Sci* 2004; **21**: 453–463.

19. Bilia AR, Bergonzi MC, Mazzi G, Vincieri FF. Analysis of plant complex matrices by nuclear magnetic resonance spectroscopy: St. John's wort extract. *J Agric Food Chem* 2001; **49**: 2115–2124.

20. Rager I, Ross G, Schmidt PC, Kovar KA. Rapid quantification of constituents in St. John's wort extracts by NIR spectroscopy. *J Pharm Biomed Anal* 2002; **28**: 439–446.

21. Mai I, Bauer S, Perloff ES, Johne A, Uehleke B, Frank B, Budde K, Roots I. Hyperforin content determines the magnitude of St. John's wort–cyclosporine drug interactions. *Clin Pharmacol Ther* 2004; **76**: 330–340.

22. Zanoli P. Role of hyperforin in the pharmacological activities of St. John's wort. *CNS Drug Rev* 2004; **10**: 203–218.

23. Mannel M. Drug interactions with St. John's wort: mechanism and clinical implications. *Drug Saf* 2004; **27**: 773–797.

24. Cui Y, Gurley B, Ang CYW, Leakey J. Determination of hyperforin in human plasma using solid-phase extraction and high performance liquid chromatography with ultraviolet detection. *J Chromatogr B* 2002; **780**: 129–135.

25. Bauer S, Stormer E, Graubaum HJ, Roots I. Determination of hyperforin, hypericin and pseudohypericin in human plasma using high-performance liquid chromatography analysis with fluorescence and ultraviolet detection. *J Chromatogr B* 2001; **765**: 29–35.

26. Pirker R, Huck CW, Bonn GK. Simultaneous determination of hypericin and hyperforin in human plasma and serum using liquid–liquid extraction, high performance liquid chromatography–tandem mass spectrometry. *J Chromatogr B* 2002; **777**: 147–153.

27. RiedelKD, Rieger K, Martin-Facklam M, Mikus G, Haefeli WE, Burhenne J. Simultaneous determination of hypericin and hyperforin in human plasma with liquid chromatography–tandem mass spectrometry. *J Chromatogr B* 2004; **813**: 27–33.

28. Ko R, Greenwald M, Loscutoff S, Au A, *et al.* Lethal ingestion of Chinese tea containing chan su. *West J Med* 1996: **164**; 71–75.

29. Fushimi R, Amino N. Digoxin concentration in blood. *Rinsho Byori* 1995: **43**; 34–40 [in Japanese with English abstract].

30. Panesar NS. Bufalin and unidentified substances in traditional Chinese medicine cross react in commercial digoxin assay. *Clin Chem* 1992: **38**; 2155–2156.

31. Dasgupta A, Biddle D, Wells A, Datta P. Positive and negative interference of Chinese medicine chan su in serum digoxin measurement: elimination of interference by using a monoclonal chemiluminescent digoxin assay or monitoring free digoxin concentration. *Am J Clin Pathol* 2000; **114**: 174–179.

32. Wahed A, Dasgupta A. Positive and negative *in vitro* interference of Chinese medicine dan shen in serum digoxin measurement: elimination of interference by monitoring free digoxin concentration. *Am J Clin Pathol* 2001; **116**; 403–408.

33. McRae S. Elevated serum digoxin levels in a patient taking digoxin and Siberian ginseng. *CMAJ* 1996; **155**: 293–295.

34. Dasgupta A, Wu S, Actor J, Olsen M, Wells A, Datta P. Effect of Asian and Siberian ginseng on serum digoxin measurement by five digoxin immunoassays: significant variation in digoxin-like immunoreactivity among commercial ginsengs. *Am J Clin Pathol* 2003; **119**: 298–303.

35. Chow L, Johnson M, Wells A, Dasgupta A. Effect of traditional Chinese medicines chan su, lu-shen-wan, danshen and Asian ginseng on serum digoxin measurement by Tina-Quant (Roche) and Synchron LX system (Beckman) digoxin immunoassays. *J Clin Lab Anal* 2003; **17**: 22–27.

36. Hong Z, Chan K, Yeung HW. Simultaneous determination of bufadienolides in traditional Chinese medicine preparations, liu-shen-wan by liquid chromatography. *J Pharm Pharmacol.* 1992: **44**; 1023–1026.

37. Chan WY, Ng TB, Yeung HW. Examination for toxicity of a Chinese drug, the total glandular secretion product chan su in pregnant mice and embryos. *Biol Neonate* 1995: **67**; 376–380.

38. Hong Z, Chan K, Yeung HW. Simultaneous determination of bufadienolides in the traditional Chinese medicine preparation, liu-shen-wan by liquid chromatography. *J Pharm Pharmacol* 1992; **44**: 1023–1026.

39. Li SQ. Separation and purification of an endogenous inhibitor of sodium pump from chansu by thin layer chromatography and reverse phase high performance liquid chromatography. *Se Pu* 2001; **19**: 555–557 [in Chinese].

40. Wang Z, Wen J, Zhang J, Ye M, Guo D. Simultaneous determination of four bufadienolides in human liver by high performance liquid chromatography. *Biomed Chromatogr* 2004; **18**: 318–322.

41. Shi Z, He J, Yao T, Chang W, Zhao M. Simultaneous determination of cryptotanshinone, tanshinone I and tanshinone IIA in traditional Chinese medicinal preparations containing *Radix salvia* miltiorrhiza by HPLC. *J Pharm Biomed Anal* 2005; **37**: 481–486.

42. Hu P, Luo GA, Zhao ZZ, Jiang ZH. Quantitative determination of four diterpenoids in *Radix salvia* miltiorrhizae using LC-MS-MS. *Chem Pharm Bull (Tokyo)* 2005; **53**: 705–709.

43. Zhang J, He Y, Cui M, Li L, Yu H, Zhang G, Guo D. Metabolic studies on the total phenolic acids from roots of Salvia miltiorrhizae in rats. *Biomed Chromatogr* 2005; **19**: 51–59.

44. Harkey MR, Henderson GL, Gershwin ME, Stern ME, Stern JS, Hackman RM. Variability in commercial ginseng products: an analysis of 25 preparations. *Am J Clin Nutr* 2001; **73**: 1101–1106.

45. Bonfill M, Casals I, Palazon J, Mallol A, Morales C. Improved high performance liquid chromatographic determination of ginsenosides in *Panax ginseng* based pharmaceuticals using a diol column. *Biomed Chromatogr* 2002; **16**: 68–72.

46. Luchtefeld R, Kostoryz E, Smith RE. Determination of ginsenosides Rb1, Rc, and Re in different dosage forms of ginseng by negative ion electrospray liquid chromatography–mass spectrometry. *J Agric Food Chem* 2004; **52**: 1953–1956.

47. Zhu S, Zou K, Fushimi H, Cai S, Komatsu K. Comparative study on triterpene saponins of ginseng drugs. *Planta Med* 2004; **70**: 666–677.

48. Sun J, Wang G, Haitang X, Hao L, Guoyu P, Tucker I. Simultaneous rapid quantification of ginsenoside Rg(1) and its secondary glycoside Rh(1) and aglycone protopanaxatriol in rat plasma by liquid chromatography–mass spectrometry after solid phase extraction. *J Pharm Biomed Anal* 2005; **38**: 126–132.

49. Escher M, Desmeules J. Hepatitis associated with kava, a herbal remedy. *BMJ* 2001; **322**: 139.

50. Humberston CL, Akhtar J, Krenzelok EP. Acute hepatitis induced by kava-kava. *J Toxicol Clin Toxicol* 2003; **41**: 109–113.

51. Stickel F, Baumuller HM, Seitz K, Vasilakis D, Seitz G, Seitz HK, Schuppan D. Hepatitis induced by kava (*Piper methysticum* rhizoma). *J Hepatol* 2003; **39**: 62–67.

52. Clouatre DL. Kava kava: examining new reports of toxicity. *Toxicol Lett* 2004; **150**: 85–96.

53. Alderman S, Kailas S, Goldfarb S, *et al.* Cholestatic hepatitis after ingestion of chaparral leaves: confirmation by endoscopic retrograde cholangiopancreatography and liver biopsy. *J Clin Gastroenterol* 1994; **19**: 242–247.

54. Gordon DW, Rosenthal G, Hart J, Sirota R, Baker AL. Chaparral ingestion. The broadening spectrum of liver injury caused by herbal medications. *JAMA* 1995; **273**: 489–490.

55. Abbott PJ. Comfrey: assessing the low dose health risk. *Med J Aust* 1988; **149**: 678–682.

56. Ridker PM, Ohkuma S, Mcdermott WV, Trey C, Huxtable RJ. Hepatic venocclusive disease associated with the consumption of pyrrolizidine containing dietary supplement. *Gastroenterology* 1985; **88**: 1050–1054.

57. Hirono I, Mori H, Haga M. Carcinogenic activity of *Symphytum officinale*. *J Natl Cancer Inst* 1989; **9**: 510–511.

58. Laliberte L, Villeneuve JP. Hepatitis after use of germander, a herbal remedy. *CMAJ* 1996; **154**: 1689–1692.

59. Bosisio E, Givarini F, Dell'Agli M, Galli G, Galli CL. Analysis of high-performance liquid chromatography of teucrin A in beverages flavored with an extract of *Teucrium chamaedrys* L. *Food Addit Contam* 2004; **21**: 407–414.

60. Duffield AM, Jamieson DD, Lidgard RO, Duffield PH, Bourne DJ. Identification of some urinary metabolites of the intoxicating beverage kava. *J Chromatogr* 1989; **475**: 273–281.

61. Gaub M, von Brocke A, Ross G, Kovar KA. High-performance liquid chromatography–coordination ion spray mass spectrometry (HPLC–CIS/MS): a new tool for the analysis of non polar compound classes in plant extract using the example of *Piper methysticum* Forst. *Phytochem Anal* 2004; **15**: 300–305.

62. Obermeyer WR, Musser SM, Betz JM, Casey RE, Pohland AE, Page SW. Chemical studies of phytoestrogens and related compound in dietary supplements: flax and chaparral. *Proc Soc Exp Biol Med* 1995; **208**: 6–12.

63. Mroczek T, Glowniak K, Wlaszczyk A. Simultaneous determination of *N*-oxide and free bases of pyrrolizidine alkaloids by cation-exchange solid phase extraction and ion-pair high performance liquid chromatography. *J Chromatogr A* 2002; **949**: 249–262.

64. Schaneberg BT, Molyneux RJ, Khan IA. Evaporative light scattering detection of pyrrolizidine alkaloids. *Phytochem Anal* 2004; **15**: 36–39.

65. Avula B, Manyam RB, Bedir E, Khan IA. HPLC analysis of neoclerodane diterpenoids from *Teucrium chamaedrys*. *Pharmazie* 2003; **58**: 494–496.

66. Gilbert GJ. Ginkgo biloba. *Neurology* 1997; **48**: 1137.

67. Fessenden JM, Wittenborn W, Clarke L. Gingko biloba: ; a case report of herbal medicine and bleeding postoperatively from a laparoscopic cholecystectomy. *Am Surg* 2001; **67**: 33–35.

68. Hauser D, Gayowski T, Singh N. Bleeding complications precipitated by unrecognized ginkgo biloba use after liver transplantation. *Transpl Int* 2002; **15**: 377–379.

69. Fuzzati N, Pace R, Villa F. A simple HPLC-UV method for the assay of ginkgolic acid in ginkgo biliba extract. *Fitoterapia* 2003; **74**: 247–256.

70. Tang C, Wei X, Yin C. Analysis of ginkgolides and bilobalide in *Ginkgo biloba* L extract injections by high performance liquid chromatography with evaporative light scattering detection. *J Pharm Biomed Anal* 2003; **33**: 811–817.

71. Chiu HL, Lin HY, Yang TCC. Determination of ginkgolide A, B and bilobalide in *Ginkgo biloba* L extracts by microdialysis–HPLC. *Anal Bioanal Chem* 2004; **379**: 445–448.

72. Dubber MJ, Kanfer I. High-performance liquid chromatographic determination of selected flavonols in ginkgo biloba solid oral dosage forms. *J Pharm Pharm Sci* 2004; **7**: 303–309.

73. Dubber MJ, Sewram V, Mshicileli N, Shephard GS, Kanfer I. The simultaneous determination of selected flavonol glycosides and aglycones in ginkgo biloba oral dosage forms by high performance liquid chromatography–electrospray ionization mass spectrometry. *J Pharm Biomed Anal* 2005; **37**: 723–731.

74. Wang FM, Yao TW, Zeng S. Disposition of quercetin and kaempferol in human following an oral administration of ginkgo biloba extract tablets. *Eur J Drug Metab Pharmacokinet.* 2003; **28**: 173–177.

75. Hori Y, Fujisawa M, Shimada K, Oda A, Katsuyama S, Wada K. Rapid analysis of 4-*O*-methylpyridoxine in the serum of patients with ginkgo biloba seed poisoning by ion-pair high performance liquid chromatography. *Biol Pharm Bull* 2004; **27**: 486–491.

76. Melchart WE, Linde K, Brandmaier R, Lersch C. Echinacea root extracts for the prevention of upper respiratory track infection – a double blind, placebo controlled randomized trial. *Arch Fam Med* 1998; **7**: 541–545.

77. Grimm W, Muller H. A randomized controlled trial of the effect of fluid extract of *Echinacea purpurea* on the incidence and severity of colds and respiratory infection. *Am J Med* 1999; **106**: 138–143.

78. Mullis RJ. Echinacea associated anaphylaxis. *Med J Aust* 1998; **168**: 170–171.

79. Schicke B, Hagels H, Freudenstein J, Watzig H. A sensitive TLC method to identify *Echinaceae pallidae* radix. *Pharmazie* 2004; **59**: 608–611.

80. Perry NB, Burgess EJ, Glennie VL. Echinacea standardization: analytical methods for phenolic compounds and typical levels in medicinal species. *J Agric Food Chem* 2001; **49**: 1702–1706.

81. Molgaard P, Ohnsen S, Christensen P, Cornett C. HPLC method validation for simultaneous analysis of cichoric acid and alkamides in *Echinacea purpurea* plants and products. *J Agric Food Chem* 2003; **51**: 6922–6933.

82. Pellati F, Benvenuti S, Melegari M, Lasseigne T. Variability in the composition of anti-oxidant compounds in *Echinacea* species by HPLC. *PhytochemAnal* 2005; **16**: 77–85.

83. Luo XB, Chen B, Yao SZ, Zeng JG. Simultaneous analysis of caffeic acid derivatives and alkamides in roots and extracts of *Echinacea purpurea* by high performance liquid chromatography–photodiode array detection–electrospray mass spectrometry. *J Chromatogr A* 2003; **986**: 73–81.

84. Leathwood PD, Chauffard F, Heck F, *et al.* Aqueous extract of valerian root (*Valeriana officinalis* L) improves sleep quality in man. *Pharmacol Biochem Behav* 1982; **17**: 65–71.

85. Chan TY, Tang CH, Critchley JA. Poisoning due to an over-the-counter hypnotic, Sleep-Qik (hyoscine, cyproheptadine, valerian). *Postgrad Med J* 1995; **71**: 227–228.

86. Shohet D, Wills RB, Stuart DL. Valepotriates and valerenic acids in commercial preparations of valerian available in Australia. *Pharmazie* 2001; **56**: 860–863.

87. Torrado JJ. *In vitro* release of valerenic and hydroxyvalerenic acids from valerian tablets. *Pharmazie* 2003; **58**: 636–638.

88. Marles RJ, Kaminski J, Arnason JT. A bioassay for the inhibition of serotonin release from bovine platelets. *J Nat Prod* 1992; **55**: 1044–1056.

89. Heptinstall S, Williamson L, White A, *et al.* Extracts of feverfew inhibit granule secretion in blood platelets and polymorphonuclear leukocytes. *Lancet* 1985; **1**: 1071–1073.

90. Makheja AN, Bailey JM. Platelet phospholipase inhibitor from the medicinal herb feverfew (*Tanacetum panthenium*). *Prostaglandins Leuko Med* 1982; **8**: 653–660.

91. Johnson ES. Efficacy of feverfew as a prophylactic treatment of migraine. *BMJ* 1985; **291**: 569–573.

92. Murphy JJ, Heptinstall S, Mitchell JRA. Randomized double blind placebo controlled trial of feverfew in migraine prevention. *Lancet* 1988; **2**: 189–192.

93. Zhou JZ, Kou X, Stevenson D. Rapid extraction and high performance liquid chromatographic determination of parthenolide in feverfew (*Tanacetum parthenium*). *J Agric Food Chem* 1999: **47**: 1018–1022.

94. Nelson MH, Cobb SE, Shelton J. Variations in parthenolide content and daily dose of feverfew products. *Am J Health Syst Pharm* 2003; **59**: 1527–1531.

95. Curry EA, Murry DJ, Yoder C, Fife K, *et al.* Phase I dose escalation trial of feverfew with standardized dose of parthenolide in patients with cancer. *Invest New Drugs* 2004; **22**: 299–305.

96. Silagy CA, Neil HAW. A meta-analysis of effect of garlic on blood pressure. *J Hypertens* 1994; **12**: 463–468.

97. Jabbari A, Argani H, Ghorbanihaghjo A, Mahdavi R. Comparison between swallowing and chewing garlic on levels of serum lipids, cyclosporine, creatinine and lipid peroxidation in renal transplant patients. *Lipids Health Dis* 2005; May 19; (4)1–11 [open access article].

98. Tattleman E. Health effects of garlic. *Am Fam Physician* 2005; **72**: 103–106.

99. Petry JJ. Garlic and postoperative bleeding [Letter]. *Plast Reconstr Surg* 1995; **96**: 483–484.

100. Arnault I, Christides JP, Mandon N, Haffner T, Kahane R, Auger J. High-performance ion pair chromatography method for simultaneous analysis of allin, deoxyallin, allicin and dipeptide precursors in garlic products using multiple mass spectrometry and UV detection. *J Chromatogr A* 2003; **991**: 69–75.

101. Rosen RT, Hiserodt RD, Fukuda EK, Ruiz RJ, *et al.* Determination of allicin, S-allylcysteine and volatile metabolites of garlic in breath, plasma or simulated gastric fluid. *J Nutr* 2001; **131**: 968S–971S.

102. Haller CA, Benowitz NL. Adverse and central nervous system events associated with dietary supplements containing ephedra alkaloids. *N Engl J Med* 2000; **343**: 1833–1838.

103. Ichikawa M, UdayamaM, Imamura K, Shiraishi S, Matsuura M. HPLC determination of (+)-pseudoephedrine, (−)-ephedrine in Japanese herbal medicines containing ephedra herb using solid phase extraction. *Chem Pharm Bull (Tokyo)* 2003; **51**: 635–639.

104. Li HX, Ding MY, Lv K, Yu JY. Separation and determination of ephedrine alkaloids and tetramethylpyrazine in *Ephedra sinica* Stapf by gas chromatography–mass spectrometry. *J Chromatogr Sci* 2001; **39**: 370–374.

105. Cottiglia F, Bonsignore L, Acsu L, Deidda D, Pompei R, Casu M, Floris C. New phenolic constituents from *Ephedra nebrodensis. Nat Prod Res* 2005; **19**: 117–123.

106. Shao G, Wu F, wang DS, Zhu R, Luo X. Quantitative analysis of (*l*)-ephedrine and (*d*)-pseudoephedrine in plasma by high-performance liquid chromatography with fluorescence detection. *Yao Xue Xue Bao* 1995; **30**: 384–389.

107. Jacob P, III, Haller CA, Duan M, Yu L, Peng M, Benowitz NL. Determination of ephedra alkaloid and caffeine concentrations in dietary supplements and biological fluids. *J Anal Toxicol* 2004; **28**: 152–159.

108. Huang WF, Wen KC, Hsiao ML. Adulteration by synthetic therapeutic substances of traditional Chinese medicine in Taiwan. *J Clin Pharmacol* 1997; **37**: 344–350.

109. Vander stricht BI, Parvasis OE, Vanhaelen-Fastre RJ. Remedies may contain cocktail of active drugs. *BMJ* 1994; **308**: 1162.

110. Lau G, Lo DS, Yao YJ, Leong HT, Chan CL, Chu SS. A fatal case of hepatic failure possibly induced by nitrosofenfluramine: a case report. *Med Sci Law* 2004; **44**: 252–263.

111. Cole MR, Fetrow CW. Adulteration of dietary supplements. *Am J Health Syst Pharm* 2003; **60**: 1576–1580.

112. Ko RJ. Adulterants in Asian patent medicines. *N Engl J Med* 1998; **339**: 847.

113. Saper RB, Kales SN, Paquin J, Burns MJ, *et al.* Heavy metal content of Ayurvedic herbal medicine products. *JAMA* 2004; **292**: 2868–1873.

4

Measurement of Plasma L-DOPA and L-Tyrosine by High-Performance Liquid Chromatography as a Tumor Marker in Melanoma

Thierry Le Bricon,[1] Sabine Letellier,[1,2] Konstantin Stoitchkov,[3,4] and Jean-Pierre Garnier[1,2]

[1]*Laboratoire de Biochimie A Hôpital Saint-Louis (AP-HP), 1 Avenue Claude Vellefaux, 75010 Paris, France*
[2]*Faculté des Sciences Pharmaceutiques et Biologiques, Université Paris V, 4 Avenue de l'Observatoire, 75270 Paris, France*
[3]*Department of Dermatology, National Center of Oncology, Plovdivsko Pole Street 6, Sofia 1756, Bulgaria*
[4]*EORTC DC, Avenue E. Mounier 83–11, 1200 Brussels, Belgium*

4.1 INTRODUCTION

Melanoma is the leading cause of death from skin diseases: one out of five patients diagnosed with melanoma will die from this cancer. The incidence of melanoma is increasing by 4% per year in the USA and is reaching up to 10 per 100 000 per annum in Europe (lifetime risk ~1 in 200). Melanoma is a malignancy of melanocytes, cells of neuroectodermic origin migrating essentially to the skin during embryogenesis.

The primary risk factor for melanoma is childhood sun exposure and prevention through education towards children and parents is essential. Individuals whose skin type and/or genetic background place them as high risk should also adopt sun protection behavior. Only ~10% of melanomas are probably related to a genetic context, which involves CDKN2 and/or CDK4 genes coding for cyclin-dependent kinase (CDK) inhibitors. The American Joint Committee on Cancer (AJCC) melanoma classification is based on the tumor–node–metastasis (TNM) system.[1] This defines four different stages and some sub-stages based, for example, on

Chromatographic Methods in Clinical Chemistry and Toxicology   Edited by R. L. Bertholf and R. E. Winecker

histopathological criteria of the primary tumor. Early surgical removal of the primary tumor is the best curative treatment of melanoma. Disease progression is associated with poor prognosis due to the high metastatic potential. In the treatment of metastasizing melanoma, chemotherapy is disappointing (15–20% response to dacarbazine), even with the introduction of adjuvant therapy with interleukin-2 and interferon-α.

Efforts of researchers are concentrated on early detection of tumors and recurrence, selection of high-risk patients for adjuvant therapy and monitoring of treatment efficiency. In the last 25 years, a wide range of molecules have been investigated as tumor markers in melanoma.[2] Unfortunately, most have a limited role in screening, early diagnosis and staging due to their low sensitivity. A non-specific enzyme, lactate deshydrogenase (LDH), has recently been introduced in the revised AJCC melanoma staging system[1] (stage IV, M1c: elevated serum levels at the time of staging). Available laboratory techniques include molecular biology [such as reverse transcriptase polymerase chain reaction (RT-PCR)] for tyrosinase and other melanoma-associated mRNAs such as MART-1, GalNAc-T, PAX-3 and MAGE-A3, immunoassays [such as enzyme-linked immunosorbent assays (ELISAs)] for melanoma antigens and high-performance liquid chromatography (HPLC) for melanin precursors.

The present review will focus on the analysis of L-DOPA and L-tyrosine by HPLC in plasma. Both are involved in the first step of melanogenesis controlled by a melanocyte-specific enzyme, tyrosinase. The plasma L-DOPA/L-tyrosine ratio has been evaluated as a tumor marker in melanoma during a 10-year collaboration between the Biochemistry Laboratory, the Dermatology Department of Saint-Louis Hospital (AP-HP) in Paris (France) and the Dermatology Department of the National Center of Oncology in Sofia (Bulgaria).

4.2 MELANOGENESIS

Melanocytes produce pigments of complex heterogeneous polyphenol-like structure, known collectively as melanins. In humans, melanins play a major role in photoprotection based on their radical scavenging abilities. They are responsible for the production of different colored patterns in hair and superficial epidermis.

4.2.1 Overview of the pathway

Two basic types of melanins exist in mammals: reddish brown phaeomelanins and brownish black eumelanins. Both are produced by enzymatically catalyzed and chemical reactions in specific subcellular melanogenic compartments, known as melanosomes.

H. S. Raper started the most important studies on melanogenesis in the period 1920–35. The classical Raper–Mason scheme postulated a linear pathway involving L-tyrosine, L-DOPA, L-DOPAquinone, L-DOPAchrome, dihydroxyindoles and indolequinones, leading to the end-products melanins. During the last decade, new technologies such as the molecular biology of the pigmented-related genes have substantially modified the traditional concept of melanogenesis.[3,4] For example, melanin synthesis in mammals is catalyzed by at least two other metalloenzymes, the tyrosinase-related proteins 1 and 2.

The amino acid L-tyrosine is the starting material of melanin biosynthesis. The first step of melanogenesis is centered on tyrosinase (EC 1.14.18.1), a copper-containing glycoprotein. Until recently, this enzyme was believed to catalyze the hydroxylation of L-tyrosine

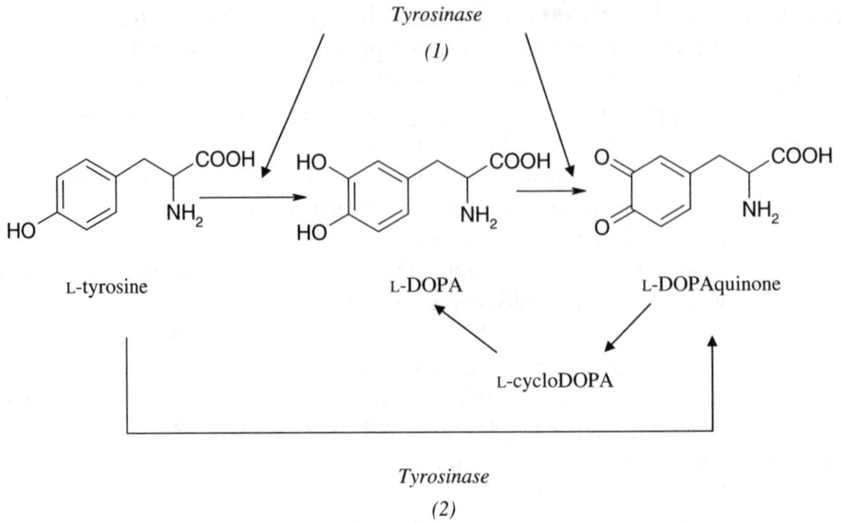

Figure 4.1 Early phases of the melanogenesis pathway. The first steps are critically regulated by the melanocyte-specific enzyme tyrosinase. L-DOPA is directly formed from L-tyrosine (1) and/or indirectly via L-DOPAquinone (2). Adapted from *Melanoma Research*, **9**, Letellier S, Garnier JP, Spy J, Stoitchkov K, Le Bricon T, Baccard M, Revol M, Kerneis Y, Bousquet B. Development of metastases in malignant melanoma is associated with an increase of plasma L-DOPA/L-tyrosine ratio, pages 389–394, 1999, with permission from Lippincott Williams & Wilkins

to 3,4-dihydroxyphenylalanine (L-DOPA) and the subsequent oxidation of L-DOPA to L-DOPAquinone (Figure 4.1). This classical role of tyrosinase has been contested, and L-DOPA could be indirectly formed from L-tyrosine via L-DOPAquinone. The catalytic activities of the human melanogenic enzymes, including tyrosinase, are still a matter of debate.

Eumelanogenesis and phaeomelanogenesis pathways diverge at the point where L-DOPAquinone, a reactive intermediate, undergoes either a reductive endocyclization yielding L-DOPAchrome, the precursor of the eumelanogenic pathway, or a reductive addition of thiols by cysteine to give the catechol 5-S-cysteinylDOPA (5-S-CD). The different 5-S-CD isomers are then oxidized to form phaeomelanins.

4.2.2 Potential tumor markers

Malignant melanocytes present defective melanosomes and tend to exhibit up-regulated melanogenesis. Melanogenuria (in the form of dark urine) is observed in some patients with widespread disease. End-product pigments, enzymes and melanin precursors or intermediates of the melanogenesis have therefore been measured in urine and blood from melanoma patients for more than 30 years.

Based on its key regulating role, the melanocyte-specific enzyme tyrosinase has been evaluated by the measurement of its activity (abandoned due to poor sensitivity), substrates and/or metabolites (L-tyrosine, L-DOPA) and, more recently, by the detection and now quantification of specific mRNAs, mostly in blood (plasma or serum). Tyrosinase appears to be the most reliable mRNA target for RT-PCR: it is a specific marker of melanocytic differentiation, expressed both in primary and metastatic melanoma. Many researchers aiming at the detection of melanoma cells in blood, bone marrow, lymph nodes and sentinel nodes have pursued this promising line of investigation. Unexpectedly, the results are so far extremely variable and sometimes negative in advanced melanoma patients.

Other evaluated serological markers from the melanogenesis pathway include the phaeomelanin metabolite 5-S-CD (see Ref. 5 for an review of 5-S-CD in melanoma with 2648 samples taken from 218 patients) and to a lesser extent the eumelanin metabolite 6-hydroxy-5-methoxyindole-2-carboxylic acid. Since 5-S-CD is very sensitive to light and oxidation, analysis requires immediate centrifugation, freezing of the samples after blood collection and exclusion of light and air-oxygen.

4.3 L-DOPA ALONE

4.3.1 Urine analysis

Blois and Bonda[6] in 1976 first detected elevated L-DOPA levels in urine of patients with metastatic disease using ion-exchange column chromatography with colorimetric detection. L-DOPA excretion levels appeared to be proportional to tumor burden and to provide earlier evidence of distant metastases than imaging techniques. This relation with disease stage was confirmed by Faraj et al.[7] using an enzyme radioimmunoassay; surgical intervention substantially decreased the urinary output of L-DOPA.

Incomplete 24-h urine collection, interfering substances, tedious sample preparation and low L-DOPA concentration in diluted samples have limited the routine use of urinary L-DOPA, even with the development HPLC coupled with electrochemical detection.[8,9] The reversed-phase HPLC method proposed by Ito et al.[10] used a simple alumina extraction, a procedure also applicable to serum (and even tissues). In addition, these sensitive and specific HPLC methods allowed the separation of both L-DOPA and 5-S-CD.[8–10]

4.3.2 Blood (plasma or serum) analysis

Using a radioenzymatic technique, Faraj et al.[11] measured plasma L-DOPA in 98 melanoma patients. For those without metastases, the mean L-DOPA plasma concentration ($1.01 \pm 0.12 \,\mu g \, L^{-1}$, $n = 21$) was not different from that of normal controls ($n = 32$). However, it was increased ($p < 0.001$) in patients with metastases to regional lymph nodes ($2.08 \pm 0.46 \,\mu g \, L^{-1}$, $n = 65$) and more distant metastases ($8.40 \pm 3.50 \,\mu g \, L^{-1}$, $n = 12$). The development from regional to distant metastases in four patients was accompanied by a 4–6-fold increase in the concentration of plasma L-DOPA. This clinical study, for the first time, suggested that measurement of L-DOPA in plasma might be useful in melanoma patients, at least in those with metastatic disease. Measuring L-DOPA by the technique of Faraj et al.[11] was, however, cumbersome: it should be enzymatically converted to dopamine and further to 3-O-[^{3}H]methyldopamine, before being extracted and characterized by radiochromatography.

4.4 L-DOPA/L-TYROSINE RATIO

Since L-tyrosine is the main precursor of melanins and a substrate of tyrosinase, the L-DOPA/L-tyrosine ratio could more accurately reflect tyrosinase activity than L-DOPA alone. In 1997,[12] we developed two reversed-phase HPLC techniques, one with electrochemical detection to measure simultaneously L-DOPA, norepinephrine (NE), epinephrine (E), dopamine (DA), and DOPAC (3,4-dihydroxyphenyl acetic acid) (all compounds easily oxidizable between +0.15 and +0.50 V), and one with fluorimetric detection to measure L-tyrosine (and phenylalanine) on the same blood sample.

4.4.1 Technical aspects

These two HPLC techniques, slightly modified since the original publication, are fully described in Table 4.1; analytical performances[12] are summarized in Table 4.2. All procedures

Table 4.1   Reversed-phase HPLC determination of L-DOPA and L-tyrosine[a]

Samples	
Plasma[b] or serum	Li heparinate[b]
Centrifugation	2000 g (10 min, 4 °C)
Storage	−80 °C
HPLC system	
Column	C_{18} reversed-phase (250 × 4 mm i.d.); integrated precolumn (4 × 4 mm); 5 µm Purospher particles (Merck)
Pump/injector	515 model/WISP 717 (Waters)
L-DOPA procedure (1 mL plasma or serum)	
Internal standard	+ 100 µL 3,4-hydroxybenzamide (DHBA, 200 nmol L^{-1})
Isolation	Alumina adsorption[b], final desorbing volume: 800 µL. 0.2 mol L^{-1} $HClO_4$; 150 µL of eluate injected
Mobile phase	0.05 mol L^{-1} phosphate buffer with 0.02 mol L^{-1} EDTA, 1.125 mmol L^{-1} sodium octanesulfonic acid (7% methanol), pH 2.65
Elution, run time	Isocratic: 1 mL min^{-1} flow-rate, 30 min
Detection	Coulometric (5100 A detector, ESA), analytical cell (5011 ESA) in oxidative mode (optimal potential: +0.35 V).
L-Tyrosine procedure (1 mL plasma or serum)	
Internal standard	+ 50 µL p-hydroxyphenylacetic acid (PHPA) (2.5 mmol L^{-1})
Deproteinization	+ 0.5 mL 1 mol L^{-1} trichloracetic acid, followed by centrifugation at 2000 g, 4 °C (10 min); 10 µL injected
Mobile phase	0.08 mol L^{-1} acetate buffer, with 0.02 mol L^{-1} sodium EDTA, 1.8 mmol L^{-1} sodium lauryl sulfate (10% methanol), pH 3.8
Elution, run time	Isocratic: 1 ml min^{-1} flow-rate, 10 min
Detection	Fluorimetric: excitation at 275 nm, emission at 305 nm (RF 535 detector, Shimadzu)

[a]All chemicals from Sigma-Aldrich (Saint-Quentin Fallavier, France). Other suppliers: Merck (Darmstadt, Germany), Waters (St Quentin en Yvelines, France), ESA (Bedford, MA, USA), Shimadzu (Kyoto, Japan).
[b]See Ref. 12 for more details.

Table 4.2 Characteristics of the L-DOPA and L-tyrosine HPLC procedures[a]

L-DOPA procedure	
Recovery	72.3% at 3 nM, 74.8% at 50 nM
Linearity	1–100 nM
Detection limit	0.25 nM
Precision	1.9–2.9% for intra-assay ($n = 6$), 3.0–3.9% for inter-assay ($n = 6$) (at 80–2.5 nM)
L-tyrosine procedure	
Recovery	87.9% at 25 μM, 88.2% at 200 μM
Linearity	10–500 μM
Detection limit	2.5 nM
Precision	1.6–2.4% for intra-assay, 3.9–4.5% for inter-assay (at 400–25 μM)

[a]All data from Ref. 12.

are carried out at room temperature. The total analysis time for L-DOPA analysis, including alumina adsorption, is ~3 h.

A few important remarks should be made concerning the L-DOPA procedure. For alumina adsorption (alumina activity grade I), glass tubes should never be used. We used 17×100 mm conical propylene tubes. To improve recovery, it is also important to desorb in a minimum volume of perchloric acid. The eluate (in propylene tubes) should then be directly injected into the column or kept at 4 °C until injection. The detailed extraction procedure (at 4 °C) is fully described in Ref. 12. Minimal changes in the experimental conditions of the L-DOPA procedure strongly influence the quality of the analytical process. These include methanol concentration (an increase will decrease the capacity factor for all compounds without affecting the order of elution), and counterion concentration (an increase will increase the capacity factor, except for DOPAC). The pH of the mobile phase is the most important factor to be controlled. As the pH increases, L-DOPA elution times decrease significantly with a risk of co-elution with DHBA, DOPAC or catecholamines (Figure 4.2). Changes in pH of less than one-tenth of a unit could produce shifts of 1 min in the L-DOPA retention time.

The high selectivity of HPLC coupled with electrochemical detection greatly reduces drug-induced interference, as tested for acetaminophen, acetylsalicylic acid, dihydroxy-caffeic acid and α-methylDOPA. None of the endogenous compounds found in plasma interfered with the retention time of L-DOPA or internal standard.

4.4.2 Clinical results

Five papers have been published on the value of the L-DOPA/L-tyrosine ratio as a marker of melanoma progression; the main biological and clinical results are summarized in Table 4.3.

The reference range was established at 4.6–9.4 nmol L^{-1} for L-DOPA, 32–88 μmol L^{-1} for L-tyrosine and $6.0–16.0 \times 10^{-5}$ for the L-DOPA/L-tyrosine ratio in a group of 35 healthy subjects.[12] No seasonal variations (winter/summer) were observed in a group

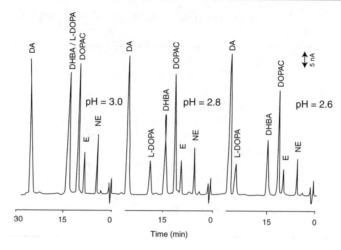

Figure 4.2 Variation of retention time of L-DOPA as a function of pH of the mobile phase. As the pH increases, the retention time of L-DOPA decreases. NE, norepinephrine; E, epinephrine; DA, dopamine; DHBA, internal standard. Reprinted from *Journal of Chromatography B*, **696**, Letellier S, Garnier JP, Spy J, Bousquet B. Determination of the L-DOPA/L-tyrosine ratio in human plasma by high-performance liquid chromatography. Usefulness as a marker in metastatic malignant melanoma, pages 9–17, Copyright 1997, with permission from Elsevier

of 11 healthy subjects;[13] additional controls consisting of 17 non-melanoma metastatic cancer patients (mostly breast cancers) did not exceed the 16.0×10^{-5} upper reference limit.[13]

Table 4.3   L-DOPA/L-tyrosine ratio in plasma/serum of melanoma patients

Study (n patients)	Biological results	Clinical results
Letellier *et al.*[12] (1997) (*n* = 46)	Normal range defined (*n* = 39)	Elevated in patients with distant metastasis vs other stages
Letellier *et al.*[13] (1999) (*n* = 90)	Specificity confirmed (in non-melanoma cancers) No seasonal variations	Elevated in distant metastasis vs other stages Relation with the number of distant metastases
Le Bricon *et al.*[14] (1999) (*n* = 50)	Poor concordance with tyrosinase mRNA	Sensitivity for distant metastasis: 41% (vs local: 30%)
Stoitchkov *et al.*[15] (2002) (*n* = 89)	No correlation with S100B (47% discordant results)	Elevated in distant metastasis Elevated in progressive patients with local metastasis (sensitivity/specificity: 78/67%)
Stoitchkov *et al.*[16] (2003) (*n* = 60)	Less variable than S100B	Decreased by chemotherapy (in responders) Prognostic value in distant metastasis

When compared with tyrosinase mRNA (by in-house nested RT-PCR),[14] and S100B (immunoluminometric LIA-mat Sangtec100; AB Sangtec Medical, Sweden),[15,16] there was no correlation ($r = 0.149$ with S100B) or little agreement (tyrosinase mRNA) with the L-DOPA/L-tyrosine ratio. Tyrosinase mRNA positivity by RT-PCR was associated with rapid disease progression and death within 2 months. By contrast, some cases of very low L-DOPA/L-tyrosine ratio levels have been reported shortly before death.[14] It was not sensitive to lymph node dissection (LND), whereas surgery decreased S100B by 52% in patients with elevated S100B.[16] These differences are probably due to the different respective mechanisms of production and release of these markers. The L-DOPA/L-tyrosine ratio reflects the overall metabolic activity of the melanocyte population (although an extracellular conversion of L-tyrosine to L-DOPA by tyrosinase during the loss of tumor cell integrity cannot be excluded). A local decrease in tyrosinase activity in lymph nodes or in some metastases might be also present.

Using the L-DOPA/L-tyrosine ratio, rather than L-DOPA alone, reduces inter-individual variability, at least in metastatic patients (K. Stoitchkov, unpublished data). A weak positive correlation was found between L-DOPA and L-tyrosine ($+0.296$, $p < 0.01$, Spearman correlation) in a group of 98 melanoma patients from the National Center of Oncology in Sofia (Bulgaria). Disease stage influenced plasma L-DOPA and L-DOPA/L-tyrosine ratio ($p < 0.001$), but not L-tyrosine concentration (data not shown). In patients with distant metastases, the 25th–75th percentile range represented 66% of the median value (L-DOPA/L-tyrosine ratio) vs 80% (L-DOPA alone). A similar effect was noted for the 5th–95th percentiles (124 vs 182%), a phenomenon also observed in patients with localized metastases.

Concerning individual variability of the L-DOPA/L-tyrosine ratio, some elements can be derived from a retrospective analysis of our database with a total of 255 patients diagnosed and/or followed in St Louis Hospital between 1994 and 2003 (S. Letellier, unpublished data). Apparent disease stability (no change to higher stage) was confirmed in 44 out of 116 patients without metastases (stage I–II) (median follow-up: 6 months from inclusion). The median difference between the first and the second samples was $+0.4 \times 10^{-5}$ (or $+2.0\%$, NS); the 75th percentile (reflecting intra-individual variability) represented 17% of the median value. Twelve patients exceeded the 75th percentile (up to $+143\%$ increase). The inter-individual variability (inclusion values) was 21%, as previously found in healthy subjects.[12]

4.4.3 Future directions

A large multi-marker study [including LDH, melanoma antigens S100B and melanoma inhibitory activity (MIA) and the L-DOPA/L-tyrosine ratio for serological tumor markers] has been initiated in Paris in a multi-center approach. The aims of this long-term (10 years) prospective study are to optimize the biological follow-up of melanoma patients and to select the most pertinent marker(s) or combination of markers. Identifying the most appropriate category of patients (AJCC stage or group defined, for example, by a given cut-off) who would benefit from the use of these markers is also of great interest. When analyzing inclusion values in our homogeneous group of 255 patients, the L-DOPA/L-tyrosine ratio was, for the first time, significantly higher in stage III (local metastases)

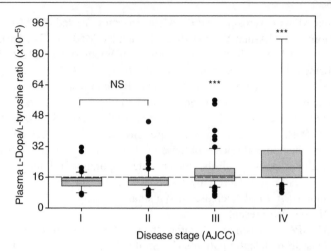

Figure 4.3 Plasma L-DOPA/L-tyrosine ratio and AJCC disease stage. A total of 255 melanoma patients were included: stage I, 54; stage II, 62; stage III, 67; and stage IV, 72 (AJCC staging using Ref. 2). *** Significantly different from other stages (Kruskal–Wallis ANOVA and Dunn's multiple comparison procedure). Box, 25th percentile/median/75th percentile; error bars, 5th percentile/95th percentile. There was 14 upper outliers: 7 in stage III, including a patient with Dubreuilh melanosis (184.8×10^{-5}), and 7 in stage IV (values up to 310×10^{-5})

than lower stages (no metastases) (Kruskal–Wallis ANOVA and Dunn's multiple comparison procedure, $p < 0.05$) (Figure 4.3). In stage III, 57% of patients were above the 16.0×10^{-5} upper cut-off[12] for the L-DOPA/L-tyrosine ratio (vs 22–24% in stage I–II). Patients with clinically localized disease, but with elevated L-DOPA/L-tyrosine ratio (above 16.0×10^{-5} or other cut-off to be defined by ROC analysis), might be considered as being at 'higher risk' for disease progression. Accordingly, they might deserve to be followed more closely and to be candidates for alternative or more aggressive therapies. An example of a longitudinal follow-up of a melanoma patient with the L-DOPA/L-tyrosine ratio is presented in Figure 4.4.

4.5 CONCLUSION

L-DOPA and L-tyrosine can be efficiently measured in human plasma by reversed-phase HPLC with electrochemical (L-DOPA) or fluorimetric (L-tyrosine) detection. Careful attention should be paid, however, to the entire analytical process. The specificity of the L-DOPA/L-tyrosine ratio for melanoma and its sensitivity (even in some patients with early disease) have been established. In advanced melanoma, it possesses a prognostic value, a relation with disease progression and treatment efficacy (chemotherapy). Studies on a large number of patients and with a long follow-up period are needed to refine the use of the L-DOPA/L-tyrosine ratio.

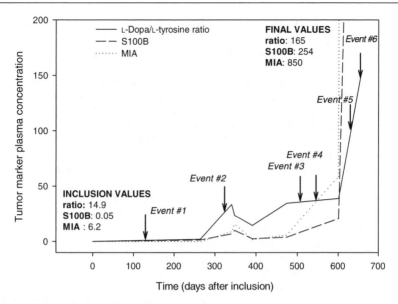

Figure 4.4 Example of longitudinal follow-up by plasma L-DOPA/L-tyrosine ratio. A 52-year-old male (initial melanoma: right arm) was monitored by the L-DOPA/L-tyrosine ratio, S100B and melanoma inhibitory activity (MIA) for 21 months. Six major clinical events are reported: 1, second local recurrence, 3 years after initial diagnosis; 2, bone and pulmonary metastases (stage IV); 3, evolution of subcutaneous metastases; 4, evolution of cutaneous and pulmonary metastases; 5, metastases to axillary nodes; 6, death

REFERENCES

1. C.M. Balch, A.C. Buzaid, S.-J. Soong, M.B. Atkins, N. Cascinelli, D.G. Coit, *et al.*, Final version of the American Joint Committee on Cancer staging system for cutaneous melanoma, *J. Clin. Oncol.*, **19**, 3635–3648 (2001).

2. L. Brochez, J.M. Naeyaert, Serological markers for melanoma, *Br. J. Dermatol.*, **143**, 256–268 (2000).

3. G. Prota, Recent advances in the chemistry of melanogenesis in mammals, *J. Invest. Dermatol.*, **75**, 122–127 (1980).

4. P.A. Riley, Melanogenesis and melanoma, *Pigment Cell Res.*, **16**, 548–552, (2003).

5. K. Wakamatsu, T. Kageshita, M. Furue, N. Hatta, Y. Kiyohara, J. Nakayama, *et al.*, Evaluation of 5-*S*-cysteinyldopa as a marker of melanoma progression: 10 years' experience, *Melanoma Res.*, **12**, 245–253 (2002).

6. M.S. Blois, P.W. Banda, Detection of occult metastatic melanoma by urine chromatography, *Cancer Res.*, **36**, 3317–3323 (1976).

7. B.A. Faraj, D.H. Lawson, D.W. Nixon, D.R. Murray, V.M. Camp, F.M. Ali, *et al.*, Melanoma detection by enzyme-radioimmunoassay of L-dopa, dopamine, and 3-*O*-methyldopamine in urine, *Clin. Chem.*, **27**, 108–112 (1981).

8. C. Hansson, L.E. Edholm, G. Agrup, H. Rorsman, A.M. Rosengren, E. Rosengren, The quantitative determination of 5-*S*-cysteinyldopa and dopa in normal serum and in serum from patients with malignant melanoma by means of high-pressure liquid chromatography, *Clin. Chim. Acta*, **88**, 419–427 (1978).

9. C. Hansson, G. Agrup, H. Rorsman, A.M. Rosengren, E. Rosengren, L.E. Edholm, Analysis of cysteinyldopas, dopa, dopamine, noradrenaline and adrenaline in serum and urine using high-performance liquid chromatography and electrochemical detection, *J. Chromatogr.*, **162**, 7–22 (1979).

10. S. Ito, T. Kato, K. Maruta, K. Fujita, T. Kurahashi, Determination of DOPA, dopamine, and 5-S-cysteiny-DOPA in plasma, urine, and tissue samples by high-performance liquid chromatography with electrochemical detection, *J. Chromatogr.*, **311**, 154–159 (1984).

11. B.A. Faraj, V.M. Camp, D.R. Murray, M. Kutner, J. Hearn, D. Nixon, Plasma L-dopa in the diagnosis of malignant melanoma, *Clin. Chem.*, **32**, 159–161 (1986).

12. S. Letellier, J.-P. Garnier, J. Spy, B. Bousquet, Determination of the L-Dopa/L-tyrosine ratio in human plasma by high performance liquid chromatography. Usefulness as a marker in metastatic malignant melanoma, *J. Chromatogr. B.*, **696**, 9–17 (1997).

13. S. Letellier, J.-P. Garnier, J. Spy, K. Stoitchkov, T. Le Bricon, M. Baccard, *et al.*, Development of metastases in malignant melanoma is associated with an increase of plasma L-Dopa/L-tyrosine ratio, *Melanoma Res.*, **9**, 389–394 (1999).

14. T. Le Bricon, K. Stoitchkov, S. Letellier, F. Guibal, J. Spy, J.-P. Garnier, *et al.*, Simultaneous analysis of tyrosinase mRNA and markers of tyrosinase activity in the blood of patients with metastatic melanoma, *Clin. Chim. Acta*, **282**, 101–113 (1999).

15. K. Stoitchkov, S. Letellier, J.-P. Garnier, B. Bousquet, N. Tsankov, P. Morel, *et al.*, Melanoma progression and serum L-Dopa/L-tyrosine ratio: a comparison with S100B, *Melanoma Res.*, **12**, 255–262 (2002).

16. K. Stoitchkov, S. Letellier, J.-P. Garnier, B. Bousquet, N. Tsankov, P. Morel, *et al.*, Evaluation of the serum L-Dopa/L-tyrosine ratio as a melanoma marker, *Melanoma Res.*, **13**, 287–293 (2003).

5

Hypersensitive Measurement of Proteins by Capillary Isoelectric Focusing and Liquid Chromatography-Mass Spectrometry

Feng Zhou and Murray Johnston

Department of Chemistry and Biochemistry, University of Delaware, Newark, DE 19716, USA

5.1 INTRODUCTION

In the post-genomic era, research in proteomics has moved to the forefront of attention as proteins play a central role in biological processes. The ability to profile proteins and peptides in cells and tissues is an important step in understanding how a biological activity progresses.[1–5] In a proteomic sample, there are usually a large number of proteins that vary widely in physicochemical properties, including molecular weight (MW), isoelectric point (pI), solubility, acidity/basicity and hydrophobicity/hydrophilicity. Moreover, the protein concentrations can extend over six orders of magnitude, and it is often the least abundant proteins that are of the greatest interest in systems biology.[6] For these reasons, there is a continuing need to develop and evaluate new platforms for protein separation and analysis. Two-dimensional polyacrylamide gel electrophoresis (2D-PAGE) is the core technology for profiling protein expression.[7,8] This method can separate proteins by their pI in the first dimension and eletrophoretic mobility in a polyacrylamide gel in the second dimension. As a multidimensional separation method, 2D-PAGE can resolve >1000 proteins in a single run.[8] However, there are two main drawbacks, sensitivity and throughput, that ultimately limit its use in proteomics research. 2D-PAGE usually employs Coomassie Brilliant Blue or silver staining to visualize separated proteins in the gel. Silver staining is the more sensitive of

Chromatographic Methods in Clinical Chemistry and Toxicology Edited by R. L. Bertholf and R. E. Winecker
© 2007 John Wiley & Sons, Ltd

these, detecting 0.1–10 ng amounts of protein, which correspond to 20–200 fmol for a 50-kDa protein.[7] The maximum dynamic range of 2D-PAGE is 10^4, making it inadequate for the study of proteins in low abundance.[9] Throughput is another limitation, since the entire process of gel preparation to protein separation and analysis is slow and difficult to automate fully. More importantly, the reproducibility of both the position and intensity of a protein spot are limited, which makes it difficult to compare the results from different runs.[1]

In principle, multidimensional liquid-based separations could overcome these problems. Liquid-based methods facilitate direct coupling with mass spectrometry (MS) for protein characterization. If the separation mode is analogous to 2D-PAGE, then comprehensive and powerful databases built from 2D-PAGE data can be effectively utilized.[10] Several liquid-based pH fractionation methods have been developed to replace the first dimension of 2D-PAGE. By coupling these with reversed-phase liquid chromatography (RPLC) and MS, an equivalent multidimensional separation to 2D-PAGE (pI vs MW under denaturing conditions) can be achieved. Lubman and co-workers reported the coupling of a Rotofor with non-porous RPLC for intact protein profiling.[11–13] The Rotofor is a relatively simple device that is the liquid analog of an isoelectric focusing (IEF) gel. The chamber is filled with liquid containing protein and ampholyte. A high voltage is then applied across the chamber and the proteins migrate to locations based on their pI. After separation, each protein fraction is transferred from chamber into a small vial for further analysis by RPLC-MS. Shang et al.[14] used a Rotofor system for intact protein separation in which the chambers were separated by Immobiline-impregnated polyacrylamide gel membranes. With this modification, the use of ampholyte was not necessary, which eliminated the possibility of ampholyte interference in MS analysis. Speicher and co-workers reported a similar liquid-based isoelectrofocusing system to fractionate protein mixtures.[15–18] Lubman and co-workers have developed a system consisting of chromatofocusing (CF) coupled with RPLC-MS.[19–21] CF is an ion-exchange-based separation in which proteins are bound to an anion exchanger and then eluted by a continuous decrease of the buffer pH.[22–24] CF can easily be coupled with RPLC-MS to build an automated two-dimensional separation system.

Capillary isoelectric focusing (CIEF) coupled with MS is another promising approach to sensitive protein separation and analysis.[25–33] In CIEF, the fused-silica capillary contains both the protein mixture and ampholyte. When an electric potential is applied, the negatively charged acidic components in the ampholyte migrate toward the anode and decrease the pH in that end of the capillary, whereas the positively charged basic constituents migrate toward the cathode and increase the pH. These pH changes continue until each ampholyte species reaches its pI, at which point its charge is neutral and it ceases to migrate. The ampholyte buffering capacity provides a continuous pH gradient along the capillary. Protein analytes are amphoteric macromolecules that also focus in narrow zones based on their pI values in the same way as the individual ampholytes. The focusing effect of CIEF permits the analysis of dilute protein samples with a typical concentration factor of 50–100-fold.[6,34,35] CIEF provides a high-resolution separation based on pI, whereas subsequent MS analysis provides a mass measurement with high precision and accuracy. A display of pI vs MW can be obtained from CIEF-MS data with higher throughput and sensitivity than 2D-PAGE.[27,28]

Unfortunately, substantial problems remain with the CIEF-MS approach. Most importantly, the ampholyte used in CIEF co-elutes with the analyte and perturbs the electrospray ionization (ESI) process in a manner that suppresses analyte signal intensity and degrades mass resolution.[36,37] In addition, the resolution of CIEF alone is not sufficient to separate proteins completely in complex samples. When proteins co-elute from an ESI source, some high-concentration species may quench the ionization of other low-concentration species. In an extreme case, the signal of these low concentration species may be lost.[38] Finally, it is difficult to add reagents that enhance the solubility of hydrophobic proteins because their presence, like ampholyte, degrades ESI performance. For these reasons, it is desirable to insert an additional separation step between CIEF and MS such as RPLC. RPLC is easily coupled with ESI-MS and provides the opportunity to remove ampholyte and further separate proteins that are not sufficiently resolved by CIEF. A CIEF-RPLC-MS system for peptides has been reported by Lee and co-workers.[39,40] In this system, an injection loop attached to a microselection valve is placed between the anodic and cathodic cells to transfer peptide fractions into the RPLC system. Ampholyte is removed from the peptide fraction with a C_{18} trap column prior to RPLC. This system is capable of detecting thousands of peptides in a single run with high sensitivity. This system was also applied recently for intact yeast protein profiling.[41]

In our work, an improved CIEF-RPLC-MS system has been developed for protein separation and characterization.[42] A microdialysis membrane separates the cathodic cell from the capillary column. During the focusing step, catholyte is able to traverse the membrane but protein molecules cannot. After focusing, proteins are hydrodynamically pushed past the membrane to a microselection valve that collects and transfers fractions to the RPLC system for further separation. This configuration permits a linear pH gradient to be maintained in the separation capillary with no voltage applied to the microselsection valve. The cathodic cell and microselection valve can easily and safely be retrofitted into commercial instruments.

5.2 A ROBUST CIEF-RPLC INTERFACE

Figure 5.1 shows a schematic of the microdialysis membrane-based cathodic cell for CIEF. The cathodic cell is built around a PEEK tee (Valco, Houston, TX, USA). One collinear port is connected to a 55-cm CIEF capillary (50 μm i.d., 375 μm o.d.). The opposite collinear port is connected to a 20-cm fused-silica capillary that leads to a waste vial. The perpendicular

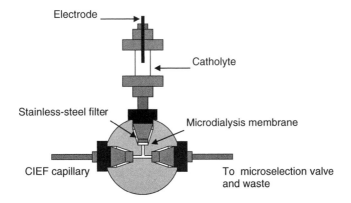

Figure 5.1 Diagram of the microdialysis membrane-based cathodic cell. Reprinted with permission from F. Zhou and M. Johnston, *Analytical Chemistry*, **76**, 2734–2740, Copyright 2004 American Chemical Society

port is connected to the cathodic vial. The cathodic vial is built from two reducing unions and a 5-cm length of Dayco polyethylene tubing (1/4 in o.d., 3/8 in i.d.). The first reducing union connects the polyethylene tubing to the PEEK tee through a 3-cm length of PEEK tubing (1/16 in o.d., 400 μm i.d.) that attaches to the perpendicular port. A microdialysis membrane (MW cutoff 3500 Da) cut to a 1/16-in diameter is inserted into the perpendicular port and supported by a 0.5-μm stainless-steel filter. The second reducing union connects the other end of polyethylene tubing to another 3-cm length of PEEK tubing (1/16 in o.d., 400 μm i.d.) containing a 6-cm Pt wire as the cathodic electrode. This cathodic 'vial' is filled with ~1.5 mL of catholyte and any air bubbles are carefully removed.

CIEF is performed with the microdialysis membrane-based cathodic cell mounted inside a commercially available automated capillary electrophoresis system (Beckman P/ACE MDQ with a UV detector). The CIEF capillary is filled with ampholyte solution (1:80 ampholye:water). Protein and peptide test mixtures are injected from the anodic side for 2 min by pressure. Focusing is performed in an electric field of 300 V cm^{-1}; the anolyte is 91 mM H$_3$PO$_4$ and the catholyte (sealed in the cathodic cell) is 20 mM NaOH. The platinum wire in cathodic cell is connected to the negative electrode of the instrument and set to 0 V. After a focusing period of 4 min, a small pressure is applied to the anodic end of the capillary to push the focused bands hydrodynamically through a UV absorption detector located 10 cm upstream of the cathodic cell. The 300 V cm^{-1} electric field is maintained across the capillary during hydrodynamic mobilization. Absorption is monitored at 280 nm. Sample losses to the cathodic cell are studied by repositioning the UV detector 10 cm downstream of the cathodic cell. All capillary lengths and sizes are identical for each detector location. Figure 5.2a shows the separation achieved with the microdialysis membrane-based cathodic

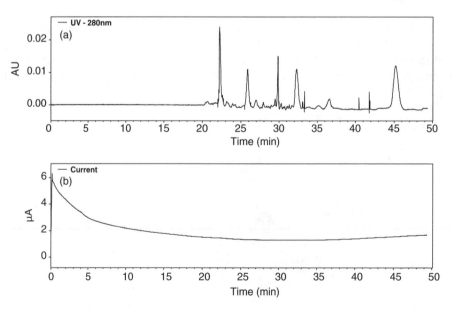

Figure 5.2   CIEF separation with the microdialysis membrane-based cathodic cell. (a) UV absorption (280 nm) vs time; (b) current profile. Reprinted with permission from F. Zhou and M. Johnston, *Analytical Chemistry*, **76**, 2734–2740, Copyright 2004 American Chemical Society

cell for a sample containing 0.1 nmol of ribonuclease A, myoglobin, carbonic anhydrase II, β-lactoglobulin, trypsin inhibitor and 1 nmol of CCK flanking peptide. In this experiment, the absorbance detector was placed upstream from the cathodic cell as described above. The current profile in Figure 5.2b shows a rapid decrease in the first few minutes with little change afterwards. A slight upturn in current is observed at very long times and is caused by hydrodynamic movement of phosphoric acid into the capillary. The current profile in Figure 5.2b is virtually identical with that obtained from a separation using the conventional cathodic cell, which indicates free movement of hydroxide ions across the microdialysis membrane. A plot of pI vs migration time with the membrane-based cathodic cell is also identical with that obtained with the commercial cathodic cell. In both cases, a slight non-linearity is observed, the origin of which origin has been discussed in detail elsewhere.[43] Although hydroxide ions can move across the microdialysis membrane by both diffusion and electrophoresis, diffusion is slow and a single fill of the cathodic cell can be used for over 1 month. Sample loss to the membrane-based cathodic cell was studied by comparing the peak area of ribonuclease A when the UV detector was located upstream vs downstream of the cell. Within experimental error, no change in peak area was observed between the two detector locations.

5.3 FIRST-GENERATION CIEF-RPLC-MS SYSTEM FOR PROTEINS

A diagram of the first-generation CIEF-RPLC-MS system is shown in Figure 5.3. The membrane-based cathodic cell assembly is connected to a six-port electronically controlled microselection valve with a 5-cm length of fused-silica capillary (50 μm i.d., 375 μm o.d.). The ampholyte solution (1:80 ampholye:water) and protein–peptide test mixtures are injected and separated by CIEF as described above. The focused bands are then hydro-dynamically pushed past the cathodic cell into a 1-μL stainless-steel injection loop attached to the microselection valve. The valve position is then switched, and proteins fractioned by the loop are transferred to a C_{18} reversed-phase trap column (5 mm × 300 μm i.d.) attached to a second six-port valve located on a capillary LC system. Figure 5.3 illustrates the valve positions during the transfer step. After trapping, both six-port valves are switched back to their original positions. Fractionated proteins are eluted from the trap column and separated by RPLC. Meanwhile, another fraction is collected on the 1-μL loop and the sequence is repeated. During the entire time, the 300 V cm^{-1} electric field is applied to the CIEF capillary to maintain the pH gradient. Typically, some electroosmotic flow occurs from the anodic end of the CIEF capillary during the course of the experiment. This flow is counterbalanced hydrodynamically by placing the waste vial ∼10 cm above the anodic end of CIEF capillary.

In order to collect and analyze protein fractions sequentially, hydrodynamic flow must be cycled on and off. In the 'go' mode, pressure is applied to the anodic end to push focused bands into the injection loop. In the 'stop' mode, the pressure is removed so that the analytes remaining in the capillary do not move. After the collected fraction has been transferred to the trapping column and washed, the microselection valve switches back to its original configuration and pressure is reapplied to the anodic end to collect a new fraction. While the new fraction is being collected, the fraction loaded on the trap column is eluted and analyzed by RPLC-MS.

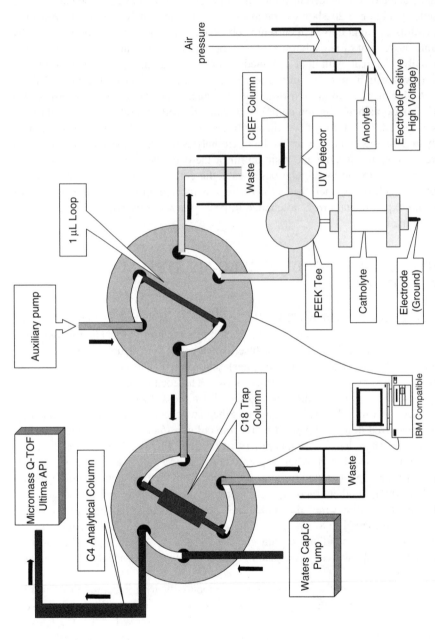

Figure 5.3 Diagram of the first-generation CIEF-RPLC-MS system. Reprinted with permission from F. Zhou and M. Johnston, *Analytical Chemistry*, **76**, 2734–2740, Copyright 2004 American Chemical Society

The protein fraction sampled by the injection loop is loaded on the trap column and washed with a solution of water–acetonitrile–acetic acid (94.9:5.0:0.1 v/v/v) at a flow-rate of $20\,\mu L\,min^{-1}$ for 3 min. Most ampholyte is removed during this time. The protein fraction is then eluted from the trap column with mobile phase and further separated on a C_4 reversed-phase column (5 cm × 300 μm i.d.). Mobile phase A [water–acetonitrile–acetic acid (94.9:5.0:0.1)] and mobile phase B [acetonitrile–water–acetic acid (94.9:5.0:0.1)] are delivered at a flow-rate of $20\,\mu L\,min^{-1}$ using a two-step gradient of 5 min (phase B from 25 to 60%) and 2 min (phase B from 60 to 90%). Proteins are mass analyzed with an ESI-QTOF mass spectrometer (Waters-Micromass, Beverly, MA, USA). Data processing is performed with MaxEnt1, which converts the raw spectrum of multiply charged ions into a deconvoluted display of singly charged ions.

Figure 5.4 shows how ampholyte can significantly degrade protein detection by MS. Figure 5.4a and b show the mass spectrum of 2 μM cytochrome c in H_2O, water–methanol–acetic acid buffer (50:49:1) with and without 1:80 ampholyte. In the presence of ampholyte, no cytochrome c signal is observed, highlighting the need to remove ampholyte prior to mass analysis. Figure 5.4c and d show the mass spectra of 2 pmol and 2 fmol, respectively, of cytochrome c loaded on the capillary and analyzed by CIEF-RPLC-MS. In each case, mass spectra were obtained with little interference from ampholyte. The high sensitivity demonstrated in Figure 5.4d may be further improved by optimizing the RPLC step.

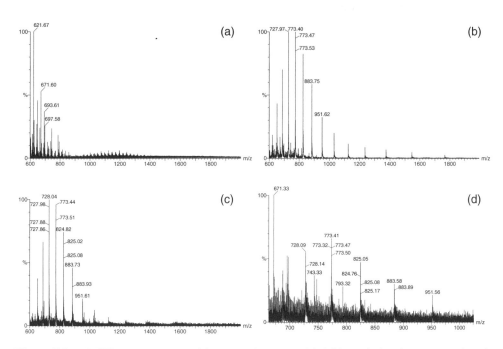

Figure 5.4 (a) ESI mass spectrum of 2 μM cytochrome c with 1:80 ampholyte in water–methanol–acetic acid (50:49:1); (b) ESI mass spectrum of 2 μM cytochrome c without ampholyte in the same buffer as in (a); (c) ESI mass spectrum of 2 pmol cytochrome c injected and analyzed by CIEF-RPLC-MS; (d) ESI mass spectra of 2 fmol cytochrome c injected and analyzed by CIEF-RPLC-MS. Reprinted with permission from F. Zhou and M. Johnston, *Analytical Chemistry*, **76**, 2734–2740, Copyright 2004 American Chemical Society

This system is then tested with a standard protein mixture containing 0.1 pmol of ribonuclease A, cytochrome *c*, myoglobin, insulin and β-lactoglobulin, 20 pmol of carbonic anhydrase II and bovine serum albumin and 1 pmol of CCK flanking peptide. The proteins and peptide are focused by CIEF and then divided into seven p*I* fractions by stop and go operation. Figure 5.5 shows the RPLC-MS analysis of these fractions. The labels A–G

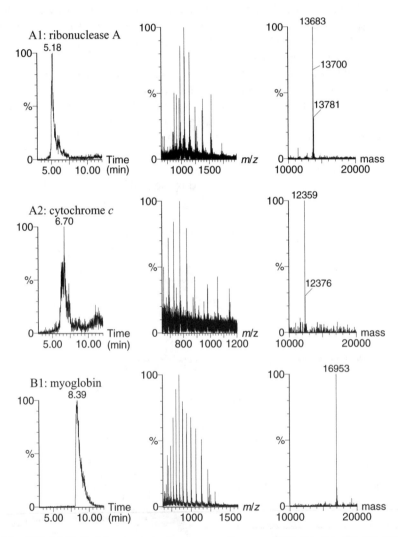

Figure 5.5   RPLC-MS of p*I* fractions. A–G represent p*I* ranges of 9.2–8.1, 8.1–7.0, 7.0–5.9, 5.9–4.8, 4.8–3.7, 3.7–2.6 and 2.6–1.5, respectively. Left panel, extracted mass chromatogram of each protein in the fraction; middle panel, raw spectrum of each protein averaged over the top 80% of the LC peak; right panel, deconvoluted spectrum (singly charged) by MaxEnt1. Only the total ion chromatogram is shown for fraction F; no deconvoluted spectrum is shown for fraction G. Reprinted with permission from F. Zhou and M. Johnston, *Analytical Chemistry*, **76**, 2734–2740, Copyright 2004 American Chemical Society

represent the seven protein fractions analyzed. Two fractions (A and D) contained two proteins, one fraction (F) contained no proteins or peptides and the remaining fractions contained just one protein/peptide. Three separate plots are shown for each protein/peptide: an extracted ion chromatogram, a raw mass spectrum averaged over the top 80% of the peak in the extracted ion chromatogram and the Maxent1 deconvoluted spectrum. For fraction F, just a total ion chromatogram is shown as the only signal observed is due to residual ampholyte. For proteins of molecular mass <20 kDa, 100 fmol are generally sufficient to obtain a high-quality spectrum. Larger proteins require more sample because the ion signal is distributed over many m/z values.

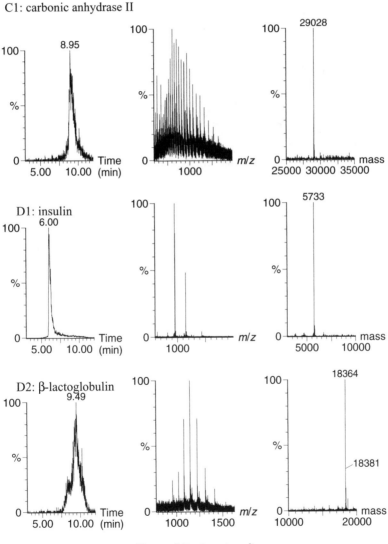

Figure 5.5 (*continued*)

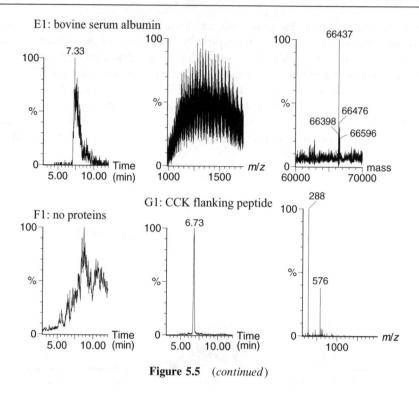

Figure 5.5   (*continued*)

5.4 SECOND-GENERATION CIEF-RPLC-MS SYSTEM

Although the first-generation system demonstrates the potential use of CIEF-RPLC-MS for protein separation and identification, it also has a disadvantage: the time allowed for RPLC is so short that the separation has to be run with a very fast gradient, giving poor resolving power. An improved CIEF-RPLC-MS system has been designed and constructed to solve this problem.[44] A schematic diagram of this second-generation system is shown in Figure 5.6. Compared with the system described above, the major difference is that there is an array of 10 storage loop located between the CIEF and RPLC instruments, and these loops retain the protein fractions, allowing more time for the RPLC step. With this setup, a shallower gradient is possible for RPLC separation, resulting in higher resolving power. In addition, a wider bore CIEF capillary (100 μm i.d.) is used so that more sample can be loaded. This system is used to profile proteins from a complex yeast enzyme concentrate, 11 μg μL^{-1}, in ampholyte solution (eCAP cIEF 3-10; diluted 1:100, ampholye:water) that also contains 6 M urea and 100 mM dithiothreitol (DTT). A total of 50 μg of protein is loaded in each run. The sample solution is injected from the anodic side by pressure. Focusing is performed in an electric field of 500 V cm^{-1}; the anolyte is 0.4 M acetic acid and the catholyte (sealed in the cathodic cell) is 0.5% NH$_4$OH. The CIEF capillary is contained within a continuously flowing coolant at 20 °C to remove the heat generated during focusing. The initial current is around 6 μA and after ∼6 min of focusing it drops to a steady-state value around 2 μA. Since the maximum sample volume that can be loaded onto the CIEF capillary is limited (only 4.3 μL in this experiment), the sample concentration is relatively

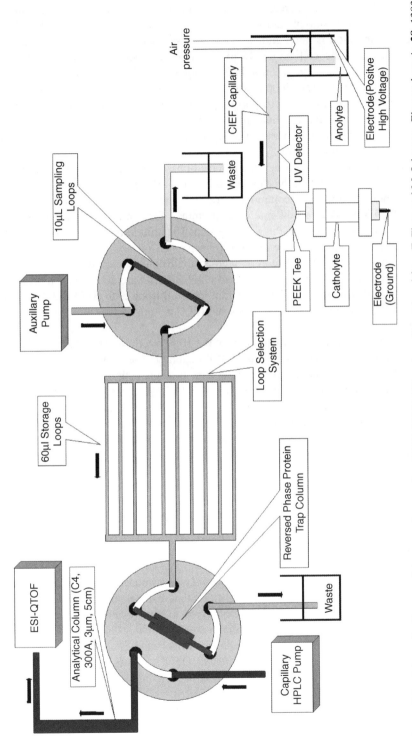

Figure 5.6 Schematic diagram of the second-generation CIEF-RPLC-MS system. Reproduced from F. Zhou and M. Johnston, *Electrophoresis*, **28**, 1383–1388 (2005), with permission from Wiley-VCH

high ($11\,\mu g\,\mu L^{-1}$) in order to increase the chance of detecting low abundance proteins. The current vs time profile is smooth and spikes in the UV trace are rarely observed, indicating that protein precipitation is not significant. The focused bands are then hydrodynamically pushed past the cathodic cell into a 10-μL stainless-steel sampling loop attached to a microselection valve. The valve position is then switched, and the proteins in the sampling loop are transferred to a storage loop constructed with 20 cm of PEEK tubing (1/16 in o.d.; $360\,\mu$m i.d.; Alltech) attached to a 10-port loop selection system. During the sampling and storage steps, other proteins are held in the CIEF capillary under voltage to maintain focusing. The microselection valve position cycles back and forth to sequentially load 10 fractions into the 10 storage loops.

Each stored sample is then loaded on a reversed-phase protein trap column attached to a second six-port valve located on the capillary LC system. Proteins and peptides in the stored sample are captured on the trap column and washed with a solution of water–acetonitrile–fomic acid (94.9:5.0:0.1) at a flow-rate of $20\,\mu L\,min^{-1}$ for 3 min. Most of the ampholyte, urea and DTT are removed during this washing step. The protein fraction is then eluted from the trap column with mobile phase and further separated on a C_4 reversed-phase column (5 cm × 300 μm i.d.). Mobile phase A [water–acetonitrile–fomic acid (94.9:5.0:0.1)] and mobile phase B [acetonitrile–water–acetic acid (94.9:5.0:0.1)] are delivered at a flow-rate of $10\,\mu L\,min^{-1}$ using a two-step gradient of 40 min (phase B from 20 to 60%) and 2 min (phase B from 60 to 90%). The sample eluted from the RPLC column is sent directly into the ESI-QTOF mass spectrometer.

Figure 5.7 shows total ion chromatograms of the 10 pI fractions. The pI ranges shown are based on a linear gradient assumed in the CIEF capillary. The protein bands can become wider in the low pH range and it is possible for some proteins with higher pI to diffuse into a lower pI fraction. Each chromatogram in Figure 5.7 results from the co-elution of many proteins/peptides. This aspect is illustrated in Figure 5.8, where mass spectra from three different time segments in the chromatogram of fraction 8 (pI 5.1–4.4) are shown. Each mass spectrum displayed in Figure 5.8 is the combination of 25 successive scans from the time period shown the chromatogram. The mass spectra are clearly different and indicate the presence of numerous proteins/peptides. The insets of the mass spectra in Figure 5.8 show expanded regions of individual peaks. These particular peaks exhibit isotopically resolved ions that allow the precise monoisotopic masses (i.e. masses for molecules containing exclusively ^{12}C, ^{1}H, ^{14}N, ^{16}O and ^{32}S) to be determined directly from the raw mass spectra. However, many peaks in the mass spectra are not isotopically resolved as the molecular mass is above the resolving power of the mass spectrometer. Therefore, the deconvolution software MaxEnt1, supplied by the mass spectrometer vendor (Waters), is used to determine the molecular masses that give rise to the various m/z values observed in the mass spectrum. MaxEnt1 is based on a maximum entropy algorithm that assumes that if a molecular mass is observed in charge state z, it also should be observed to some extent in charge states $z+1$ and $z-1$, $z+2$ and $z-2$, etc.[45–47] The main parameters input by the user are the expected range of molecular masses (3000–30 000 Da for this work) and peak width at half-height (i.e. mass spectrometer resolving power; 0.15 Da for the QTOF spectrometer used in this work). The output is an isotopically averaged mass spectrum which contains only singly charged peaks ($z = 1$) of the molecular masses observed.

The raw and deconvoluted mass spectra from scans 2000–2025 of the eighth protein fraction are shown in Figure 5.9a and b, as an example of proteins that are not isotopically resolved. The most intense feature of the deconvoluted spectrum in Figure 5.9b, 16 293 Da,

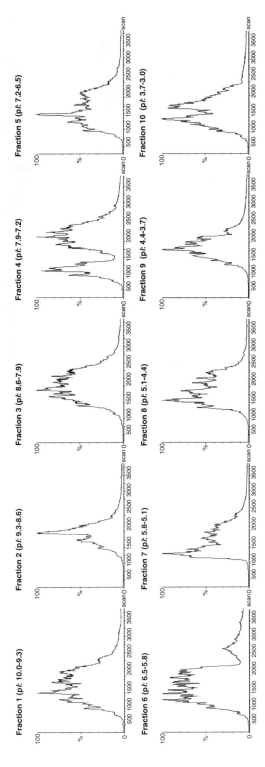

Figure 5.7 Total ion chromatograms of 10 p*I* fractions of soluble yeast proteins. Reproduced from F. Zhou and M. Johnston, *Electrophoresis*, **28**, 1383–1388 (2005), with permission from Wiley-VCH

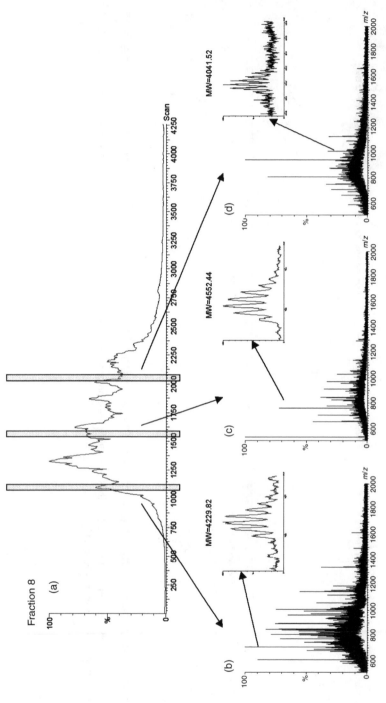

Figure 5.8 (a) Total ion chromatogram of the eighth fraction. Mass spectra corresponding to three segments of the chromatogram are shown in (b), (c) and (d). (b) Mass spectrum from scans 1150–1175. The +6 peak of a protein (monoisotopic mass = 4229.82) is shown in the inset. (c) Mass spectrum from scans 1510–1535. The +6 peak of a protein (monoisotopic mass = 4552.44) is shown in the inset. (d) Mass spectrum from scans 2050–2075. The +4 peak of a protein (monoisotopic mass = 4041.52) is shown in the inset. Reproduced from F. Zhou and M. Johnston, *Electrophoresis*, **28**, 1383–1388 (2005), with permission from Wiley-VCH

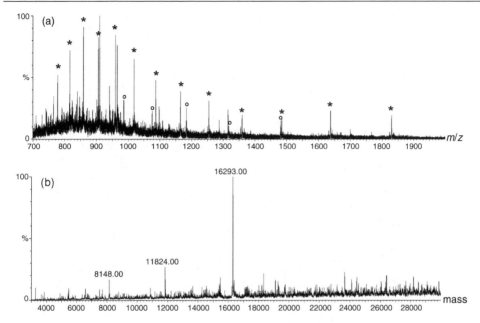

Figure 5.9 (a) Raw mass spectrum from scans 2000–2025 of the eighth fraction. Stars indicate the charge state distribution of 16 293 Da. Circles indicate the charge distribution of 11 824 Da. (b) The deconvoluted mass spectrum of (a) obtained by MaxEnt1. Reproduced from F. Zhou and M. Johnston, *Electrophoresis*, **28**, 1383–1388 (2005), with permission from Wiley-VCH

corresponds to the +9 to +21 charge state distribution indicated by stars in the raw spectrum in Figure 5.9a. The second most intense feature of the deconvoluted spectrum, 11 824 Da, corresponds to the +8 to +14 charge state distribution indicated by circles in the raw spectrum. Other features in the deconvoluted spectrum are not as easily matched with extended charge state distributions in the raw spectrum and may or may not represent real proteins in the sample. The baseline in Figure 5.9b increases with increasing mass, and it is possible that the increasing baseline and the greater number of low-intensity peaks in the higher mass range of the deconvoluted spectrum simply reflect the elevated baseline and background peaks (from artifacts associated with MaxEnt1 processing and also signal from residual additives, cone-induced fragmentation of protein molecular ions, etc.) in the low m/z range of Figure 5.9a. Also, the prominent peak at 8148 Da in the deconvoluted spectrum is most likely an artifact of the 16 293 Da peak – the MaxEnt1 algorithm is known to produce artifact peaks at half and twice the mass of the main peak,[10] in this case 32 586 Da (beyond the range of deconvolution) and 8148 Da.

The large number of proteins and peptides observed in Figures 5.8 and 5.9 is also characteristic of other spectra in this fraction and of most spectra in the remaining nine fractions. In all, there are over 1600 spectra in the dataset, each representing the average of 25 individual scans within the chromatogram of a single fraction. In the present work, manual interpretation (similar to the preceding paragraphs) was required for each combination of raw and processed spectra as automation software was not available. To shorten the analysis time, spectra corresponding to low total ion current (defined as less than 20% of the

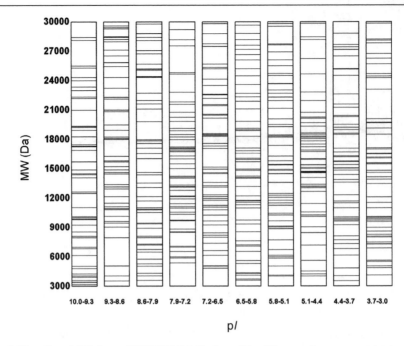

Figure 5.10 p*I* vs MW from CIEF-RPLC-MS data. The 50 most intense proteins/peptides as determined by MaxEnt1 are shown for each fraction. Reproduced from F. Zhou and M. Johnston, *Electrophoresis*, **28**, 1383–1388 (2005), with permission from Wiley-VCH

highest signal in the chromatogram) were excluded, leaving about 600 spectra for detailed interpretation. To simplify the analysis further, deconvolution was performed only over a mass range of 3000–30 000 Da. This mass range was appropriate for the present work as 2D-PAGE analysis of the yeast enzyme concentrate showed very few proteins larger than 30 kDa. Furthermore, the background and spurious peaks illustrated in Figure 5.9b increased substantially above 30 000 Da.

Figure 5.10 summarizes the results of this analysis in part by displaying the 50 most intense peaks derived by MaxEnt1 from each fraction, 500 protein 'bands' in total, as a plot of p*I* vs MW. This figure shows only the p*I* vs MW values, not the relative abundances, because the relationship between peak intensity in the raw mass spectrum and relative height in the MaxEnt1 output is still under investigation. The MW values displayed in Figure 5.10 were selected in the following way. First, the five most intense peaks in each spectrum were tabulated, taking care to exclude artifact peaks at half or twice the mass of intense peaks. Using Figure 5.9b as an example, the peaks at 16 293 and 11 824 Da were included among the five most intense peaks but that at 8148 Da was excluded. Next, the combined list of five peaks from each spectrum in a single fraction (on the order of 300 in total) was reduced to the 50 most intense peaks while removing redundant values (i.e. the same protein in adjacent spectra). Again using Figure 5.9b as an example, 16 293 Da was included in the list of 50 peaks, but 11 824 Da was not. This procedure eliminated the spurious peaks from Figure 5.10, but it also excluded many proteins. For this reason, the number of 'bands' shown in Figure 5.10 does not reflect the ultimate capability of the technique, nor does it reflect the fact that more proteins are present in some p*I* fractions than others. Several proteins are observed in adjacent p*I* fractions, presumably because the p*I* is at the interface between the two or because several isoforms exist with similar molecular masses.

5.5 FUTURE IMPROVEMENTS

Two specific experimental improvements are on the horizon. First, a two-loop set-up will replace the existing one-loop set-up for the collection of p*I* fractions. In the new set-up, one loop will collect the fraction from the CIEF capillary while the contents of the second loop are transferred to the storage loop. In this way, CIEF can run in a continuous mode with improved reproducibility of p*I* measurement over the stop and go mode. Second, a monolithic column will replace the standard packed column in RPLC. Since monolithic columns typically achieve efficiencies of 25 000, chromatographic resolution can be increased while significantly shortening the analysis time.[48,49]

Finally, an important need is to identify the separated proteins 'on the fly'. The CIEF-RPLC-MS approach potentially provides a seamless interface to tandem mass spectrometry for protein identification using top-down approaches to proteomics such as ECD (electron-capture dissociation),[50–52] ETD (electron-transfer dissociation)[53,54] and the pseudo-MS^3.[55] In top-down proteomics, individual proteins are identified through a sequence tag derived from molecular ion fragmentation. The coupling of one or more of these top-down techniques with a multidimensional separation that is based in part on p*I* will provide a fully automated alternative to 2D-PAGE that has the ability to increase significantly the throughput of proteomics investigations.

ACKNOWLEDGMENT

The work described in this chapter was supported by a grant from the National Science Foundation (DBI-0096578).

REFERENCES

1. D.M. Lubman, M.T. Kachman, H.X. Wang, S.Y. Gong, F. Yan, R.L. Hamler, K.A. O'Neil, K. Zhu, N.S. Buchanan and T.J. Barder, Two-dimensional liquid separation–mass mapping of proteins from human cancer cell lysates, *Journal of Chromatography B*, **782**, 183–196 (2002).
2. N. Masumori, T.Z. Thomas, P. Chaurand, T. Case, M. Paul, S. Kasper, R.M. Caprioli, T. Tsukamoto, S.B. Shappell and R.J. Matusik, A probasin–large T antigen transgenic mouse line develops prostate adenocarcinoma and neuroendocrine carcinoma with metastatic potential, *Cancer Research*, **61**, 2239–2249 (2001).
3. J.D. Wulfkuhle, K. McLean, D. Sgroi, A. Sahin, E. Petricoin and P. Steeg, Proteomic analysis of breast cancer progression, *Clinical Cancer Research*, **7**, 3684s–3684s (2001).
4. C.S. Giometti, S.L. Tollaksen, C. Chubb, C. Williams and E. Huberman, Analysis of proteins from human breast epithelial-cells using 2-dimensional gel-electrophoresis, *Electrophoresis*, **16**, 1215–1224 (1995).
5. C.S. Giometti, S.L. Tollaksen, S. Mukund, Z.H. Zhou, K.R. Ma, X.H. Mai and M.W.W. Adams, 2-dimensional gel-electrophoresis mapping of proteins isolated from the hyperthermophile *Pyrococcus furiosus*, *Journal of Chromatography A*, **698**, 341–349 (1995).

6. Y.F. Shen and R.D. Smith, Proteomics based on high-efficiency capillary separations, *Electrophoresis*, **23**, 3106–3124 (2002).

7. A. Gorg, C. Obermaier, G. Boguth, A. Harder, B. Scheibe, R. Wildgruber and W. Weiss, The current state of two-dimensional electrophoresis with immobilized pH gradients, *Electrophoresis*, **21**, 1037–1053 (2000).

8. P.H. O'Farrell, High-resolution 2-dimensional electrophoresis of proteins, *Journal of Biological Chemistry*, **250**, 4007–4021 (1975).

9. G.L. Corthals, V.C. Wasinger, D.F. Hochstrasser and J.C. Sanchez, The dynamic range of protein expression: a challenge for proteomic research, *Electrophoresis*, **21**, 1104–1115 (2000).

10. C.S. Giometti, K. Williams and S.L. Tollaksen, A two-dimensional electrophoresis database of human breast epithelial cell proteins, *Electrophoresis*, **18**, 573–581 (1997).

11. D.B. Wall, M.T. Kachman, S.Y. Gong, R. Hinderer, S. Parus, D.E. Misek, S.M. Hanash and D.M. Lubman, Isoelectric focusing nonporous RP HPLC: a two-dimensional liquid-phase separation method for mapping of cellular proteins with identification using MALDI-TOF mass spectrometry, *Analytical Chemistry*, **72**, 1099–1111 (2000).

12. K. Zhu, J. Kim, C. Yoo, F.R. Miller and D.M. Lubman, High sequence coverage of proteins isolated from liquid separations of breast cancer cells using capillary electrophoresis–time-of-flight MS and MALDI-TOF MS mapping, *Analytical Chemistry*, **75**, 6209–6217 (2003).

13. H.X. Wang, M.T. Kachman, D.R. Schwartz, K.R. Cho and D.M. Lubman, Comprehensive proteome analysis of ovarian cancers using liquid phase separation, mass mapping and tandem mass spectrometry: a strategy for identification of candidate cancer biomarkers, *Proteomics*, **4**, 2476–2495 (2004).

14. T.Q. Shang, J.M. Ginter, M.V. Johnston, B.S. Larsen and C.N. McEwen, Carrier ampholyte-free solution isoelectric focusing as a prefractionation method for the proteomic analysis of complex protein mixtures, *Electrophoresis*, **24**, 2359–2368 (2003).

15. X. Zuo and D.W. Speicher, A method for global analysis of complex proteomes using sample prefractionation by solution isoelectrofocusing prior to two-dimensional electrophoresis, *Analytical Biochemistry*, **284**, 266–278 (2000).

16. X. Zuo, L. Echan, P. Hembach, H.Y. Tang, K.D. Speicher, D. Santoli and D.W. Speicher, Towards global analysis of mammalian proteomes using sample prefractionation prior to narrow pH range two-dimensional gels and using one-dimensional gels for insoluble and large proteins, *Electrophoresis*, **22**, 1603–1615 (2001).

17. X. Zuo and D.W. Speicher, Comprehensive analysis of complex proteomes using microscale solution isoelectrofocusing prior to narrow pH range two-dimensional electrophoresis, *Proteomics*, **2**, 58–68 (2002).

18. X. Zuo, P. Hembach, L. Echan, and D.W. Speicher, Enhanced analysis of human breast cancer proteomes using micro-scale solution isoelectrofocusing combined with high resolution 1-D and 2-D gels, *Journal of Chromatography B*, **782**, 253–265 (2002).

19. B.E. Chong, F. Yan, D.M. Lubman, and F.R. Miller, Chromatofocusing nonporous reversed-phase high-performance liquid chromatography/electrospray ionization time-of-flight mass spectrometry of proteins from human breast cancer whole cell lysates: a novel two-dimensional liquid chromatography/mass spectrometry method, *Rapid Communications in Mass Spectrometry*, **15**, 291–296 (2001).

20. S.P. Zheng, K.A. Schneider, T.J. Barder, and D.M. Lubman, Two-dimensional liquid chromatography protein expression mapping for differential proteomic analysis of normal and O157:H7 *Escherichia coli*, *Biotechniques*, **35**, 1202–1212 (2003).

21. K. Zhu, M.T. Kachman, F.R. Miller, D.M. Lubman and R. Zand, Use of two-dimensional liquid fractionation for separation of proteins from cell lysates without the presence of methionine oxidation, *Journal of Chromatography A*, **1053**, 133–142 (2004).

22. Y.S. Liu and D.J. Anderson, Gradient chromatofocusing high-performance liquid chromatography. 1. Practical aspects, *Journal of Chromatography A*, **762**, 207–217 (1997).

23. Y.S. Liu and D.J. Anderson, Gradient chromatofocusing high-performance liquid chromatography. 2. Theoretical aspects, *Journal of Chromatography A*, **762**, 47–54 (1997).

24. Y.M. Li, J.L. Liao, R. Zhang, H. Henriksson, and S. Hjerten, Continuous beds for microchromatography: chromatofocusing and anion exchange chromatography, *Analytical Biochemistry*, **267**, 121–124 (1999).

25. Q. Tang, A.K. Harrata and C.S. Lee, Capillary isoelectric-focusing electrospray mass-spectrometry for protein analysis, *Analytical Chemistry*, **67**, 3515–3519 (1995).

26. L.Y. Yang, C.S. Lee, S.A. Hofstadler, L. Pasa-Tolic and R.D. Smith, Capillary isoelectric focusing–electrospray ionization Fourier transform ion cyclotron resonance mass spectrometry for protein characterization, *Analytical Chemistry*, **70**, 3235–3241 (1998).

27. P.K. Jensen, L. Pasa-Tolic, G.A. Anderson, J.A. Horner, M.S. Lipton, J.E. Bruce and R.D. Smith, Probing proteomes using capillary isoelectric focusing–electrospray ionization Fourier transform ion cyclotron resonance mass spectrometry, *Analytical–Chemistry*, **71**, 2076–2084 (1999).

28. P.K. Jensen, L. Pasa-Tolic, K.K. Peden, S. Martinovic, M.S. Lipton, G.A. Anderson, N. Tolic, K.K. Wong and R.D. Smith, Mass spectrometic detection for capillary isoelectric focusing separations of complex protein mixtures, *Electrophoresis*, **21**, 1372–1380 (2000).

29. L. Pasa-Tolic, P.K. Jensen, G.A. Anderson, M.S. Lipton, K.K. Peden, S. Martinovic, N. Tolic, J.E. Bruce and R.D. Smith, High throughput proteome-wide precision measurements of protein expression using mass spectrometry, *Journal of the American Chemical Society*, **121**, 7949–7950 (1999).

30. S. Martinovic, S.J. Berger, L. Pasa-Tolic and R.D. Smith, Separation and detection of intact noncovalent protein complexes from mixtures by on-line capillary isoelectric focusing-mass spectrometry, *Analytical Chemistry*, **72**, 5356–5360 (2000).

31. S. Martinovic, L. Pasa-Tolic, C. Masselon, P.K. Jensen, C.L. Stone and R.D. Smith, Characterization of human alcohol dehydrogenase isoenzymes by capillary isoelectric focusing–mass spectrometry, *Electrophoresis*, **21**, 2368–2375 (2000).

32. T.S. Liu, X. Xiao, R. Zeng and Q. Xia, Protein isoforms observed by ultrahigh resolution capillary isoelectric focusing–electrospray ionization mass spectrometry, *Shengwu Huaxue Yu Shengwu Wuli Xuebao*, **34**, 423–432 (2002).

33. N.J. Clarke and S. Naylor, Capillary isoelectric focusing-mass spectrometry: analysis of protein mixtures from human body fluids, *Biomedical Chromatography*, **16**, 287–297 (2002).

34. W. Thormann, A. Tsai, J.P. Michaud, R.A. Mosher, and M. Bier, Capillary isoelectric-focusing – effects of capillary geometry, voltage gradient and addition of linear polymer, *Journal of Chromatography*, **389**, 75–86 (1987).

35. Y.F. Shen, S.J. Berger and R.D. Smith, Capillary isoelectric focusing of yeast cells, *Analytical Chemistry*, **72**, 4603–4607 (2000).

36. Q. Tang, A.K. Harrata and C.S. Lee, High-resolution capillary isoelectric focusing-electrospray ionization mass spectrometry for hemoglobin variants analysis, *Analytical Chemistry*, **68**, 2482–2487 (1996).

37. Q. Tang, A.K. Harrata and C.S. Lee, Two-dimensional analysis of recombinant *E. coli* proteins using capillary isoelectric focusing electrospray ionization mass spectrometry, *Analytical Chemistry*, **69**, 3177–3182 (1997).

38. J.L. Sterner, M.V. Johnston, G.R. Nicol and D.P. Ridge, Signal suppression in electrospray ionization Fourier transform mass spectrometry of multi-component samples, *Journal of Mass Spectrometry*, **35**, 385–391 (2000).

39. J.Z. Chen, C.S. Lee, Y.F. Shen, R.D. Smith and E.H. Baehrecke, Integration of capillary isoelectric focusing with capillary reversed-phase liquid chromatography for two-dimensional proteomics separation, *Electrophoresis*, **23**, 3143–3148 (2002).

40. J.Z. Chen, B.M. Balgley, D.L. DeVoe and C.S. Lee, Capillary isoelectric focusing-based multi-dimensional concentration/separation platform for proteome analysis, *Analytical Chemistry*, **75**, 3145–3152 (2003).

41. Y.J. Wang, B.M. Balgley, P.A. Rudnick, E.L. Evans, D.L. DeVoe and C.S. Lee, Integrated capillary isoelectric focusing/nano-reversed phase liquid chromatography coupled with ESI-MS for characterization of intact yeast proteins, *Journal of Proteome Research*, **4**, 36–42 (2005).

42. F. Zhou and M.V. Johnston, Protein characterization by on-line capillary isoelectric focusing, reversed-phase liquid chromatography, and mass spectrometry, *Analytical Chemistry*, **76**, 2734–2740 (2004).

43. K. Shimura, W. Zhi, H. Matsumoto and K. Kasai, Accuracy in the determination of isoelectric points of some proteins and a peptide by capillary isoelectric focusing: utility of synthetic peptides as isoelectric point markers, *Analytical Chemistry*, **72**, 4747–4757 (2000).

44. F. Zhou and M.V. Johnston, Protein profiling by capillary isoelectric focusing, reversed-phase liquid chromatography, and mass spectrometry, *Electrophoresis*, **26**, 1383–1388 (2005).

45. A.G. Ferrige, M.J. Seddon and S. Jarvis, Maximum-entropy deconvolution in electrospray mass-spectrometry. *Rapid Communications in Mass Spectrometry*, **5**, 374–377 (1991).

46. A.G. Ferrige, M.J. Seddon, B.N. Green, S.A. Jarvis and J. Skilling, Disentangling electrospray spectra with maximum entropy. *Rapid Communications in Mass Spectrometry*, **6**, 707–711 (1992).

47. A.G. Ferrige, M.J. Seddon, J. Skilling and N. Ordsmith, The application of Maxent to high-resolution mass-spectrometry. *Rapid Communications in Mass Spectrometry*, **6**, 765–770 (1992).

48. A. Premstaller, H. Oberacher, W. Walcher, A.M. Timperio, L. Zolla, J.P. Chervet, N. Cavusoglu, A. van Dorsselaer and C.G. Huber, High-performance liquid chromatography–electrospray ionization mass spectrometry using monolithic capillary columns for proteomic studies, *Analytical Chemistry*, **73**, 2390–2396 (2001).

49. W. Walcher, H. Oberacher, S. Troiani, G. Holzl, P. Oefner, L. Zolla and C.G. Huber, Monolithic capillary columns for liquid chromatography–electrospray ionization mass spectrometry in proteomic and genomic research, *Journal of Chromatography B*, **782**, 111–125 (2002).

50. M.W. Senko, S.C. Beu, and F.W. McLafferty, High-resolution tandem mass-spectrometry of carbonic-anhydrase, *Analytical Chemistry*, **66**, 415–417 (1994).

51. D.M. Horn, Y. Ge and F.W. McLafferty, Activated ion electron capture dissociation for mass spectral sequencing of larger (42 kDa) proteins, *Analytical Chemistry*, **72**, 4778–4784 (2000).

52. F.Y. Meng, B.J. Cargile, L.M. Miller, A.J. Forbes, J.R. Johnson and N.L. Kelleher, Informatics and multiplexing of intact protein identification in bacteria and the archaea, *Nature Biotechnology*, **19**, 952–957 (2001).

53. J.E.P. Syka, J.J. Coon, M.J. Schroeder, J. Shabanowitz and D.F. Hunt, Peptide and protein sequence analysis by electron transfer dissociation mass spectrometry, *Proceedings of the National Academy of Sciences of the United States of America*, 101, 9528–9533 (2004).

54. J.J. Coon, J.E.P. Syka, J. Shabanowitz and D.F. Hunt, Tandem mass spectrometry for peptide and protein sequence analysis, *Biotechniques*, **38**, 519–523 (2005).

55. J.M. Ginter, F. Zhou and M.V. Johnston, Generating protein sequence tags by combining cone and conventional collision induced dissociation in a quadrupole time-of-flight mass spectrometer, *Journal of the American Society for Mass Spectrometry*, **15**, 1478–1486 (2004).

6

Chromatographic Measurement of Transferrin Glycoforms for Detecting Alcohol Abuse and Congenital Disorders of Glycosylation

Anders Helander

Department of Clinical Neuroscience, Karolinska Institute and Karolinska University Hospital, SE-171/76 Stockholm, Sweden

6.1 INTRODUCTION

Transferrin is a glycoprotein with an M_r of about 80 000 that is synthesized mainly in the hepatocytes and secreted into the blood. Transferrin is the major iron-transport protein and plays an important physiological role in iron homeostasis and also in the regulation of cellular growth.[1] Transferrin exists in many different forms, which are determined by the glycosylation pattern, genetic polymorphism and the degree of iron saturation.[2] The total transferrin concentration in human serum is normally in the range 2.0–3.5 g L^{-1}. Changes in the transferrin concentration commonly occur in a number of pathological conditions, for example, increased serum levels are seen in iron deficiency anemia whereas malnutrition and infections cause decreased levels.

Qualitative modifications in the glycosylation of this protein occur after prolonged heavy alcohol consumption and also in patients with rare congenital disorders of glycosylation (CDG).[3] Therefore, measurement of transferrin microheterogeneity has been used both as a biomarker for detection and follow-up of alcohol abuse [known as carbohydrate-deficient transferrin (CDT)][4] and for diagnosis of CDG.[5] This chapter will review the microheterogeneity of human transferrin and the methods used for the measurement of transferrin

Chromatographic Methods in Clinical Chemistry and Toxicology   Edited by R. L. Bertholf and R. E. Winecker
© 2007 John Wiley & Sons, Ltd

glycoforms to detect alcohol abuse (CDT) and CDG, with the primary focus on chromatographic methods.

6.2 TRANSFERRIN MICROHETEROGENEITY

The transferrin molecule consists of three subunits: a single polypeptide chain containing 679
amino acid residues, a maximum of two asparagine-linked oligosaccharide units (N-glycans)
of complex structure in the C-terminal domain at positions 413 and 611, and two binding sites
for iron, one each within the N- and C-terminal domain.[1] Normally, serum transferrin is only
partially (~30%) iron-saturated. The N-glycans may be biantennary, triantennary or even
tetraantennary in structure. The different transferrin isoforms, or glycoforms, traditionally
have been named depending on the number of terminal, negatively charged sialic acid
residues on the glycans. The major glycoform under normal conditions, accounting for about
75–80% of total serum transferrin,[6,7] contains two disialylated biantennary glycans (i.e. a
total of four terminal sialic acid residues) and was accordingly named tetrasialotransferrin
(Figure 6.1). Other glycoforms that can be detected in blood from healthy individuals are
pentasialo- (~15% of serum transferrin), trisialo- (~5%), hexasialo- (~2%) and disialotransferrin (<2%). The remaining glycoforms, asialo-, monosialo-, heptasialo- and octasia-
lotransferrin, normally occur at <1% or in only trace amounts.[6,7]

Human transferrin displays genetic polymorphism with transferrin C being the most
common phenotype in all populations, whereas allelic B (lower isoelectric point, pI) and D
(higher pI) variants, with a different primary structure but a normal set of glycans, occur at
lower frequencies.[8]

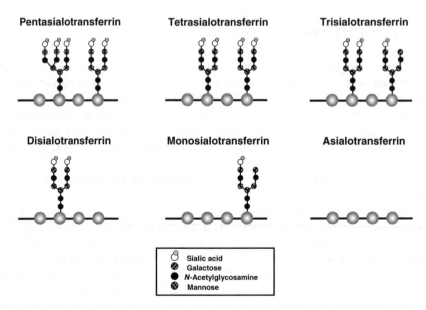

Figure 6.1 Structural illustration of the major normal glycoforms of human serum transferrin and
those being influenced by chronic alcohol abuse (carbohydrate-deficient transferrin; CDT) and in
congenital disorders of glycosylation (CDG)

6.3 CARBOHYDRATE-DEFICIENT TRANSFERRIN (CDT)

The abnormal glycoform pattern of serum transferrin that occurs as a result of prolonged heavy alcohol consumption, CDT, has emerged as a useful biochemical marker for identifying persons chronically abusing alcohol and for monitoring abstinence from alcohol during treatment.[4,9] Regular high alcohol intake averaging at least 50–80 g of ethanol per day for two or more weeks typically results in an altered transferrin glycoform profile. The biological mechanism by which alcohol causes changes in the carbohydrate composition of the molecule has not yet been conclusively identified, but likely involves interferences with the systems responsible for glycosyl transfer.[4,10,11] When drinking is discontinued, the glycoform pattern of serum transferrin slowly normalizes with a half-life of ~1.5–2 weeks,[4,12] and the time to reach a stable baseline level may require abstinence for 1 month or longer.[13]

The transferrin glycoforms were originally separated, identified and quantified by iso-electric focusing (IEF) based on the differences in net charge of the molecule. The glycoforms with isoelectric points at or above pH 5.7 after complete iron saturation, corresponding to disialo-, monosialo- and asialotransferrin, were collectively named CDT.[4] These glycoforms were first believed to contain truncated glycans deficient only in the terminal saccharide(s), but more recent studies have demonstrated that the major modification observed is instead a loss of one or both of the entire glycans (i.e. disialo- and asialotransferrin, respectively) and, consequently, also of the terminal negatively charged sialic acid residues.[14–18]

The major benefit of CDT compared with other laboratory tests used in routine clinical medicine to indicate chronic alcohol abuse, such as the liver enzyme γ-glutamyltransferase (GGT) and the mean corpuscular volume of erythrocytes (MCV), is its high specificity for alcohol and resulting low rate of false-positive results.[9,19,20] The limitations of CDT as a clinical marker include the lack of standardized procedures for the measurement and definition of CDT. Several different causes for falsely high or low CDT values have been reported over the years, mostly resulting from the use of inaccurate analytical methods, and/or expressing the CDT concentration in absolute (e.g. units or mg L^{-1}) instead of relative (%CDT) terms.[2,21,22] These limitations prompted the International Federation of Clinical Chemistry and Laboratory Medicine (IFCC) to convene a working group for the standardization of CDT measurement that addressed such issues as definition of the analyte, establishment of reference methodology and standardized calibration.

6.4 CONGENITAL DISORDERS OF GLYCOSYLATION (CDG)

Congenital disorders of glycosylation (CDG), formerly named carbohydrate-deficient glycoprotein syndrome, are a rare but increasingly recognized group of inherited multi-system disorders caused by mutations in the genes encoding the enzymes involved in the glycosylation of proteins and other glycoconjugates.[5,23,24] Glycosylation is a ubiquitous form of post-translational modification that has considerable impact on numerous biological processes.[25] CDG are divided into two subgroups, CDG-I and CDG-II. CDG-I comprises defects in the assembly and transfer of the dolichol-linked oligosaccharides (*N*-glycans) to the protein, whereas CDG-II refers to defects in the downstream processing of the protein-linked glycans. As a result, glycans are either completely missing (CDG-I) or become structurally abnormal (CDG-II). Cases in which the underlying defect has not been identified are called CDG-x.

The clinical symptoms of CDG are highly variable between individuals, and also within individual subtypes, but typically include psychomotor, growth and mental retardation from early childhood. CDG-Ia is by far the most frequent subtype with roughly 500 patients reported worldwide to date, while <100 have been diagnosed with type Ib-I*L* and only a few with type II (IIa–IIe).[24,26,27] However, it is generally assumed that CDG is largely under-diagnosed, partly due to the limited awareness of these disorders, but also to the variable and non-specific clinical presentation.[28–30] Another reason for underdiagnosis of this disorder may be that methods for CDG testing are not generally available.[31]

The diagnosis of CDG is based on a combination of clinical symptoms and biochemical tests, the latter including the presence of a defective glycosylation pattern of glycoproteins and enzymatic assays.[5,32,33] Confirmation and subtyping of CDG are established by molecular identification of the underlying genetic defects (i.e. detection of specific mutations). If CDG is suspected, initial diagnosis is currently based on the identification of alterations of the *N*-linked glycosylation of serum transferrin. Transferrin is a very convenient marker for this purpose, partly because of its relative abundance in serum, and also because assays for transferrin microheterogeneity are routinely available (i.e. the CDT assays for detection of alcohol misuse).

6.5 ANALYTICAL METHODS FOR TRANSFERRIN MICROHETEROGENEITY

Over the years, many different methods for carbohydrate-modified transferrin have been used for the routine determination of CDT. This has sometimes created confusion, related to discrepancies between methods (e.g. test values are given in various absolute and relative amounts), and also their ability to distinguish between normal and elevated values (i.e. the sensitivity and specificity for heavy alcohol consumption). For example, a very high total transferrin concentration might render falsely high results, when expressing the CDT content as an absolute amount and not in relation to total transferrin.[21,34,35] Moreover, changes in the definition of the analyte (e.g. heterogeneous mixtures of the asialo-, monosialo-, disialo-, and/or trisialotransferrin glycoforms in the 'CDT' fraction)[2,21,36] and the lack of traceability of reference intervals have hampered the clinical implementation and acceptance of this laboratory test.

Separation of transferrin glycoforms is based on variations in the carbohydrate structure and mainly the number of terminal, negatively charged, sialic acid residues, yielding molecules with different net charges. Originally this was achieved by IEF, but several other laboratory techniques have also been used for the investigation and diagnosis of CDT and CDG, including high-performance liquid chromatography (HPLC), capillary electrophoresis (CE), mass spectrometry (MS), western blotting, lectin analysis and immunoassays.[2,7,12,37–43] So far, the most common methods for detection of these analytes have been IEF and immunoassays. A limitation of IEF is that there is not a linear relationship between the concentration and staining intensity of tetrasialotransferrin, the major band as determined by densitometric analysis. This limitation makes accurate relative quantification of the different glycoforms impossible. Many immunoassays for CDT (CDTect, %CDT TIA and %CDT) are limited because they require an initial minicolumn separation of the CDT fraction from the non-CDT glycoforms prior to quantification, and genetic transferrin variants may cause falsely high or low values.

6.6 CHROMATOGRAPHIC METHODS FOR CDT

HPLC measurement of CDT is routinely used in hospital and research laboratories as a way to detect chronic alcohol misuse. The first HPLC method for the quantification of iron-saturated transferrin glycoforms was introduced in 1993,[12] and was based on anion-exchange chromatographic separation of the different glycoforms by salt gradient elution followed by photometric detection. Since then, several variations of the original method have been published[7,44–48] and a number of commercial HPLC assays for CDT have been introduced.[49]

6.6.1 HPLC conditions and potential interferences

A common feature of the HPLC methods for transferrin glycoforms is that quantification relies on the selective absorbance of the iron–transferrin complex at $\sim$460–470 nm.[50] This represents an analytical advantage over CE methods, which use detection at $\sim$214 nm where all proteins absorb. For example, C-Reactive Protein (CRP) has been demonstrated to interfere with the CE measurements by co-eluting with monosialotransferrin in the electropherograms.[51] In contrast, CRP does not interfere with the quantification of transferrin glycoforms by HPLC, even at very high concentrations ($>$200 mg L^{-1}).[7] Likewise, bilirubin and hemoglobin added in concentrations up to 150 μmol L^{-1} and 0.7 g L^{-1}, respectively, did not cause any significant interference with the quantification of transferrin glycoforms.

Using HPLC, the relative amount of any single transferrin glycoform, or combination of glycoforms, to total transferrin (all glycoforms) can be calculated from the area under the curve (%AUC). In recent studies, baseline or valley-to-baseline integration for all transferrin peaks has replaced the original valley-to-valley integration mode, because the results obtained in this way were found to be less sensitive to variations in peak resolution.[7] Another improvement is the use of ferric nitrilotriacetate (FeNTA), a well-known transferrin iron donor,[52] instead of FeCl$_2$ for complete saturation of serum transferrin. Because the FeNTA complex gives almost instant iron saturation, no additional incubation time is necessary. Preparation of serum without addition of FeNTA results in a very different HPLC profile (Figure 6.2a and b). This occurs because the net charge of the transferrin molecule depends on the level of iron saturation, and the non-saturated (Fe$_0$) or partially saturated [transferrin saturated at the N-terminal (Fe$_{1N}$) or C-terminal (Fe$_{1C}$) domain] glycoforms will thus elute differently from the completely iron-saturated (Fe$_2$) forms.[2] However, the high reproducibility of test results observed under standard HPLC conditions[7] indicate that the transferrin iron load is complete and stable, with no risk for loss of iron during the chromatographic separation.

Both serum and the pretreated samples are relatively stable upon storage, and also during normal handling in the laboratory. In addition to serum, the HPLC method has also been tested for use with plasma.[7] With lithium heparin and sodium heparin–fluoride plasma, the method allowed the separation and measurement of all transferrin glycoforms similarly to those for serum. However, with EDTA plasma, a high, asymmetric peak was always observed with a similar retention time as asialotransferrin (Figure 6.2c). This peak was also observed with blank samples (i.e. EDTA tube containing water; Figure 6.2c, inset), confirming that it is an artefact peak and not the result of iron removal from transferrin by complexation with EDTA. With citrated plasma, a shoulder can be observed on the tetrasialotransferrin peak (Figure 6.2d), but this was not observed with blank citrate samples.

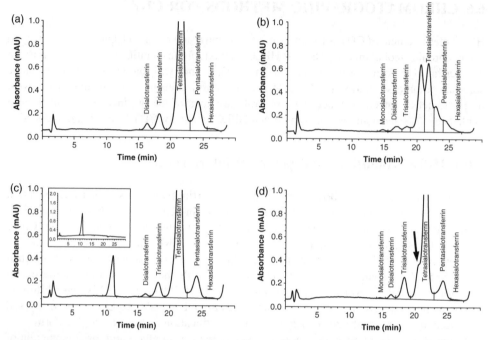

Figure 6.2 Transferrin glycoform patterns for serum and plasma samples with the HPLC method. HPLC traces for (a) a serum sample of the predominant transferrin C homozygote variant from a heavy drinker showing a slightly elevated disialotransferrin (2.1% of total AUC for transferrin), (b) the same serum sample but without treatment with FeNTA (i.e. no complete iron saturation of transferrin), (c) EDTA plasma from a light drinker showing an asymmetric EDTA peak eluting at about 11 min (inset: EDTA blank sample) and (d) citrated plasma from a light drinker showing an additional peak (at arrow) partially co-eluting with tetrasialotransferrin. Reproduced from A. Helander, A. Husa and J.-O. Jeppsson, *Clin. Chem.*, **49**, 1881–1890 (2003), by permission of the American Association of Clinical Chemistry

6.6.2 Chromatographic separation of transferrin glycoforms

The improved HPLC methods provide good separation between the different iron-saturated glycoforms of serum transferrin. A typical HPLC elution profile obtained for a serum sample from a control subject by a candidate HPLC reference method for CDT,[7] using a SOURCE 15Q PE 4.6/100 anion-exchange chromatographic column (Amersham Biosciences), is shown in Figure 6.3a. The peaks representing disialo-, trisialo-, tetrasialo-, pentasialo- and hexasialotransferrin were readily identified from their characteristic positions in the chromatogram. Most importantly, disialo- and trisialotransferrin are almost baseline separated. Because the trisialotransferrin level normally exceeds the disialotransferrin level by ~4-fold,[6,7] partial co-elution of these neighboring peaks may otherwise cause overestimation of disialotransferrin, the major CDT glycoform, and hence increase the risk of false-positive identifications of alcohol abuse.[2] In sera showing a high relative amount of trisialotransferrin (>6%), a peak representing monosialotransferrin is sometimes measurable usually at <0.5% of total transferrin (Figure 6.3c), eluting immediately ahead of disialotransferrin.

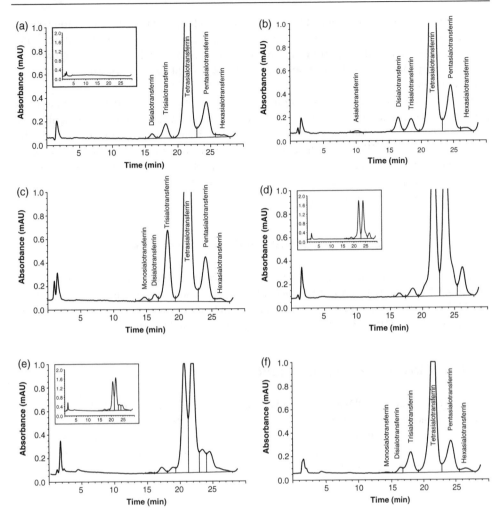

Figure 6.3 Transferrin glycoform patterns for different serum samples with the HPLC method. HPLC traces for (a) a transferrin C serum from a light drinker (inset: the same sample after immunosubtraction of transferrin), (b) serum from a heavy drinker with elevated disialotransferrin (3.2%) and a detectable (~0.4%) asialotransferrin, (c) serum from a person with high trisialotransferrin (8.4%) and a measurable (~0.5%) monosialotransferrin, (d) serum from a transferrin BC heterozygote (inset: the same chromatogram at full scale), (e) serum from a transferrin CD heterozygote (inset: the same chromatogram at full scale) and (f) serum from transferrin variant tentatively identified as a 'C2C3' heterozygote. Reproduced from A. Helander, A. Husa and J.-O. Jeppsson, *Clin. Chem.*, **49**, 1881–1890 (2003), by permission of the American Association of Clinical Chemistry

In healthy individuals, tetrasialotransferrin normally accounts for about 75–80% of serum transferrin, whereas trisialo- and pentasialotransferrin make up ~5 and ~10–15%, respectively, and disialotransferrin ~0.5–2% (mean 1.16%).[7] Patients who admitted drinking alcohol in the previous week (range 15–780 g day^{-1}) typically showed elevated values of disialotransferrin (mean 4.25%). Asialotransferrin, which is not detectable in serum from social drinkers, was also detected at 0.15–4.38% of total transferrin, in 57% of chronic heavy drinkers and in 62% of the sera with disialotransferrin >2%.[7] A typical HPLC elution profile for a serum sample collected from a chronic alcohol consumer is shown in Figure 6.3b. Small gender differences have been observed for the higher glycoforms but not for disialotransferrin.[7]

When transferrin in serum was adsorbed with rabbit anti-human transferrin antibody, no peaks were detectable in the HPLC trace (Figure 6.3a, inset), thus confirming that all peaks observed in the chromatogram at 460–470 nm are transferrin glycoforms.[7] Using the proposed candidate HPLC reference method,[7] the separation of transferrin glycoforms was achieved within 30 min but, for routine applications, the total run time can be reduced markedly by applying a steeper salt gradient. A recent commercial HPLC assay allows reproducible separation within only about 6 min.[49]

6.6.3 Genetic transferrin variants and glycoform types

HPLC traces of sera for some of the most common genetic transferrin variants, the transferrin BC and CD heterozygotes, and a transferrin C heterozygote tentatively identified as 'C2C3',[22] are shown in Figure 6.3d–f. In addition to these, a number of very rare variants, including the transferrin B homozygote, have been identified (unpublished work).

6.6.4 Sensitivity and reproducibility

The limit of detection (LOD) of the candidate reference HPLC method for CDT[7] in %AUC for asialo- and monosialotransferrin was approximately 0.05% of total serum transferrin in the normal concentration range ($\sim$2.0–3.5 g L^{-1}). For routine use, a limit of quantification (LOQ) of $\sim$0.1% is applied. It should be noted that the LOD and LOQ depend in part on the transferrin concentration in the serum sample. Furthermore, as all glycoforms except tetrasialotransferrin occur at low concentrations,[6] and the absorbance of the transferrin–iron complex at 460–470 nm is only $\sim$10% compared with its absorbance at 280 nm,[12] the performance of the photometric detector is very important. However, in dilution studies, using a serum sample containing a very high transferrin concentration and water as diluent (tested range 5–0.5 g L^{-1}), the AUC for total transferrin was linear and the calculated relative amount of disialotransferrin was not markedly changed at the lower total transferrin concentrations (CV 6.0%).[7] The intra- and inter-assay imprecision for serum samples containing normal and elevated disialotransferrin levels (range 1–5.6%) were below 5% (CV) and the retention times for the transferrin glycoforms were also stable over time.

6.7 CHROMATOGRAPHIC METHODS FOR CDG

IEF has long been the most common and reference procedure for CDG, but methods based on CE,[53] HPLC[31] and LC–MS[39] have also been introduced. The CDG-I pattern is characterized by an increase in transferrin glycoforms missing one or both of the entire N-glycans (i.e. disialo- and asialotransferrin, respectively),[39,54] whereas CDG-II is associated with an increase in trisialo- and monosialotransferrin, indicating the presence of truncated glycans.[55–57] A limitation of the HPLC methods has been that they require large sample volumes (0.1–1.0 mL) compared with IEF (<10 μL) and were therefore less suitable to screen for CDG in children, where the amount of sample available for determination is often limited. However, an improved method allowed for reproducible detection and relative quantification of the different transferrin glycoforms in approximately 10 μL of serum or plasma.[31]

6.7.1 HPLC testing for CDG

Using the improved HPLC method,[31] sera from CDG-I patients showed increased relative amounts of disialo- and asialotransferrin and concomitant reductions in tetrasialotransferrin (Figure 6.4). Patients with the CDG-Ia subtype showed the highest amounts of disialo- (range, 19–42%) and asialotransferrin (3.4–26%), whereas the patients with type Ib and Ig had less asialotransferrin (<3%). After the CDG-Ib patient had been treated with low-dose mannose,[58] the abnormal transferrin pattern improved, as indicated by marked reductions in disialo- (from 18 to 6%) and asialotransferrin (from 2.6 to 0.7%) (Figure 6.4e).

Sera from patients with undefined CDG-II defects showed the type II pattern with typical increases in trisialo- (range, 7.1–32%) and monosialotransferrin (5.7–15%) (Figure 6.5), indicating the presence of truncated glycans. By comparison, trisialotransferrin levels were normally <8% in controls and heavy drinkers, and monosialotransferrin was measurable only in individuals with a high trisialotransferrin level.[22] In all four CDG-II sera, two unidentified peaks were also observed (labeled A and B in Figure 6.5), and these peaks most likely represent other truncated variants.[31] This was supported by the observation that they showed similar retention times to two of the peaks that appeared after control serum was treated with neuraminidase (Figure 6.5d), which removes the terminal sialic acids of the N-glycans.[16,51] These observations suggest that the HPLC method allows for the separation of transferrin glycoforms not only based on the net charge of the molecule (i.e. sialic acid content), but also on structural differences of the glycans.

The CDG-I pattern resembles that observed after chronic alcohol consumption, albeit the relative increases were much higher in CDG-Ia (Figure 6.4). However, because CDG testing is typically performed at young age, there is usually no risk of false-positive results due to alcohol misuse. Furthermore, the CDG patterns are clearly distinguishable from those observed with the most common genetic transferrin variants,[22] which may cause falsely high or low values with the minicolumn immunoassays for CDT.

6.7.2 LC-MS testing for CDG

In addition to HPLC with photometric detection, an automated immunoaffinity LC method with electrospray MS detection of the transferrin glycoforms has been published.[39] This method focuses on identifying the glycoforms lacking one or both of the N-glycans (called 'mono-oligosaccharide' and 'a-oligosaccharide',[39] respectively, but corresponding to disialo- and asialotransferrin) and calculating relative ratios between the different forms (mono-/di-oligo and a-/di-oligo; i.e. disialo-/tetrasialotransferrin and asialo-/tetrasialotransferrin). Because increases in asialo- and disialotransferrin are characteristic for the type I pattern (Figure 6.4), this method should primarily be useful for the diagnosis of patients with CDG-I, which is the most common variant of the disease.

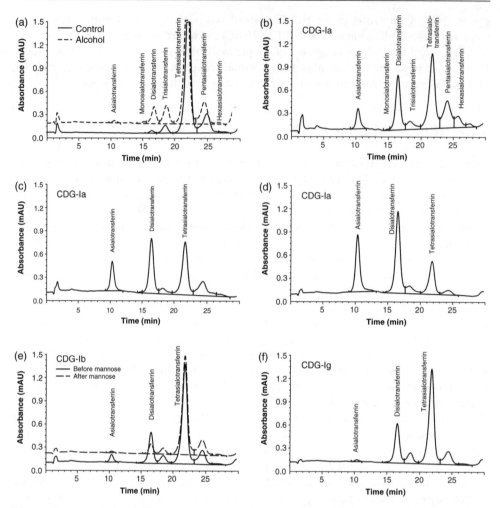

Figure 6.4 Transferrin glycoform patterns for serum samples from CDG-I patients by HPLC. HPLC traces for (a) a control serum sample from a light drinker (solid line) and from a heavy drinker with elevated asialo- and disialotransferrin and a measurable monosialotransferrin (broken line), (b–d) serum samples from three CDG-Ia patients showing elevated asialo- and disialotransferrin and reduced tetrasialotransferrin, (e) a serum sample from a CDG-Ib patient showing elevated asialo- and disialotransferrin (solid line) and another sample from the same patient after mannose therapy (broken line) and (f) a serum sample from a CDG-Ig patient showing elevated asialo- and disialotransferrin. Parts (a), (d), (e) and (f) reproduced from A. Helander, J. Bergström and H.H. Freeze, *Clin. Chem.*, **50**, 954–958 (2004), by permission of the American Association of Clinical Chemistry

6.8 SUMMARY AND CONCLUSIONS

Improved HPLC methods for the separation and quantification of transferrin glycoforms, based on anion-exchange chromatographic separation followed by selective photometric detection at 460–470 nm or MS detection, are useful for the detection and monitoring of

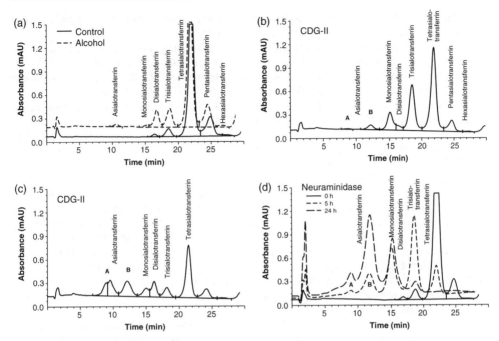

Figure 6.5 Transferrin glycoform patterns for serum samples from CDG-II patients by HPLC. HPLC traces for (a) a control serum sample from a light drinker (solid line) and from a heavy drinker with elevated asialo- and disialotransferrin and a measurable monosialotransferrin (broken line), (b, c) serum samples from two CDG-II patients with uncharacterized defects showing elevated mono- and trisialotransferrin and a reduced tetrasialotransferrin, and also two unknown peaks (A and B) and (e) a control serum before, and at different times after, treatment with neuraminidase. Parts (a), (b) and (d) reproduced from A. Helander, J. Bergström and H.H. Freeze, *Clin. Chem.*, **50**, 954–958 (2004), by permission of the American Association of Clinical Chemistry

alcohol abuse. These methods can also be used for preliminary diagnosis of CDG and for distinguishing between CDG types I and II. For CDT standardization, HPLC has been proposed as a possible reference method since it allows for reproducible and visible detection of the different glycoforms, and genetic variants and glycoform types that may cause incorrect determination of CDT with some assays are readily identified from a unique chromatographic pattern. In this respect, HPLC should be applicable to any combination of transferrin glycoforms, making it a useful diagnostic tool for clinical applications to CDT and CDG.

REFERENCES

1. G. de Jong, J.P. van Dijk and H.G. van Eijk, The biology of transferrin, *Clin. Chim. Acta*, **190**, 1–46 (1990).
2. T. Arndt, Carbohydrate-deficient transferrin as a marker of chronic alcohol abuse: a critical review of preanalysis, analysis, and interpretation, *Clin. Chem.*, **47**, 13–27 (2001).
3. G. de Jong, R. Feelders, W.L. van Noort and H.G. van Eijk, Transferrin microheterogeneity as a probe in normal and disease states, *Glycoconj. J.*, **12**, 219–226 (1995).

4. H. Stibler, Carbohydrate-deficient transferrin in serum: a new marker of potentially harmful alcohol consumption reviewed, *Clin. Chem.*, **37**, 2029–2037 (1991).

5. T. Marquardt and J. Denecke, Congenital disorders of glycosylation: review of their molecular bases, clinical presentations and specific therapies, *Eur. J. Pediatr.*, **162**, 359–379 (2003).

6. O. Mårtensson, A. Härlin, R. Brandt, K. Seppä and P. Sillanaukee, Transferrin isoform distribution: gender and alcohol consumption, *Alcohol. Clin. Exp. Res.*, **21**, 1710–1715 (1997).

7. A. Helander, A. Husa and J.-O. Jeppsson, Improved HPLC method for carbohydrate-deficient transferrin in serum, *Clin. Chem.*, **49**, 1881–1890 (2003).

8. M.I. Kamboh and R.E. Ferrell, Human transferrin polymorphism, *Hum. Hered.*, **37**, 65–81 (1987).

9. A. Helander, Biological markers in alcoholism, *J. Neural Transm. Suppl.*, 15–32 (2003).

10. C.S. Lieber, Carbohydrate deficient transferrin in alcoholic liver disease: mechanisms and clinical implications, *Alcohol*, **19**, 249–254 (1999).

11. P. Sillanaukee, N. Strid, J.P. Allen and R.Z. Litten, Possible reasons why heavy drinking increases carbohydrate-deficient transferrin, *Alcohol. Clin. Exp. Res.*, **25**, 34–40 (2001).

12. J.O. Jeppsson, H. Kristensson and C. Fimiani, Carbohydrate-deficient transferrin quantified by HPLC to determine heavy consumption of alcohol, *Clin. Chem.*, **39**, 2115–2120 (1993).

13. A. Helander and S. Carlsson, Carbohydrate-deficient transferrin and gamma-glutamyl transferase levels during disulfiram therapy, *Alcohol. Clin. Exp. Res.*, **20**, 1202–1205 (1996).

14. E. Landberg, P. Påhlsson, A. Lundblad, A. Arnetorp and J.-O. Jeppsson, Carbohydrate composition of serum transferrin isoforms from patients with high alcohol consumption, *Biochem. Biophys. Res. Commun.*, **210**, 267–274 (1995).

15. J. Peter, C. Unverzagt, W.D. Engel, D. Renauer, C. Seidel and W. Hösel, Identification of carbohydrate deficient transferrin forms by MALDI-TOF mass spectrometry and lectin ELISA, *Biochim. Biophys. Acta*, **1380**, 93–101 (1998).

16. H. Henry, F. Froehlich, R. Perret, J.D. Tissot, B. Eilers-Messerli, D. Lavanchy, C. Dionisi-Vici, J.J. Gonvers and C. Bachmann, Microheterogeneity of serum glycoproteins in patients with chronic alcohol abuse compared with carbohydrate-deficient glycoprotein syndrome type I, *Clin. Chem.*, **45**, 1408–1413 (1999).

17. C. Flahaut, J.C. Michalski, T. Danel, M.H. Humbert and A. Klein, The effects of ethanol on the glycosylation of human transferrin, *Glycobiology*, **13**, 191–198 (2003).

18. P. Kleinert, T. Kuster, S. Durka, D. Ballhausen, N.U. Bosshard, B. Steinmann, E. Hanseler, J. Jaeken, C.W. Heizmann and H. Troxler, Mass spectrometric analysis of human transferrin in different body fluids, *Clin. Chem. Lab. Med.*, **41**, 1580–1588 (2003).

19. H. Stibler, S. Borg and M. Joustra, Micro anion exchange chromatography of carbohydrate-deficient transferrin in serum in relation to alcohol consumption (Swedish Patent 8400587-5), *Alcohol. Clin. Exp. Res.*, **10**, 535–544 (1986).

20. G.J. Meerkerk, K.H. Njoo, I.M. Bongers, P. Trienekens and J.A. van Oers, Comparing the diagnostic accuracy of carbohydrate-deficient transferrin, gamma-glutamyltransferase, and mean cell volume in a general practice population, *Alcohol Clin. Exp. Res.*, **23**, 1052–1059 (1999).

21. A. Helander, Absolute or relative measurement of carbohydrate-deficient transferrin in serum? Experiences with three immunological assays, *Clin. Chem.*, **45**, 131–135 (1999).

22. A. Helander, G. Eriksson, H. Stibler and J.-O. Jeppsson, Interference of transferrin isoform types with carbohydrate-deficient transferrin quantification in the identification of alcohol abuse, *Clin. Chem.*, **47**, 1225–1233 (2001).

23. H.H. Freeze, Human disorders in *N*-glycosylation and animal models, *Biochim. Biophys. Acta*, **1573**, 388–393 (2002).

24. J. Jaeken, Komrower Lecture. Congenital disorders of glycosylation (CDG): it's all in it!, *J. Inherit. Metab. Dis.*, **26**, 99–118 (2003).

25. C.R. Bertozzi and L.L. Kiessling, Chemical glycobiology, *Science*, **291**, 2357–2364 (2001).

26. J. Jaeken and H. Carchon, Congenital disorders of glycosylation: a booming chapter of pediatrics, *Curr. Opin. Pediatr.*, **16**, 434–439 (2004).

27. C.G. Frank, C.E. Grubenmann, W. Eyaid, E.G. Berger, M. Aebi and T. Hennet, Identification and functional analysis of a defect in the human ALG9 gene: definition of congenital disorder of glycosylation type IL, *Am. J. Hum. Genet.*, **75**, 146–150 (2004).

28. H.H. Freeze, Congenital disorders of glycosylation and the pediatric liver, *Semin. Liver Dis.*, **21**, 501–515 (2001).

29. P. Briones, M.A. Vilaseca, M.T. Garcia-Silva, M. Pineda, J. Colomer, I. Ferrer, J. Artigas, J. Jaeken and A. Chabas, Congenital disorders of glycosylation (CDG) may be underdiagnosed when mimicking mitochondrial disease, *Eur. J. Paediatr. Neurol.*, **5**, 127–131 (2001).

30. G.M. Enns, R.D. Steiner, N. Buist, C. Cowan, K.A. Leppig, M.F. McCracken, V. Westphal, H.H. Freeze, F. O'Brien J, J. Jaeken, G. Matthijs, S. Behera and L. Hudgins, Clinical and molecular features of congenital disorder of glycosylation in patients with type 1 sialotransferrin pattern and diverse ethnic origins, *J. Pediatr.*, **141**, 695–700 (2002).

31. A. Helander, J. Bergström and H.H. Freeze, Testing for congenital disorders of glycosylation by HPLC measurement of serum transferrin glycoforms, *Clin. Chem.*, **50**, 954–958 (2004).

32. J. Fang, V. Peters, B. Assmann, C. Korner and G.F. Hoffmann, Improvement of CDG diagnosis by combined examination of several glycoproteins, *J. Inherit. Metab. Dis.*, **27**, 581–590 (2004).

33. G. Matthijs, E. Schollen and E. Van Schaftingen, The prenatal diagnosis of congenital disorders of glycosylation (CDG), *Prenat. Diagn.*, **24**, 114–116 (2004).

34. K. Sorvajärvi, J.E. Blake, Y. Israel and O. Niemelä, Sensitivity and specificity of carbohydrate-deficient transferrin as a marker of alcohol abuse are significantly influenced by alterations in serum transferrin: comparison of two methods, *Alcohol. Clin. Exp. Res.*, **20**, 449–454 (1996).

35. J. Keating, C. Cheung, T.J. Peters and R.A. Sherwood, Carbohydrate deficient transferrin in the assessment of alcohol misuse: absolute or relative measurements? A comparison of two methods with regard to total transferrin concentration, *Clin. Chim. Acta*, **272**, 159–169 (1998).

36. L. Dibbelt, Does trisialo-transferrin provide valuable information for the laboratory diagnosis of chronically increased alcohol consumption by determination of carbohydrate-deficient transferrin?, *Clin. Chem.*, **46**, 1203–1205 (2000).

37. R.P. Oda, R. Prasad, R.L. Stout, D. Coffin, W.P. Patton, D.L. Kraft, J.F. O'Brien and J.P. Landers, Capillary electrophoresis-based separation of transferrin sialoforms in patients with carbohydrate-deficient glycoprotein syndrome, *Electrophoresis*, **18**, 1819–1826 (1997).

38. C. Colome, I. Ferrer, R. Artuch, M.A. Vilaseca, M. Pineda and P. Briones, Personal experience with the application of carbohydrate-deficient transferrin (CDT) assays to the detection of congenital disorders of glycosylation, *Clin. Chem. Lab. Med.*, **38**, 965–969 (2000).

39. J.M. Lacey, H.R. Bergen, M.J. Magera, S. Naylor and J.F. O'Brien, Rapid determination of transferrin isoforms by immunoaffinity liquid chromatography and electrospray mass spectrometry, *Clin. Chem.*, **47**, 513–518 (2001).

40. R. Artuch, I. Ferrer, J. Pineda, J. Moreno, C. Busquets, P. Briones and M.A. Vilaseca, Western blotting with diaminobenzidine detection for the diagnosis of congenital disorders of glycosylation, *J. Neurosci. Methods*, **125**, 167–171 (2003).

41. K. Yamashita, T. Ohkura, H. Ideo, K. Ohno and M. Kanai, Electrospray ionization-mass spectrometric analysis of serum transferrin isoforms in patients with carbohydrate-deficient glycoprotein syndrome, *J. Biochem. (Tokyo)*, **114**, 766–769 (1993).

42. M.C. Ferrari, R. Parini, M.D. Di Rocco, G. Radetti, P. Beck-Peccoz and L. Persani, Lectin analyses of glycoprotein hormones in patients with congenital disorders of glycosylation, *Eur. J. Endocrinol.*, **144**, 409–416 (2001).

43. A. Helander, M. Fors and B. Zakrisson, Study of Axis-Shield new %CDT immunoassay for quantification of carbohydrate-deficient transferrin (CDT) in serum, *Alcohol Alcohol.*, **36**, 406–412 (2001).

44. P. Simonsson, S. Lindberg and C. Alling, Carbohydrate-deficient transferrin measured by high-performance liquid chromatography and CDTect immunoassay, *Alcohol Alcohol.*, **31**, 397–402 (1996).

45. F. Renner, K. Stratmann, R.D. Kanitz and T. Wetterling, Determination of carbohydrate-deficient transferrin and total transferrin by HPLC: diagnostic evaluation, *Clin. Lab.*, **43**, 955–964 (1997).
46. F. Renner and R.D. Kanitz, Quantification of carbohydrate-deficient transferrin by ion-exchange chromatography with an enzymatically prepared calibrator, *Clin. Chem.*, **43**, 485–490 (1997).
47. E. Werle, G.E. Seitz, B. Kohl, W. Fiehn and H.K. Seitz, High-performance liquid chromatography improves diagnostic efficiency of carbohydrate-deficient transferrin, *Alcohol Alcohol.*, **32**, 71–77 (1997).
48. U. Turpeinen, T. Methuen, H. Alfthan, K. Laitinen, M. Salaspuro and U.H. Stenman, Comparison of HPLC and small column (CDTect) methods for disialotransferrin, *Clin. Chem.*, **47**, 1782–1787 (2001).
49. A. Helander, J.P. Bergström, Determination of carbohydrate-deficient transferrin in human serum using the Bio-Rad% CDT by HPLC test, *Clin Chim. Acta*, 187–190 (2006).
50. J.O. Jeppsson, Isolation and partial characterization of three human transferrin variants, *Biochim. Biophys. Acta*, **140**, 468–476 (1967).
51. F.J. Legros, V. Nuyens, E. Minet, P. Emonts, K.Z. Boudjeltia, A. Courbe, J.L. Ruelle, J. Colicis, F. de L'Escaille and J.P. Henry, Carbohydrate-deficient transferrin isoforms measured by capillary zone electrophoresis for detection of alcohol abuse, *Clin. Chem.*, **48**, 2177–2186 (2002).
52. H.G. van Eijk, W.L. van Noort, M.J. Kroos and C. van der Heul, Analysis of the iron-binding sites of transferrin by isoelectric focussing, *J. Clin. Chem. Clin. Biochem.*, **16**, 557–560 (1978).
53. H.A. Carchon, R. Chevigne, J.B. Falmagne and J. Jaeken, Diagnosis of congenital disorders of glycosylation by capillary zone electrophoresis of serum transferrin, *Clin. Chem.*, **50**, 101–111 (2004).
54. N. Callewaert, E. Schollen, A. Vanhecke, J. Jaeken, G. Matthijs and R. Contreras, Increased fucosylation and reduced branching of serum glycoprotein *N*-glycans in all known subtypes of congenital disorder of glycosylation I, *Glycobiology*, **13**, 367–375 (2003).
55. J. Jaeken, H. Schachter, H. Carchon, P. De Cock, B. Coddeville and G. Spik, Carbohydrate deficient glycoprotein syndrome type II: a deficiency in Golgi localised N-acetyl-glucosaminyltransferase II, *Arch. Dis. Child.*, **71**, 123–127 (1994).
56. B. Hansske, C. Thiel, T. Lubke, M. Hasilik, S. Honing, V. Peters, P.H. Heidemann, G.F. Hoffmann, E.G. Berger, K. von Figura and C. Korner, Deficiency of UDP-galactose:*N*-acetylglucosamine beta-1,4-galactosyltransferase I causes the congenital disorder of glycosylation type IId, *J. Clin. Invest.*, **109**, 725–733 (2002).
57. V. Peters, J.M. Penzien, G. Reiter, C. Korner, R. Hackler, B. Assmann, J. Fang, J.R. Schaefer, G.F. Hoffmann and P.H. Heidemann, Congenital disorder of glycosylation IId (CDG-IId) – a new entity: clinical presentation with Dandy–Walker malformation and myopathy, *Neuropediatrics*, **33**, 27–32 (2002).
58. V. Westphal, S. Kjaergaard, J.A. Davis, S.M. Peterson, F. Skovby and H.H. Freeze, Genetic and metabolic analysis of the first adult with congenital disorder of glycosylation type Ib: long-term outcome and effects of mannose supplementation, *Mol. Genet. Metab.*, **73**, 77–85 (2001).

7

Chromatographic Measurements of Catecholamines and Metanephrines

Eric C. Y. Chan and Paul C. L. Ho

Department of Pharmacy, National University of Singapore, 18 Science Drive 4, Singapore 117543

7.1 BACKGROUND

Pheochromocytoma is a tumor of neuroectodermal origin that arises from the chromaffin cells (pheochromocytes) of the sympathoadrenal system. The term pheochromocytoma is derived from the chemical and pathological characteristics of a cell; 'pheochromocyte' means a cell that takes on a 'dusky' color when exposed to chromium salts (Greek: *phaios and chroma* mean dusky and color, respectively), and 'cytoma' means tumor. Almost 90% of these tumors are found in one or both adrenal glands, but they may be located anywhere along the sympathetic chain and, rarely, in aberrant sites. The 'rule of 10' has been used to describe pheochromocytoma: 10% are extraadrenal and, of those, 10% are extra-abdominal; 10% are malignant; 10% are found in patients who do not have hypertension and, finally, 10% are hereditary.[1] Those functioning tumors arising outside the adrenal medulla are termed extraadrenal pheochromocytomas, whereas those non-secreting extra-adrenal tumors are called paragangliomas.[2]

Many patients do not spontaneously report certain symptoms and signs that are particularly valuable clues to the physician in suggesting the diagnosis of pheochromocytoma. It is therefore important to obtain a detailed history from any patients suspected of having this tumor. However, routine screening of pheochromocytoma in the workup of every hypertensive is not recommended because this tumor is involved in only 0.1% of cases of hypertension. Testing should be reserved for those patients with features suggestive of the tumor, or those with incidentally discovered adrenal masses. The diagnosis of pheochromocytoma is established by the demonstration of elevated 24-h urinary excretion of free

Chromatographic Methods in Clinical Chemistry and Toxicology Edited by R. L. Bertholf and R. E. Winecker

catecholamines and total metanephrines. Surgical resection of the tumor is the standard treatment.[3] If the primary tumor is localized to the adrenal gland and is benign, then survival is the same as that of a normal age-matched population.[2] In patients with unresectable, recurrent or metastatic disease, long-term survival is possible; however, the overall 5-year survival is less than 50%. Pharmacological treatment of the catecholamine excess is essential and surgery, radiation therapy or chemotherapy may provide palliative benefit.

The catecholamines are biosynthesized from the amino acid L-tyrosine as outlined in Figure 7.1. The known biochemical events leading to catecholamine metabolism and urinary excretion are also summarized in Figure 7.1. There is evidence that functioning pheochromocytomas have the various enzymes necessary for the conversion of tyrosine to catecholamines, and the level of activity of some of these enzymes (tyrosine hydroxylase, Dopa decarboxylase and dopamine β-hydroxylase) is higher than in normal adrenal tissue.[4]

In order to explain the enormous capacity of some pheochromocytomas to synthesize catecholamines in the presence of high tyrosine concentrations in many of these tumors, it has been postulated that feedback inhibition of tyrosine hydroxylase may be lacking or inefficient.[5] Most pheochromocytomas secrete a combination of norepinephrine (NE) and epinephrine (E); however, NE is usually liberated in a considerably larger concentration than E. Of diagnostic importance is the fact that pheochromocytomas that secrete at least some E are located in the adrenal gland.[6] The enzyme phenylethanolamine-N-methyltransferase (PNMT) was only detected in tumors when E was one of the catecholamines being secreted; non-functioning pheochromocytomas did not contain any catecholamine-synthesizing enzymes.[4] Eisenhofer *et al.*[7] reported the presence of catechol-O-methyltransferase (COMT), the enzyme responsible for conversion of catecholamines to metanephrines in pheochromocytomas. These results indicated that the elevated plasma levels of free metanephrines in pheochromocytoma patients are derived from catecholamines that are produced and metabolized within the tumors. Only rarely do pheochromocytomas secrete dopamine (DA), Dopa and serotonin (5-HT). It has been suggested that the elaboration of Dopa can occur only in malignant pheochromocytoma;[8] however, benign pheochromocytomas have also been reported to secrete both catecholamines and Dopa.[9] John *et al.* evaluated 86 patients with pheochromocytoma, and reported that the preoperative 24-h urinary DA was in the normal range for benign pheochromocytomas but increased in malignant pheochromocytomas.[10] They concluded that high preoperative 24-h urinary DA levels, among some other factors, might be used to predict the malignant potential of pheochromocytoma. It is noteworthy that, although increased urinary excretion of DA and its metabolite 4- hydroxy-3-methoxyphenylacetic acid [homovanillic acid (HVA)] rarely occurs in patients with pheochromocytoma, a rise in excretion of these substances occurs frequently in patients with other neural crest tumors (e.g. neuroblastoma and melanoma).

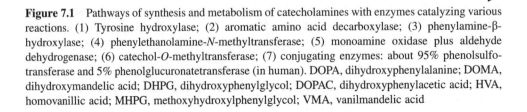

Figure 7.1 Pathways of synthesis and metabolism of catecholamines with enzymes catalyzing various reactions. (1) Tyrosine hydroxylase; (2) aromatic amino acid decarboxylase; (3) phenylamine-β-hydroxylase; (4) phenylethanolamine-N-methyltransferase; (5) monoamine oxidase plus aldehyde dehydrogenase; (6) catechol-O-methyltransferase; (7) conjugating enzymes: about 95% phenolsulfotransferase and 5% phenolglucuronatetransferase (in human). DOPA, dihydroxyphenylalanine; DOMA, dihydroxymandelic acid; DHPG, dihydroxyphenylglycol; DOPAC, dihydroxyphenylacetic acid; HVA, homovanillic acid; MHPG, methoxyhydroxylphenylglycol; VMA, vanilmandelic acid

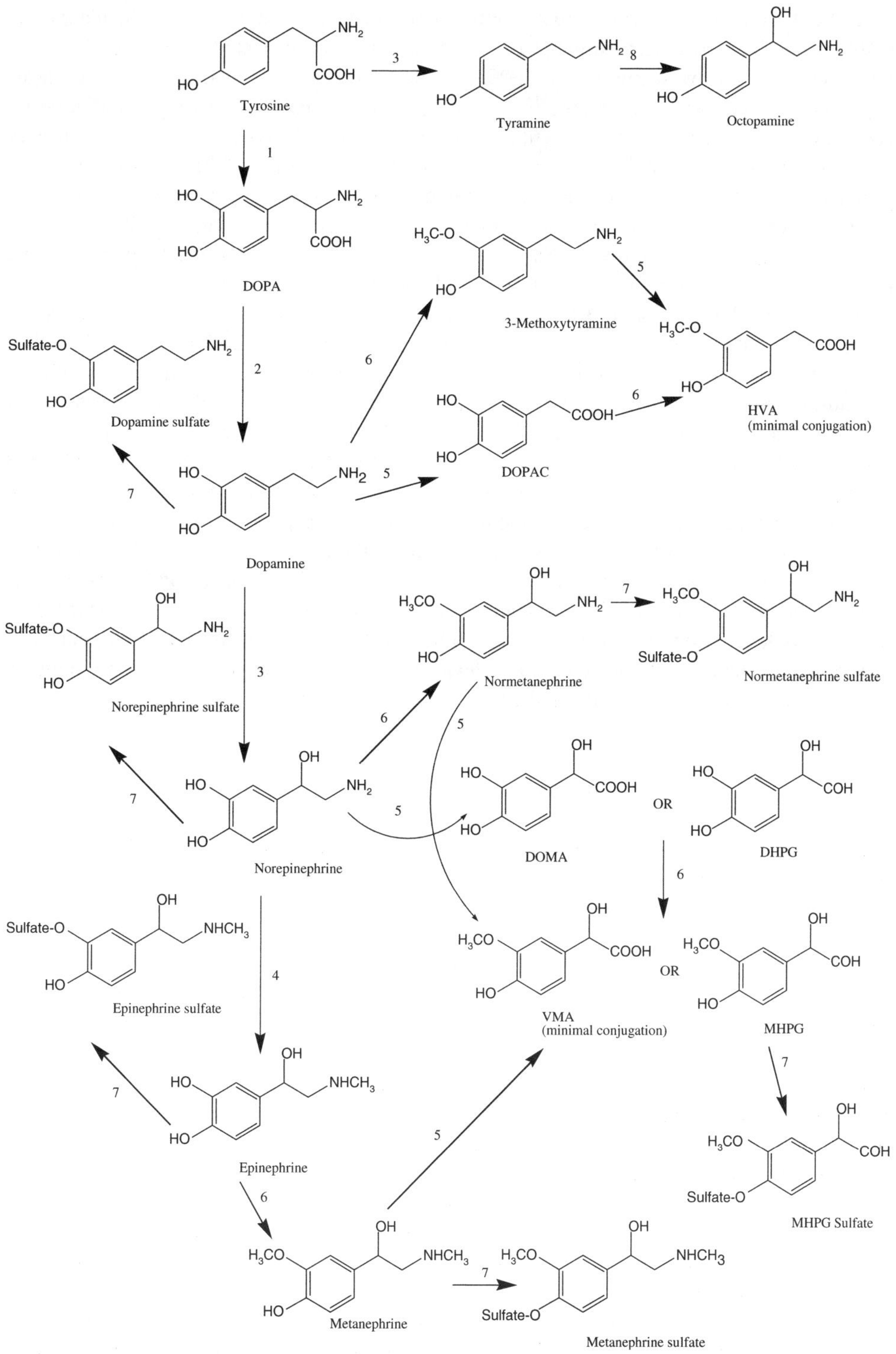

Tyrosine
Tyramine
Octopamine
DOPA
Dopamine sulfate
3-Methoxytyramine
HVA
(minimal conjugation)
DOPAC
Dopamine
Norepinephrine sulfate
Normetanephrine
Normetanephrine sulfate
Norepinephrine
DOMA
OR
DHPG
Epinephrine sulfate
VMA
(minimal conjugation)
MHPG
MHPG Sulfate
Epinephrine
Metanephrine
Metanephrine sulfate

The determination of urinary metanephrines [normetanephrine (NM) and metanephrine (MN)] is the best urine test available for screening patients suspected of having pheochromocytoma. This view seems to be shared by most investigators who have had the most experience in studying such patients. A study by Rosano et al., assessing the diagnostic accuracy of identifying patients with pheochromocytoma by determining 24-h urinary catecholamines [using high-performance liquid chromatography (HPLC)], total metanephrines and VMA, revealed that measurement of total metanephrines was the best discriminator between essential hypertension and pheochromocytoma.[11] These results were in agreement with those of Samaan et al., who found urinary metanephrines to be the most sensitive (97% positive) test for pheochromocytoma, whereas increased 24-h urinary catecholamines and VMA were detected in 76 and 88% of these patients, respectively.[12] Lucon et al. reviewed case histories of 50 patients in Brazil with pheochromocytoma and reported that urinary metanephrines were the best indicators with a diagnostic sensitivity of 97%.[13] Levels of NE in the urine and blood and urinary VMA showed sensitivities of 93, 92 and 90%, respectively. The lowest sensitivity was noted in plasma (67%) and urinary (64%) E and plasma DA (63%). In the Singapore General Hospital, the urinary free catecholamine and catecholamine metabolite levels of four pheochromocytoma patients were compared with the corresponding levels from 12 non-pheochromocytoma patients.[14] It was found that although the urinary VMA test was sensitive (100%), its specificity was only 31%, whereas the urinary total metanephrine test was both sensitive (100%) and specific (100%). As mentioned earlier, increased 24-h urinary DA level is one of the factors that indicates the likelihood of malignant pheochromocytoma.[10] Hence, patients with this elevated urinary DA should have more frequent follow-up evaluations to identify malignancy at earlier stages.

7.1.1 Total or individual assays

It has been suggested that evaluating M and NM excretions separately could increase the diagnostic sensitivity of these tests, because in some cases pheochromocytoma can secrete predominantly E, whereas in others the predominant catecholamine is NE, DA or Dopa.[15] In a retrospective study of 372 patients with sustained hypertension, for whom the determination of 24-h urinary metanephrines or VMA was ordered, Kairisto et al. found that evaluating NM excretion has better diagnostic specificity at 100% diagnostic sensitivity than did the sum of NM and M excretions. They also found that the determination of VMA in addition to NM and MN did not yield additional diagnostic information for any of the patients.[16] Similar findings were reported by Gerlo and Sevens after reviewing the measurement of catecholamines and their metabolites in 19 patients with pheochromocytoma.[17] They commented that assays of total catecholamines and total metanephrines are diagnostically inadequate, and that NE/E and NM/MN should be measured separately. Specific assays for E or MN are essential for identifying pure E-secreting tumors. Combined assays diminish the impact of E or MN and may cause false negatives. The specific measurement of E or MN is also valuable in the diagnosis of multiple endocrine neoplasia type 2 (MEN-II)-associated pheochromocytoma. It is also important to note that hypertension may not be manifested in patients harboring predominantly E-secreting tumors.[18] Therefore, measuring NM and MN with a specific procedure is the single most reliable diagnostic test for all patients suspected of having a pheochromocytoma, for patients with suggestive symptoms and for asymptomatic patients in whom a 'clinically silent' unsuspected adrenal mass has been incidentally detected. The assay of urinary VMA does not yield additional diagnostic information to urinary metanephrines and may fail to detect predominantly E-secreting tumors.[13,17,19]

This brief review of the literature pertaining to the diagnosis of pheochromocytoma indicates that 24-h urinary measurements of individual NE, E, NM and MN might be most useful for the diagnosis of pheochromocytoma in patients with suggestive symptoms such as headache, perspiration and palpitations. Urinary DA level could also be measured if malignancy is suspected. The traditional assay of urinary VMA may not be necessary if metanephrines are quantitated. There had also been much discussion with regard to the use of plasma metanephrines for the diagnosis of pheochromocytoma.[7,20–23] These studies reported that tests for plasma metanephrines are more sensitive than tests for plasma catecholamines or urinary metanephrines for the diagnosis of this tumor, particularly in paroxystic forms between crises and in patients with suspected MEN-II. Eisenhofer *et al.* demonstrated that measurement of plasma free metanephrines not only provided information about the likely presence or absence of a pheochromocytoma, but also, when a tumor was present, helped predict tumor size and location.[24]

7.2 ANALYTICAL MEASUREMENTS OF CATECHOLAMINES AND METANEPHRINES

Owing to the immense importance of catecholamines and metanephrines for the diagnosis and follow-up of pathologies such as pheochromocyotma, researchers have endeavored to improve the analytical methods used for their measurements, and substantial progress has been made. Nowadays, many of the old techniques of measuring catecholamines and metanephrines are obsolete, as new methods have greatly improved the analytical sensitivity, accuracy and clinical suitability for measuring these biogenic amines in biological fluids and tissues. To appreciate the developments of these analytical techniques, some of the earlier and current methods developed for the measurement of urinary catecholamines and metanephrines will be reviewed.

7.3 EARLY METHODS

7.3.1 Catecholamines

Traditional methods for the quantitation of catecholamines [norepinephrine (NE) and epinephrine (E)] relied on the production of detectable fluorophores. In the early fluorometric methods, NE and E were oxidized and rearranged under alkaline conditions to fluorescent trihydroxyindole products called adrenolutins.[25–27] Inclusion of ascorbic acid, along with rapid measurement of fluorescence, was required because the adrenolutins are easily oxidized. Drugs such as α-methyldopa, isoproterenol, quinidine, propranolol, labetalol and tricyclic depressants, which also show fluorescent properties, may interfere with this test. Modifying the original methods to include iodine as the oxidizing agent allows for the additional measurement of DA.[28] Although they are still used in some clinical laboratories, fluorescent methods have been mostly replaced by chromatographic methods.

7.3.2 Metanephrines

As with the catecholamines, fluorescence methods have also been reported for urinary metanephrine analysis. Fluorescent derivatization of the metanephrines NM and MN by chemical oxidation was based on modification of the trihydroxyindole reaction used for catecholamines. The individual metanephrines were measured following chromatographic separation and fluorescent derivatization[29,30] or through the formation of differential fluorescent compounds by oxidation at different pH levels.[31–34] Since the stability of the fluorescent products was variable, with some products decomposing within 10 min,[29] this method has limited application in current practice. Other early methods for analysis of NM and MN included electrophoresis[35,36] and paper[37,38] and thin-layer chromatography.[39] These assays were technically complex and had poor analytical sensitivity.

Measurement of total urinary metanephrines using spectrophotometric methods has also been reported.[40,41] After sample hydrolysis and isolation of metanephrines by ion-exchange chromatography, the hydroxyl- and amino-containing side-chains were oxidatively cleaved by periodate. Oxidation of both NM and MN resulted in the formation of a common end product, vanillin, which was measured spectrophotometrically at 360 nm. This method was, however, susceptible to interferences from other compounds such as methylglucamine[42] and the 4-hydroxylated metabolite of propranolol.[43] Furthermore, the method did not differentiate between NM and MN, and results were only semi-quantitative for grossly increased concentrations of total metanephrines.[44]

Radioenzymatic methods had also been used to quantify urinary NM.[45,46] Phenylethanoleamine-N-methyltransferase and [^{3}H]S-adenosylmethionine convert NM to its [^{3}H]N-methylated derivative, [^{3}H]MN. The main advantages of this method were its high sensitivity and freedom from interference by common antihypertensive drugs such as propranolol, guanethidine, reserpine, hydrochlorothiazide, labetalol and methyldopa. Since measurements of both NM and MN are imperative for the diagnosis of pheochromocytoma, this radioenzymatic assay for NM is not commonly used for that purpose.

7.4 CURRENT CHROMATOGRAPHIC METHODS

The use of high-performance liquid chromatography (HPLC) as an analytical tool for the life sciences has become increasingly important and the techniques and instrumentation involved are being developed to higher levels of sophistication. Significant improvements in the specificity of the measurements of catecholamines and metanephrines have resulted from the coupling of HPLC separation with fluorescence, electrochemical, chemiluminescence or mass spectromectric detection.

7.4.1 Chemistry of catecholamines

The catecholamines all contain a 3,4-dihydroxyphenyl group (catechol unit) as common structure element, but differ with respect to the 1-substituent (Figure 7.1). The enzyme phenylamine-β-hydroxylase or dopamine-β-hydroxylase introduces the hydroxyl group stereospecifically and converts the achiral DA into chiral NE in the R stereoisomeric configuration. The same configuration is present in E, as the enzyme phenylethanolamine-*N*-methyltransferase is only involved in the transformation of the primary amino group of NE into a secondary amine by methylation.[47] Based on their chemical structures, it is evident that the catecholamines and metanephrines possess some common physical and chemical properties. Their highly polar nature and amphoteric character due to the presence of amino and phenolic groups limit the use of conventional extraction procedures for isolation purposes. Another complicating factor is the ease with which all catechol-type compounds are oxidized in air to form quinones and secondary products, especially in alkaline media. These properties have to be considered in all work-up and handling procedures, which may include the use of antioxidants as protecting agents and avoidance of high pH or exposure to light. The solution chemistry of the catecholamines is largely dominated by the nucleophilic amino group that serves as an excellent target for various derivatization reagents. As will be discussed later, however, some important derivatives used in fluorescence detection techniques require a primary amino group and therefore cannot be applied to E and MN. The ease of oxidation of the catechol ring is another property that has also been utilized for analytical purposes. The technique of direct electrooxidation at an anodic surface (electrochemical detection) is well established for the analysis of these compounds.

7.4.2 Specimen preparation

Only a few methods for urinary catecholamines have been published that do not require pre-extraction prior to analysis.[48,49] These methods minimized sample preparation by making use of different precolumn derivatization procedures. The selection of a suitable method for sample preparation prior to analysis by HPLC depends on a number of factors, such as the biological source, the type of column used and the selectivity of the detection method. In cases where the analyte concentration is very low and the analyte is present in a complex matrix (urine or plasma) with interfering compounds, an exhaustive pretreatment may be unavoidable. Sample pretreatment is also essential to ensure the sensitivity and specificity of the assay and protect the analytical column from contamination.

Liquid- or solid-phase extraction methods have been adopted for the isolation of catecholamines and their metabolites from urine samples. The liquid extraction system is ordinarily based on the formation of a complex, in alkaline medium, between diphenylborate and the diol group in the catecholamines. However, the liquid extraction methods reported in the literature are relatively tedious and often involved multiple extraction steps.[50–52] For the more widely used solid-phase extraction methods, catecholamines may be selectively isolated from the urine sample by adsorption with activated alumina,[53,54] phenylboronic acid[55,56] or cation-exchange resins.[57–65] All the specimen preparative procedures are specific for the free catecholamines, i.e. the extracted catecholamines do not include the conjugated fraction.

The procedure for extraction of catecholamines with activated alumina was developed by Anton and Sayre,[66] and has subsequently been used in a number of studies. Alumina extraction has not been popular, although automated purification with alumina micro-columns was studied closely by Tsuchiya *et al.*[67] The sample preparation scheme includes increasing the pH of the alumina to >8.5 and vigorous shaking of the sample with the alumina, resulting in adsorption of the catecholamines by attraction of the hydroxyl groups of the catechol nucleus. The alumina can then be washed with water or buffer, and finally the catecholamines are eluted with acid, such as 0.3 M acetic acid.[53] Since catecholamines are unstable at high pH, the sample should not stay in the Tris–EDTA (pH 8.5) buffer for long periods. Tsuchiya *et al.* also pointed out that attention must be paid to catecholamine instability in alkaline solutions, especially when using an alumina procedure.[67] The alumina extraction scheme has also been applied to in-line sample pretreatment using small alumina precolumns. Using this method, the recovery of E, NE and DA was found to be approximately 100%, and for 3,4-dihydroxybenzylamine (DHBA, internal standard) the recovery was 76%.[54] Despite the strong affinity of alumina for catecholamines (NE, E and DA), the adsorption or retention of the metanephrines (NM and MN) on alumina was found to be negligible.[54] Therefore, solid-phase extraction with alumina could not be applied for the simultaneous extraction of catecholamines and metanephrines from urine. This limitation, and the fact that alumina needs chemical activation (which can be difficult to reproduce), may be reasons why alumina extraction has not been widely used.

In 1977, an affinity gel with boric acid covalently bound to a resin matrix was developed for the adsorption of catecholamines via the formation of a boric acid complex with the vicinal hydroxyl groups at increased pH.[68] The adsorbed compounds can be eluted by decreasing the pH. Determination of catecholamines in urine and plasma by in-line sample pretreatment using an internal surface boronic acid gel has been reported.[56] Phenylboronic acid columns are commercially available, with specified adsorbing properties, simplifying the specimen clean-up procedure and improving column-to-column reproducibility, The catecholamines were selectively adsorbed on the material and subsequently separated (in-line or off-line) on a reversed-phase or cation-exchange column. As with alumina extraction, a drawback of the boric acid gel is that it cannot selectively retain metanephrines since they lack the vicinal hydroxyl groups.

Cation-exchange resins have also been used to extract catecholamines from urine. One type of resin that is commonly adopted is Bio-Rex 70. Acidic metabolites were not retained by Bio-Rex 70 cation-exchange resin and therefore did not interfere with the assay. Similar results were obtained for the alcoholic metabolites and free amino acids. Under the conditions used in one study, both catecholamines and metanephrines were retained by the resin. However, only those compounds that contained a vicinal hydroxyl group (NE, E and DA) could be eluted with boric acid.[58] It was subsequently reported that formic acid at high concentration could be used to elute both catecholamines and metanephrines adsorbed on Bio-Rex 70 cation-exchange resin, thereby allowing the simultaneous extraction of catecholamines and metanephrines for HPLC analysis.[59] Using a solid-phase extraction (SPE) copolymer sorbent of *N*-divinylpyrrolidone and divinylbenzene, Vuorensola *et al.*[69] reported the extraction of dopamine and metanephrines from urine, and Sabbioni *et al.*[70] extracted 4-hydroxy-3-methoxyphenylethylene glycol and catecholamines from plasma. However, the simultaneous extraction of both catecholamines and metanephrines was not demonstrated in these studies. In the development of methods to extract both catecholamines and metanephrines, the cation-exchange cartridges described by Chan and co-workers provide a helpful starting point.[49,71–73] They reported several studies that evaluated measurements of catecholamines and metanephrines in urine samples using simultaneous extraction with a weak cation-exchange resin.

In general, minimal sample handling offers many advantages and if the HPLC-conditions are well chosen for a particular analytical problem, very crude samples can often be analyzed without off-line manipulation. Assays involving direct injection of the specimen may eliminate the necessity for an internal standard if recoveries are high and reproducible. In some cases of catecholamine analysis, the compounds may also be injected directly for

HPLC analysis without prior sample clean-up.[74-76] Gamache et al.[74] utilized the specificity of the coulometric array detection method for the analysis of urinary NM and MN without prior extraction. The urine sample was simply acid hydrolyzed, diluted and filtered prior to isocratic separation on a reversed-phase ion-pair chromatographic column. The assay was specific for the metanephrines with negligible interference from anti-hypertensive drugs. Yamaguchi et al.[75] derivatized NE, E, serotonin and 5-hydroxyindoleamines with benzylamine in the presence of potassium hexacyanoferrate(III) under mild conditions. The resulting fluorescent derivatives were then injected directly for HPLC with fluorescence detection. Panholzer et al.[76] utilized a column-switching HPLC system for the simultaneous analysis of E, NE, DA, serotonin, MN, NM, DOPAC, HVA and 5-hydroxyindoleacetic acid (5-HIAA) in human urine. The sample was injected directly on to a C_{18}-alkyl-diol silica precolumn that extracted the analytes from the urine matrix. The analytes were then eluted from the precolumn on to the analytical column by the use of column-switching techniques and were separated by ion-pair reversed-phase HPLC. The analytes were then oxidized to the corresponding quinones and converted into fluorescent derivatives by reaction with meso-1,2-diphenylethylenediamine.

7.4.3 Fluorescence detection

The catecholamines have native fluorescence, therefore it is possible to detect these biogenic amines by measuring the fluorescence of the effluent from an HPLC separation. HPLC with detection of native fluorescence provides sufficient sensitivity ($5 \, \mu g \, L^{-1}$) after catecholamine or metanephrine isolation from urine by cation-exchange chromatography[77-79] or concentration with alumina.[80] Many of the drug interferences in the traditional fluorescence procedures are eliminated by the selectivity of the chromatographic separation, but some compounds such as the α-methyldopa metabolite α-methylnorepinephrine may elute close to NM and MN, depending on the chromatographic conditions.[77] This technique has not been widely used in bioanalytical work.[47] Derivatization may also be used to enhance further the sensitivity and specificity. Postcolumn derivatization methods include ethylenediamine condensation,[81] trihydroxyindole (THI),[53,82] glycylglycine[83] and 2-cyanoacetamide.[84] The THI method is the most sensitive for HPLC measurements of NE and E but is insensitive with regard to DA and requires rather complicated instrumentation for the postcolumn derivatization reaction.[82]

Precolumn derivatization methods include 1,2-diphenylethylenediamine treatment,[51,85] dansylation of E, NE and DA,[86] derivatization of NE and DA by o-phthalaldehyde and mercaptoethanol[87] and derivatization of catecholamines with 9-fluorenylmethyloxycarbonyl chloride (FMOC-Cl).[88] Derivatization with o-phthalaldehyde increases the sensitivity of NE and DA, but E is not measured because only primary amines are derivatized. Co-analysis of catecholamines, metanephrines and other related compounds by combined electrochemical oxidation and fluorescence derivatization had also been reported.[60,89] This approach involves sequential chromatographic separation, coulometric oxidation and final chemical derivatization with 1,2-diphenylethylenediamine to fluorescent products.

Chan et al.[49] developed an assay for the simultaneous determination of catecholamines and metanephrines using FMOC derivatization. The assay is convenient for the simultaneous analysis of NM, MN, E and DA in human urine sample without prior extraction procedures. In this study, urine was directly derivatized and subjected to a simple extraction step with

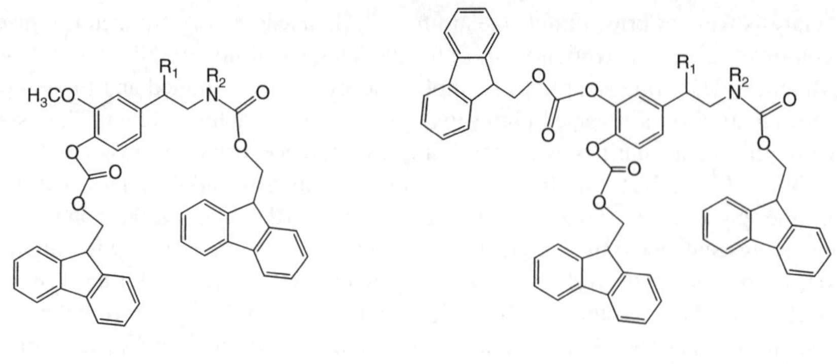

(a)   (b)

Derivative	Structure	R_1	R_2	Exact mass
FMOC-NM	a	OH	H	627.68
FMOC-MN	a	OH	CH_3	641.24
FMOC-NE	b	OH	H	835.28
FMOC-E	b	OH	CH_3	849.29
FMOC-DA	b	H	H	819.28

Figure 7.2 Chemical structures and molecular masses of the FMOC derivatives of catecholamines and metanephrines

chloroform prior to injection. With this assay, the requirement for time-consuming solid-phase extraction of these compounds in urine samples is therefore circumvented. The potential errors involved during extraction and recovery would also be reduced. After derivatization, the samples were found to be stable at ambient temperature for at least 3 days. The enhanced stability of these FMOC derivatives is imparted by the reaction of the unstable phenol or catechol functional groups with FMOC-Cl (Figure 7.2). The latter reaction was confirmed in this study using atmospheric pressure chemical ionization mass spectrometry (APCI-MS). However, this assay does not measure urinary NE accurately due to poor resolution.

In summary, these derivatization methods coupled with fluorescence detection have not gained widespread use in clinical laboratories. One possible reason for their limited use may be the preference of clinical laboratory analysts towards the use of the electrochemical detection, since it is highly selective and sensitive for these catecholic compounds and does not require pre- or postcolumn derivatization.

7.4.4 Electrochemical detection

Electrochemical detection is based on the principle that electroactive compounds are oxidized or reduced at a certain potential and the resulting flow of electrons creates a

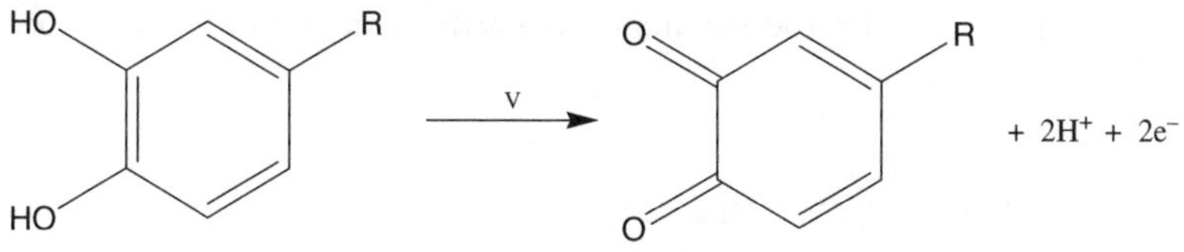

Figure 7.3 Principle of electrochemical oxidation of catecholamines

measurable current. Catecholamines and metanephrines are easily oxidized to form quinones, as illustrated in Figure 7.3, and therefore are suitable for electrochemical detection. Because these biogenic amines are more easily oxidized than many other compounds, it is possible to obtain a certain degree of selectivity with electrochemical detection if the oxidizing potential is kept as low as possible. Two types of electrochemical detectors exist. The coulometric detector oxidizes 100% of the catecholamine molecules in the effluent and therefore creates a larger current than the amperometric detector, which oxidizes less than 10% of these molecules. It has been reported that the higher yield of the coulometric detector does not seem to increase its sensitivity compared with the amperometric detector because the background noise level increases in proportion to the increase in signal, so the two detectors have a similar signal-to-noise ratio.[90] Nonetheless, the wide use of the coulometric detector for the trace analysis of free catecholamines and/or metanephrines in urine reflects its high sensitivity in practical applications.[62,74,91,92] This high sensitivity and selectivity of coulometric detection may be a result of the oxidation of the analytes in the guard cell followed by reduction at successively higher reduction potentials in the analytical cell.[93] The response of the electrochemical detector is influenced by ionic strength and pH of the mobile phase, which must be taken into account when optimizing the HPLC assay. For all their simplicity, though, electrochemical detectors require considerable attention to troubleshooting and minimization of disturbances in order to obtain optimal sensitivity.[94]

Electrochemical detection with ion-pairing adaptations of reversed-phase chromatography are the most common methodologies, and many techniques for measuring urinary catecholamines and metanephrines have been published.[57,59,65,71,72,74,95–108]

Ion-pairing with alkyl sulfonates or sulfates enhances the retention of cationic amine moieties on the lipophilic stationary phase. The effects of mobile phase composition (ion-pairing agents, ionic strength or organic components) on the chromatographic and electrochemical behavior of catecholamines and their metabolites has been reviewed previously.[93,109]

The coupling of ion-exchange chromatography with electrochemical detection has been applied to the measurement of urinary catecholamines.[62,110–113] Early applications of ion-exchange chromatography were hampered by the lack of column reproducibility, resulting in the loss of catecholamines during the analytical process.[110] Mefford reported that semi-irreversible loading of C_{18} columns with lauroylsarcosine resulted in separations that are dependent on ion-exchange mechanisms.[112] Subsequently, N-methyloleoyl taurate was applied for the ion-exchange chromatographic separation of catecholamines. This latter resin facilitated the elution of E prior to NE and, therefore, allowed enhanced sensitivity for the normally lower concentration of E.[112] This enhanced sensitivity may be of particular value in detecting small tumors that preferentially secrete E. Sarzanini et al. reported an ion chromatographic procedure for the simultaneous purification and determination of urinary NE, E and DA.[62]

7.4.5 Chemiluminescence detection

Chemiluminescence occurs when an energy-releasing reaction produces a molecule in an electronically excited state and that molecule, as it returns to the ground state, releases its energy as a photon of light. The chemiluminescent process is distinctly different from the photoluminescent processes such as fluorescence and phosphorescence. The latter processes occur after the excited state of a molecule is produced from its ground state by the absorption of light energy. Thus fluorescent species emit light only while being irradiated. Phosphorescent species may appear to emit light without being irradiated, but this emitted energy had to have been absorbed at an earlier time. Chemiluminescent reactions, however, produce light without any prior absorption of radiant energy. As such, chemiluminescence requires no excitation source (as does fluorescence and phosphorescence) but only a single light detector such as a photomultiplier tube. Furthermore, a monochromator or filter is often not needed. Most chemiluminescence methods involve only a few chemical components to generate light. Luminol chemiluminescence,[114] which has been extensively investigated, and peroxyoxalate chemiluminescence[115] are both used in bioanalytical methods and have been adapted to the analysis of catecholamines and their metabolites.

Bis(2,4,6-trichlorophenyl) oxalate (TCPO) is an example of the peroxyoxalate reaction sequence. It involves the fuel (TCPO) plus the oxidant (H_2O_2) reacting to produce an intermediate, which in this example is shown as a dioxetane (Figure 7.4) although the reaction probably produces many intermediates, and others, such as hydroperoxyoxalate,

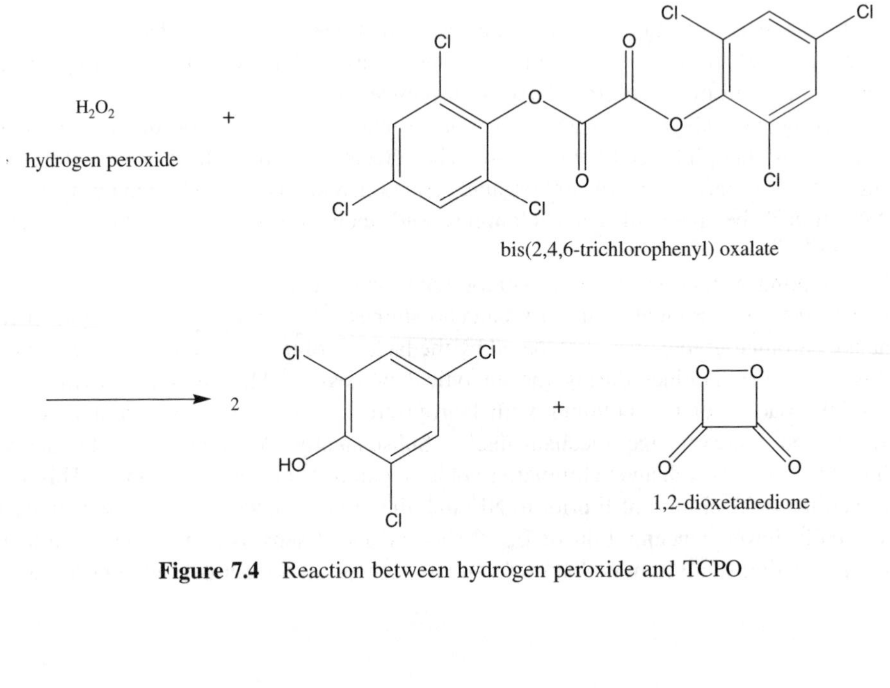

H_2O_2

hydrogen peroxide

bis(2,4,6-trichlorophenyl) oxalate

1,2-dioxetanedione

Figure 7.4   Reaction between hydrogen peroxide and TCPO

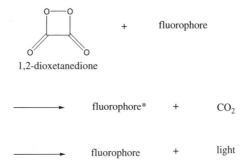

Figure 7.5 Excitation of a fluorophore by intermediate produced by the reaction between hydrogen peroxide and TCPO

have been proposed.[116] The intermediate, 1,2-dioxetanedione, excites a fluorophore, 9,10-diphenylanthracene (DPA). The process of transferring the energy of the initial reaction, the chemical reaction of hydrogen peroxide with TCPO, to light emission from the excited-state fluorophore (fluorophore*) facilitates the analytical determination of these compounds (derivatized as flurophores) (Figure 7.5).

Chemiluminescence detection following HPLC separation of derivatized catecholamines (NE, E and DA) and isoproterenol using the peroxyoxalate reaction has been reported.[117] The amines and isoproterenol, derivatized with 1,2-diarylethylenediamines, were separated on a reversed-phase HPLC column with isocratic elution with imidazole buffer (pH 5.8, 120 mM)–methanol–acetonitrile (6:2:9, v/v/v). The eluate was detected by a post-column chemiluminescence reaction using bis[4-nitro-2-(3,6,9-trioxadecyloxycarbonyl)phenyl] oxalate and hydrogen peroxide. The chromatographic detection limits for catecholamines were approximately 40–120 amol for an injection volume of 100 μL (signal-to-noise ratio of 3). A highly specific and sensitive automated HPLC–peroxyoxalate chemiluminescence method for the simultaneous determination of catecholamines (NE, E and DA) and their 3-*O*-methyl metabolites (NM, MN and 3-methoxytyramine) has also been reported.[118] Automated precolumn ion-exchange extraction of diluted plasma is coupled with HPLC separation of these compounds on an octyldecylsilica (ODS) column, postcolumn coulometric oxidation, fluorescence derivatization with ethylenediamine and finally peroxyoxalate_chemilumines-cence reaction detection. Although these methods may be applied for the quantitation of catecholamines and metanephrines, the complexity associated with the multiple steps such as the pre- or postcolumn derivatization of catecholamines and the postcolumn chemilumi-nescent reaction needs to be considered. All these steps have to be manipulated and controlled prudently to ensure the accurate measurement of these analytes.

Luminol is also widely used as a chemiluminescent reagent, but unlike the peroxyoxalate systems, it does not require an organic/mixed solvent system. The chemiluminescent emitter is a 'direct descendent' of the oxidation of luminol (or an isomer such as isoluminol) by an oxidant in alkaline solution. Probably the most useful oxidant is hydrogen peroxide, which is also used in peroxyoxalate chemiluminescence. If the fuel is luminol, the emitting species is 3-aminophthalate (Figure 7.6).

The use of luminol as a chemiluminescent reagent for the detection of catecholamines has been reported.[119–121] In a typical scheme (Figure 7.7), imidazole was used to catalyze the

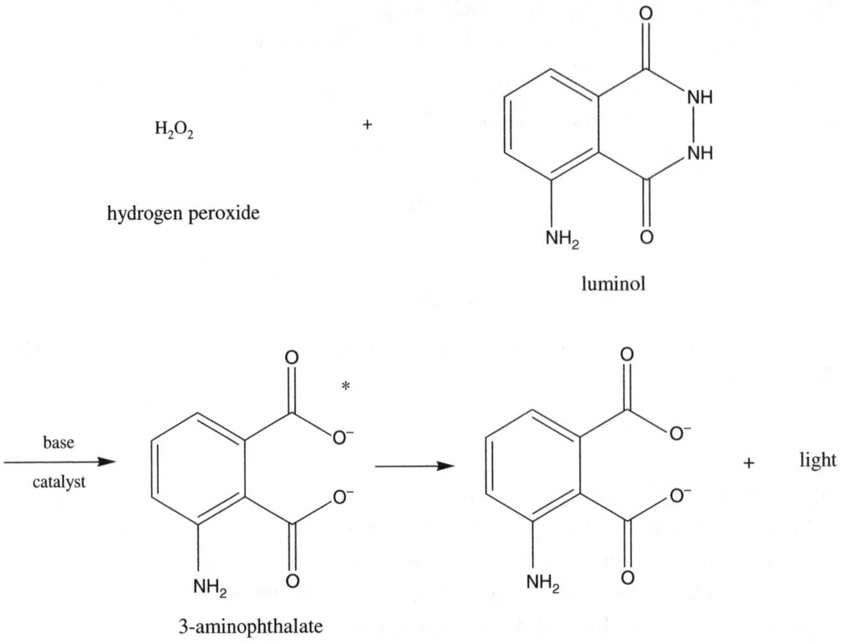

Figure 7.6 Reaction between hydrogen peroxide and luminol in the presence of peroxidase producing the excited 3-aminophthalate and the emitted light at a wavelength of 430 nm

decomposition of catecholamines to generate hydrogen peroxide, then the hydrogen peroxide was detected by luminol chemiluminescence with horseradish peroxidase as the catalyst.[119,120] However, as these assays were not coupled to HPLC separation, only total free catecholamines were measured. Israel and Tomasi[121] developed a chemiluminescent procedure for measuring catecholamines based on the observation that lactoperoxidase catalyzes both the oxidation of catecholamines and the chemiluminescent reaction of luminol with their oxidation product (hydrogen peroxide). The assay had been adapted for continuously monitoring the release of catecholamines from adrenergic tissues, from cell suspensions and from cells loaded in culture with dopamine. Nonetheless, this assay measured total catecholamines and could not be applied for measuring individual NE, E and DA.

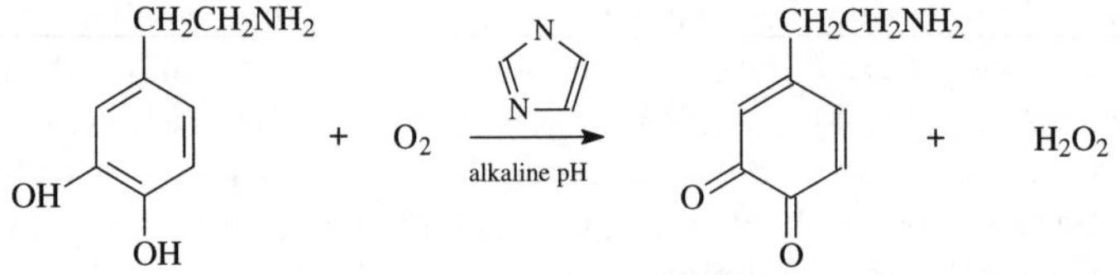

Figure 7.7 The phenolic hydroxyl group of catecholamine reacts with oxygen in aqueous solution. The reaction was catalyzed by imidazole to produce hydrogen peroxide at alkaline pH

7.4.6 Mass spectrometry

Tandem mass spectrometry is becoming an important analytical technology in the clinical laboratory environment. Applications already exist for tandem mass spectrometry in toxicology and therapeutic drug monitoring and many new applications are being developed. The combination of tandem mass spectrometry with sample introduction techniques employing atmospheric pressure ionization has enabled this technology to be readily implemented in the clinical laboratory. Although mass spectrometry has been traditionally used in other industries and applications, it is now being used for a great variety of analytes, including steroids, peptides and catecholamines.

As with other analytes, the earlier mass spectrometric measurements of urinary amines, particularly the metanephrines, were developed using gas chromatography–mass spectrometry (GC-MS), as the coupling of GC to MS technology was technically more straightforward than LC coupling to MS.[122,123] Nonetheless, the number of published GC-MS methods developed for biogenic amines is limited, probably because of the chemical manipulations required to produce volatile derivatives suitable for GC. Crockett *et al.* reported a rapid GC-MS assay for urinary metanephrine and normetanephrine using simultaneous derivatization with *N*-methyl-*N*-(trimethylsilyl)trifluoroacetamide and *N*-methylbis(heptafluoro)butryamide at room temperature.[124] The method performance compares well with that of a standard HPLC assay and avoids drug interferences that commonly affect HPLC assays for urine metanephrines.

The simultaneous LC-MS analysis of urinary catecholamines and metanephrines was first reported by Chan and Ho.[73] They developed a rapid assay (6.5 min) for the quantitation of urinary NE, E, DA, MN and NM using HPLC coupled to atmospheric pressure chemical ionization mass spectrometry (APCI-MS) (Figure 7.8). Numerous chromatographic separation techniques can be coupled with APCI-MS. The key to achieving separations and optimal MS performance in LC-MS is to find separation conditions that are compatible with atmospheric pressure ionization. Catecholamines and metanephrines are amphoteric in nature and chromatographic separations are usually performed using ion-pairing reversed-phase HPLC or weak cationic-exchange HPLC at acidic pH. However, ion-pair reagents commonly adopted for the separation of catecholamines and metanephrines, such as 1-heptanesulfonic acid and 1-octanesulfonic acid, are not suitable for APCI-MS.

In a preliminary study, a SynChropak CM 100 weak cation-exchange column was used to separate these biogenic amines in order to avoid the use of any ion-pairing reagent. One problem with this approach is that it requires the use of non-volatile buffers of higher ionic strength, such as phosphate buffer, which is not compatible with MS analysis. Subsequently, a C_{18} reversed-phase column was used and the chromatographic resolution of these biogenic amines was possible in the presence of an APCI-compatible buffer reagent consisting of ammonium formate and formic acid. Since these amines are mostly protonated at pH 3, the resolution of these compounds using reversed-phase HPLC might be attributed to the 'ion-pairing' effect of the formate ion. This finding is important because it allows the separation of biogenic amines (especially E and NM, which have the same molecular weight) within a short time of 6 min, and also allows the use of an APCI-compatible buffer reagent, both of which are important for the gas-phase ionization of the analytes in the APCI probe. Similar approaches in the LC-MS analysis of biogenic amines were reported for the screening of pheochromocytoma[125] and the analysis of catecholamines in brain tissue.[126] Tornkvist *et al.*

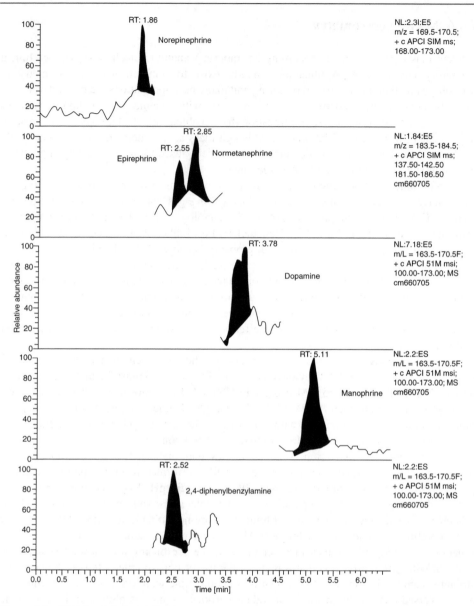

Figure 7.8 HPLC–APCI-MS SIM chromatographic profiles of an extract of a urine sample collected from a healthy male subject. From E. C. Y. Chan and P. C. Ho, *Rapid Communications in Mass Spectrometry*, **14**, 1959–1964. Copyright 2000 © John Wiley & Sons, Ltd. Reproduced with permission

pointed out that the use of mobile phase without ion-pairing agents and with high content of organic modifier facilitated the coupling of HPLC to the selective and sensitive MS detector.[126] Quantification of plasma free metanephrines is usually accomplished using HPLC with electrochemical detection, but sample preparation is labor intensive and time consuming, analytical cycle times are long and interfering substances may obscure the relevant peaks. The LC-MS-MS assay using multiple reaction monitoring is highly selective and sensitive for the measurement of trace quantities of analyte in complex matrices. Utilizing the LC-MS-MS approach and cyano analytical column chemistry, Lagerstedt *et al.* measured nanomolar concentrations of free metanephrine and normetanephrine in plasma for the diagnosis of pheochromocytoma. The analytical method demonstrated high precision and throughput.[127]

7.5 PRACTICAL CONSIDERATIONS FOR THE STABILITY OF URINARY CATECHOLAMINES AND METANEPHRINES DURING STORAGE

Catecholamines and metanephrines are prone to oxidative conversion to their corresponding quinones,[128] hence proper storage and preservation of 24-h urine samples after collection from patients is essential for accurate quantitation of these compounds. Various groups have reported stability studies of catecholamines in biological fluids.[129–133] Boomsma *et al.* studied the optimal collection and storage conditions for catecholamines in human plasma and urine,[131] and reported that catecholamines were stable at 4 °C for 1 month in unpreserved urine and for 4 months in urine preserved with EDTA and sodium metabisulfite. In acidified urine, catecholamines were also reported to be nearly unchanged after 1 year at 4 and −20 °C. Despite the claimed stability in this study, fluctuating stability profiles of the catecholamines (NE, E and DA) were observed after various lengths of storage. It was also observed that fluctuations in catecholamine concentrations occurred throughout the storage period. The authors attributed the chaotic stability profiles of these analytes in plasma and urine to the depletion of antioxidant compounds added to the biological samples. Similar fluctuations in concentrations of these biogenic amines were also observed in unpreserved urine samples. Miki and Sudo reported in their stability study that unpreserved urine samples could be used for the measurement of catecholamines, cortisol and creatinine.[133] However, they did not comment on the increases in measured concentrations of catecholamines (>10%) in acidified urine samples (pH 1 and 0.5) during storage at room temperature (15–20 °C). Metanephrines (NM and MN) were not examined in most of these stability studies.[130–133] It should be appreciated that any fluctuating concentration profiles of catecholamines and metanephrines during storage could possibly lead to an inaccurate estimation of their actual concentrations in the biological samples. This would render the test results invalid and may eventually lead to higher laboratory costs due to the need for additional tests. More importantly, such inaccurate measurements could result in the incorrect diagnosis of the disease condition. Therefore, the reason(s) for these fluctuating concentrations are of great interest. Using a stability-indicating validated amperometric HPLC method[71] for the measurements of urinary catecholamines and metanephrines, the factor(s) that account for the fluctuating stability profiles of catecholamines in biological samples during storage were identified by Chan *et al.*[72] The detected levels of catecholamines and metanephrines in biological samples were found to be affected by the simultaneous degradation of these compounds and deconjugation of their sulfoconjugates (CA-S and MN-S) when these samples were not properly stored and preserved. These two simultaneous processes most likely contributed to the chaotic stability profiles of these amines observed in some studies. Deconjugation of urinary CA-S and MN-S could occur during the first 3 days of storage at 10 and 30 °C in both unpreserved and preserved urine samples. Addition of preservatives (HCl or Na$_2$EDTA and sodium metabisulfite) was found to have no effect on the deconjugation process. Instead, deconjugation was accentuated by the presence of HCl (3 M, 5 mL L^{-1}) as a preservative in the sample and at higher temperature (30 °C). However, storing the samples at extremely low temperature minimizes the deconjugation process. Free catecholamine and metanephrine concentrations were found to be stable for at least 3 weeks in both unpreserved urine and aqueous samples stored at −80 °C.[72] From this, it was concluded that deconjugation of the sulfoconjugates and degradation of their free amines did not occur during this period in the unpreserved urine samples stored at −80 °C. It should be noted that the storage temperature of the urine during sample collection period can have a profound influence on the stability of biogenic amines. The stability of biogenic amines in biological specimens has been reviewed elsewhere.[134]

7.6 FUTURE DEVELOPMENTS

Until recently, the most common analytical technique used for the determination of catecholamines and metanephrines in urine was HPLC. Because of the minute amount of material present in the sample, methods of detection generally have to be very sensitive and selective. The detection methods used in most published studies are electrochemical[57,59,74,135] and fluorimetric.[51,60,78,79,89] Recently, chemiluminescence has also been explored as a detection method for catecholamines and their 3-O-methylated metabolites in rat plasma.[117]

Electrochemical detectors are sensitive and selective for the quantitation of urinary catecholamines and metanephrines. However, there are limitations to this mode of detection, which are seldom addressed in research papers. Based on our experience with the amperometric detector, we found that this detector is prone to malfunctioning due to a number of factors. The common malfunctions of the system include unstable baseline, excessive noise, transient spiking, noise corresponding to pump stroke, ghost peaks, high background current and sudden decrease in sensitivity. The common causes of these problems are fouling of the auxiliary electrode, the brass contact of the working electrode, and the carbon plate and/or spacer. Baseline drift and diminished sensitivity also result from carbon plate deterioration, defective contact between electrical connectors, bubbles in the flow cell, elution of metal ions leached from the injector body, ambient temperature changes and draught.

As mentioned earlier, in fluorimetric detection, the amines may be monitored either by natural fluorescence[78,79] or after derivatization reaction with either 1,2-diphenylethylenediamine (DPE),[51,85] trihydroxyindole (THI),[82,136,137] o-phthalaldehyde (OPA)[138] or fluorescamine.[138] Although the fluorescence detector is relatively easy to maintain, each of these derivatization methods faces some shortcomings. The natural fluorescence methods show low sensitivity of the catecholamines and metanephrines. DPE increases the sensitivity of all three catecholamines; however, the reaction products are stable for only 30 min after the reaction. When THI is used, HPLC requires rather complicated instrumentation for postcolumn derivatization and cannot be applied to DA. Derivatization with OPA and fluorescamine increases the sensitivity of NE and DA, but E is not measured because only primary amines are derivatized by this reaction.

These challenges observed for both electrochemical and fluorimetric modes of detection prompted the need for a more sensitive, selective and maintenance-free HPLC assay for routine, rapid, simultaneous measurements of catecholamines and metanephrines in urine. The emergence of LC-MS and LC-MS-MS[73,125−127] has provided new analytical approaches to this field of study. Although MS-MS requires expensive hardware, it is most cost effective when groups of compounds can be measured simultaneously. As the price/performance ratio of this technology improves, it will likely become more common in clinical laboratory applications.

HPLC is a proven technique that has been adopted by laboratories worldwide during the last three decades. One of the primary drivers for the growth of this technique has been the continual improvement in the packing materials (chemistry, particle size, packing technology) used to effect the separation. The underlying principles of this evolution are specified by the van Deemter equation, which relates linear velocity (flow-rate), diffusion components (both linear and eddy) and resolution [in terms of height equivalent of a theoretical plate (HETP), a measure of column efficiency]. Based on the van Deemter equation, as the particle sizes decreases to less than 2.5 μm, not only is there a significant gain in column efficiency, but also the efficiency does not diminish at increased flow-rates. By using smaller particles, speed and peak capacity (number of peak resolved per unit time in gradient separations) can be extended to greater limits, termed ultra-performance liquid chromatography (UPLC).[139] With the need of rapid turn-around times for clinical analyses and higher sensitivity and selectivity for the analysis of biogenic amines, especially in biological matrices, the fast separation, high sensitivity and resolution of UPLC are attractive features. Applications of UPLC-MS and UPLC-MS-MS in pharmaceutical R&D, food testing, environmental analysis, proteomics and life science research and clinical laboratory medicine are likely to become more common as this technology reaches its full potential.

DEDICATION

This chapter is dedicated to the late brother-in-law of P.C.L. Ho, Mr Eddie Kwok Kuen Lee, a former patient with pheochromocytoma. His medical history encouraged and stimulated the authors to pursue research on the diagnosis of the disease.

REFERENCES

1. E.L. Bravo and R.W. Gifford, Jr, Pheochromocytoma: diagnosis, localization and management, *N. Engl. J. Med.*, **311**, 1298–1303.
2. B. Shapiro and L.M. Fig, Management of pheochromocytoma, *Endocrinol. Metab. Clin. North Am.*, **18**, 443–481 (1989).
3. M.F. Brennan and H.R. Keiser, Persistent and recurrent pheochromocytoma: the role of surgery, *World J. Surg.*, **6**, 397–402 (1982).
4. N. Kimura, Y. Miura, I. Nagatsu and H. Nagura, Catecholamine synthesizing enzymes in 70 cases of functioning and non-functioning pheochromocytoma and extra-adrenal paraganglioma, *Virchows Arch. A Pathol. Anat. Histopathol.*, **421**, 25–32 (1992).
5. T. Nagatsu, T. Yamamoto and I. Nagatsu, Partial separation and properties of tyrosine hydroxylase from the human pheochromocytoma. Effect of norepinephrine, *Biochim. Biophys. Acta*, **198**, 210–218 (1970).
6. J.R. Crout and A. Sjoerdsma, Catecholamines in the localization of pheochromocytoma, *Circulation*, **22**, 516–525 (1960).
7. G. Eisenhofer, H. Keiser, P. Friberg, E. Mezey, T.T. Huynh, B. Hiremagalur, T. Ellingson, S. Duddempudi, A. Eijsbouts and J.W. Lenders, Plasma metanephrines are markers of

pheochromocytoma produced by catechol-*O*-methyltransferase within tumors, *J. Clin. Endocrinol. Metab.*, **83**, 2175–2185 (1998).

8. A.H. Anton, M. Greer, D.F. Sayre and C.M. Williams, Dihydroxyphenylalanine secretion in a malignant pheochromocytoma, *Am. J. Med.*, **42**, 469–475 (1967).

9. W.J. Louis, A.E. Doyle, W.C. Health and M.J. Robinson, Secretion of dopa in pheochromocytoma, *BMJ*, **4**, 325–327 (1972).

10. H. John, W.H. Ziegler, D. Hauri and P. Jaeger, Pheochromocytomas: can malignant potential be predicted? *Urology* **53**, 679–683 (1999).

11. T.G. Rosano, T.A. Swift and L.W. Hayes, Advances in catecholamine and metabolite measurements for diagnosis of pheochromocytoma, *Clin. Chem.*, **37**, 1854–1867 (1991).

12. N.A. Samaan, R.C. Hickey and P.E. Shutts, Diagnosis, localization and management of pheochromocytoma, *Cancer* **62**, 2451–2460 (1988).

13. A.M. Lucon, M.A.A. Pereira, B.B. Mendonca, A. Halpern, B.L. Wajchenbeg and S. Arap, Pheochromocytoma: study of 50 cases, *Clin. Urol.*, **157**, 1208–1212 (1997).

14. K.E. Tan, A.C. Fok, P.H. Engand D.H. Khoo, Diagnostic difficulties associated with phaeochromocytoma – 4 case illustrations, *Singapore Med. J.*, **38**, 493–496 (1997).

15. S.G. Sheps, N.S. Jiang and G.G. Kle, Diagnostic evaluation of pheochromocytoma, *Endocrinol. Metab. Clin. North Am.*, **17**, 397–414 (1988).

16. V. Kairisto, P. Koskinen, K. Mattila, J. Puikkonen, A. Virtanen, I. Kantola and K. Irjala, Reference intervals for 24-h urinary normetanephrine, metanephrine, and 3-methoxy-4-hydroxymandelic acid in hypertensive patients, *Clin. Chem.*, **38**, 416–420 (1992).

17. E.A.M. Gerlo and C. Sevens, Urinary and plasma catecholamines and urinary catecholamine metabolites in pheochromocytoma: diagnostic value in 19 cases, *Clin. Chem.*, **40**, 250–256 (1994).

18. T. Jan, B.E. Metzger and G. Baumann, Epinephrine-producing pheochromocytoma with hypertensive crisis after corticotrophin injection. *Am. J. Med.*, **89**, 824–825 (1990).

19. P.E. Graham, G.A. Smythe, G.A. Edwards and L. Lazarus, Laboratory diagnosis of phaeochromocytoma: which analytes should we measure? *Ann. Clin. Biochem.*, **30**, 129–134 (1993).

20. T. Oeltmann, R. Carson, J.R. Shannon, T. Ketch and D. Robertson, Assessment of *O*-methylated catecholamine levels in plasma and urine for diagnosis of autonomic disorders, *Auton. Neurosci.*, **116**, 1–10 (2004).

21. G. Eisenhofer, P. Friberg, K. Pacak, D.S. Goldstein, D.L. Murphy, C. Tsigos, A.A. Quyyumi, H.G. Brunner and J.W. Lenders, Plasma metadrenalines: do they provide useful information about sympatho-adrenal function and catecholamine metabolism? *Clin. Sci.*, **88**, 533–542 (1995).

22. J.W. Lenders, H.R. Keiser, D.S. Goldstein, J.J. Willemsen, P. Friberg, M.C. Jacobs, P.W.C. Kloppenborg, T. Thien and G. Eisenhofer, Plasma metanephrines in the diagnosis of pheochromocytoma, *Ann. Intern. Med.*, **123**, 101–109 (1995).

23. R. Mornex, L. Peyrin, R. Pagliari and J.M. Cottet-Emard, Measurement of plasma methoxyamines for the diagnosis of pheochromocytoma, *Horm. Res.*, **36**, 220–226 (1991).

24. G. Eisenhofer, J.W.M. Lenders, D.S. Goldstein, M. Mannelli, G. Csako, M.M. Walther, F.M. Brouwers and K. Pacak, Pheochromocytoma catecholamine phenotypes and prediction of tumor size and location by use of plasma free metanephrines, *Clin. Chem.*, **51**, 735–744 (2005).

25. C. Sobel and R.J. Henry, Determination of catecholamines (adrenalin and noradrenalin) in urine and tissue, *Am. J. Clin. Pathol.*, **240**, 240–245 (1957).

26. S.L. Jacobs, C. Sobel and R.J. Henry, Specificity of the trihydroxyindole method for determination of urinary catecholamines, *J. Clin. Endocrinol. Metab.*, **21**, 305–313 (1961).

27. Z. Kahane and P. Vetergaard, An improved method for the measurement of free epinephrine and norepinephrine with a phosphate–meta-phosphate buffer in the trihydroxyindole procedure, *J. Lab. Clin. Med.*, **65**, 848–858 (1965).

28. K. Oka, M. Sekiya, H. Osada, K. Fujita, T. Kato and T. Nagatsu, Simultaneous fluorometry of urinary dopamine, norepinephrine, and epinephrine compared with liquid chromatography with electrochemical detection, *Clin. Chem.*, **28**, 646–649 (1982).

29. K. Taniguchi, Y. Kakimoto and M.D. Armstrong, Quantitative determination of metanephrine and normetanephrine in urine, *J. Lab. Clin. Med.*, **64**, 469–484 (1964).

30. Z. Kahane and P. Vetergaard, Column chromatographic separation and fluorometric estimation of metanephrine and normetanephrine in urine, *Clin. Chim. Acta*, **25**, 453–458 (1969).

31. E.R.B. Smith and H. Weil-Malherbe, Metanephrine and normetanephrine in human urine: method and results, *J. Lab. Clin. Med.*, **60**, 212–223 (1962).

32. S. Brunjes, D. Wybenga and V.J. Johns, Jr, Fluorometric determination of urinary metanephrine and normetanephrine, *Clin. Chem.*, **10**, 1–12 (1964).

33. Z. Kahane and P. Vetergaard, Fluorometric assay for metanephrine and normetanephrine in urine, *J. Lab. Clin. Med.*, **70**, 333–342 (1967).

34. L.B. Biglow and H. Weil-Malherbe, A simplified method for the differential estimation of metanephrine and normetanephrine in urine, *Anal. Biochem.*, **26**, 92–103 (1968).

35. K. Yoshinaga, C. Itoh, N. Ishida, T. Sato and Y. Wada, Quantitative determination of metadrenaline and normetadrenaline in normal human urine, *Nature*, **191**, 599–600 (1961).

36. R.L. Wolf, C.E. Gherman, J.D. Lauer, H.L. Fish and B.R. Levey, A new urinary assay for separate normetanephrine and metanephrine with application for the diagnosis of phaeochromocytoma, *Clin. Sci. Mol. Med.*, **45**, 263–267 (1973).

37. R.F. Coward and P. Smith, A new screening test for phaeochromocytoma, *Clin. Chim. Acta*, **13**, 538–540 (1966).

38. R.F. Coward and P. Smith, Recovery of total metanephrines from urine and their estimation by paper chromatography, *Clin. Chim. Acta*, **14**, 672–678 (1966).

39. J.M.C. Gutteridge, Thin-layer chromatographic techniques for the investigation of abnormal urinary catecholamine metabolite patterns, *Clin. Chim. Acta*, 21, 211–216 (1968).

40. J.J. Pisano, A simple analysis for normetanephrine and metanephrine in urine, *Clin. Chim. Acta*, **5**, 406–414 (1960).

41. J.R. Crout, J.J. Pisano and A. Sjoerdsma, Urinary excretion of catecholamines and their metabolites in pheochromocytoma, *Am. Heart J.*, **61**, 375–381 (1961).

42. L.R. Johnson, M. Reese and D.H. Nelson, Interference in Pisano's urinary metanephrine assay after the use of x-ray contrast media, *Clin. Chem.*, **18**, 209–211 (1972).

43. D. Chou, M. Tsuru, J.L. Holtzman and J.H. Eckfeldt, Interference by the 4-hydroxylated metabolite of propranolol with determination of metanephrines by the Pisano method, *Clin. Chem.*, **26**, 776–777 (1980).

44. R.S. Sandhu and R.M. Freed, Catecholamines and associated metabolites in human urine, *Stand. Methods Clin. Chem.*, **7**, 231–246 (1972).

45. N.D. Vlachakis and V. DeQuattro, A simple and specific radioenzymatic assay for measurement of urinary normetanephrine, *Biochem. Med.*, **20**, 107–114 (1978).

46. K. Kobayashi, A. Foti, V. DeQuattro, R. Kolloch and L. Miano, A radioenzymatic assay for free and conjugated normetanephrine and octopamine excretion in man, *Clin. Chim. Acta*, **107**, 163–173 (1980).

47. S. Allenmark, High-performance liquid chromatography of catecholamines and their metabolites in biological material, *J. Chromatogr.*, **5**, 1–41 (1982).

48. H. Nohta, T. Yukizawa, Y. Ohkura, M. Yoshimura, J. Ishida and M. Yamaguchi, Aromatic glycinonitriles and methylamines as pre-column fluorescence derivatization reagents for catecholamines, *Anal. Chim. Acta*, **344**, 233–240 (1997).

49. E.C.Y. Chan, P.Y. Wee, P.Y. Ho and P.C. Ho, High-performance liquid chromatographic assay for catecholamines and metanephrines using fluorimetric detection with pre-column 9-fluorenylmethyloxycarbonyl chloride derivatization, *J. Chromatogr. B*, **749**, 179–189 (2000).

50. F. Smedes, J.C. Kraak and H. Poppe, Simple and fast solvent extraction system for selective and quantitative isolation of adrenaline, noradrenaline and dopamine from plasma and urine, *J. Chromatogr.*, **231**, 25–39 (1982).

51. F.A.J. van der Hoorn, F. Boomsma, A.J. Man in't Veld and M.A.D.H. Schalekamp, Improved measurement of urinary catecholamines by liquid–liquid extraction, derivatization and high-performance liquid chromatography with fluorometric detection, *J. Chromatogr.* **563**: 348–355 (1991).

52. F. Boomsma, G. Alberts, F.A. van der Hoorn, A.J. Man in't Veld and M.A.D.H. Schalekamp, Simultaneous determination of free catecholamines and epinine and estimation of total epinine and dopamine in plasma and urine by high-performance liquid chromatography with fluorimetric detection, *J. Chromatogr.*, **574**, 109–117 (1992).

53. R.C. Causon and M.E. Carruthers, Measurement of catecholamines in biological fluids by high-performance liquid chromatography: a comparison of fluorometric with electrochemical detection, *J. Chromatogr.*, **229**, 301–309 (1982).

54. J. De Jong, A.J.F. Point, U.R. Tjaden, S. Beeksma and J.C. Kraak, Determination of catecholamines in urine (and plasma) by liquid chromatography after on-line sample pretreatment on small alumina or dihydroxyborylsilica columns, *J. Chromatogr.*, **414**, 285–300 (1987).

55. C.R. Benedict, Simultaneous measurement of urinary and plasma norepinephrine, epinephrine, dopamine, dihydroxyphenylalanine, and dihydroxyphenylacetic acid by coupled-column high-performance liquid chromatography on C_8 and C_{18} stationary phases, *J. Chromatogr.*, **385**, 369–375 (1987).

56. T. Soga and Y. Inoue, Determination of catecholamines in urine and plasma by on-line sample pretreatment using an internal surface boronic acid gel, *J. Chromatogr.*, **620**, 175–181 (1993).

57. H. Weicker, M. Feraudi, H. Hagele and R. Pluto, Electrochemical detection of catecholamines in urine and plasma after separation with HPLC, *Clin. Chim. Acta*, **141**, 17–25 (1984).

58. J. Odink, H. Sandman and W.H.P. Schreurs, Determination of free and total catecholamines and salsolinol in urine by ion-pair reversed-phase liquid chromatography with electrochemical detection after a one step sample clean-up, *J. Chromatogr.*, **377**, 145–154 (1986).

59. Y.P. Chan and T.S. Siu, Simultaneous quantitation of catecholamines and *O*-methylated metabolites in urine by isocratic ion-pairing high-performance liquid chromatography with amperometric detection, *J. Chromatogr.*, **459**, 251–260 (1988).

60. H. Nohta, E. Yamaguchi, Y. Ohkura and H. Watanabe, Measurement of catecholamines, their precursor and metabolites in human urine and plasma by solid-phase extraction followed by high-performance liquid chromatography with fluorescence derivatization, *J. Chromatogr.*, **493**, 15–26 (1989).

61. H.K. Jeon, H. Nohta and Y. Ohkura, High-performance liquid chromatographic determination of catecholamines and their precursor and metabolites in human urine and plasma by postcolumn derivatization involving chemical oxidation followed by fluorescence reaction, *Anal. Biochem.*, **200**, 332–338 (1992).

62. C. Sarzanini, E. Mentasti and N. Mario, Determination of catecholamines by ion-pair chromatography and electrochemical detection, *J. Chromatogr. A*, **671**, 259–264 (1994).

63. F. Mashige, Y. Matsushima, C. Miyata, R. Yamada, H. Kanazawa, I. Sakuma, N. Takai, N. Shinozuka, A. Ohkubo and K. Nakahara, Simultaneous determination of catecholamines, their basic metabolites and serotonin in urine by high-performance liquid chromatography using a mixed-mode column and an eight-channel electrochemical detector, *Biomed. Chromatogr.*, **9**, 221–225 (1995).

64. A. Rivero-Marcotegui, A. Grijalba-Uche, M. Palacios-Sarrasqueta and S. Garcia-Merlo, Effect of the pH and the importance of the internal standard on the measurement of the urinary catecholamines by high-performance liquid chromatography, *Eur. J. Clin. Chem. Clin.*, **33**, 873–875 (1995).

65. M. Hay and P. Mormede, Determination of catecholamines and methoxycatecholamines excretion patterns in pig and rat by ion-exchange liquid chromatography with electrochemical detection, *J. Chromatogr. B*, **703**, 15–23 (1997).

66. A.H. Anton and D.F. Sayre, A study of the factors affecting the aluminium oxide–trihydroxyindole procedure for the analysis of catecholamines, *J. Pharmacol. Exp. Ther.*, **138**, 360–375 (1962).

67. H. Tsuchiya, T. Koike and T. Hayashi, On-line purification for the determination of catecholamines by liquid chromatography, *Anal. Chim. Acta*, **218**, 119–127 (1989).

68. S. Higa, T. Suzuki, A. Hayashi, I. Tsuge and Y. Yamamura, Isolation of catecholamines in biological fluids by boric acid gel, *Anal. Biochem.*, **77**, 18–24 (1977).

69. K. Vuorensola, H. Siren and U. Karjalainen, Determination of dopamine and methoxycatecholamines in patient urine by liquid chromatography with electrochemical detection and by capillary electrophoresis coupled with spectrophotometry and mass spectrometry, *J. Chromatogr. B*, **788**, 277–289 (2003).

70. C. Sabbioni, M.A. Saracino, R. Mandrioli, S. Pinzauti, S. Furlanetto, G. Gerra and A.A. Raggi, Simultaneous liquid chromatographic analysis of catecholamines and 4-hydroxy-3-methoxyphenylethylene glycol in human plasma. Comparison of amperometric and coulomteric detection, *J. Chromatogr. A*, **1032**, 65–71 (2004).

71. E.C.Y. Chan, P.Y. Wee and P.C. Ho, The value of analytical assays that are stability-indicating, *Clin. Chim. Acta*, **288**, 47–53 (1999).

72. E.C.Y. Chan, P.Y. Wee and P.C. Ho, Evaluation of degradation of urinary catecholamines and metanephrines and deconjugation of their sulfoconjugates using stability-indicating reversed-phase ion-pair HPLC with electrochemical detection, *J. Pharm. Biomed. Anal.*, **22**, 515–526 (2000).

73. E.C.Y. Chan and P.C. Ho, High-performance liquid chromatography/atmospheric pressure chemical ionization mass spectrometric method for the analysis of catecholamines and metanephrines in human urine, *Rapid Commun. Mass Spectrom.*, **14**, 1959–1964 (2000).

74. P.H. Gamache, M.L. Kingery and I.N. Acworth, Urinary metanephrine and normetanephrine determined without extraction by using liquid chromatography and coulometric array detection, *Clin. Chem.*, **39**, 1825–1830 (1993).

75. M. Yamaguchi, J. Ishida and M. Yoshimura, Simultaneous determination of urinary catecholamines and 5-hydroxyindoleamines by high-performance liquid chromatography with fluorescence detection, *Analyst*, **123**, 307–311 (1998).

76. T.J. Panholzer, J. Beyer and K. Lichtwald, Coupled-column liquid chromatographic analysis of catecholamines, serotonin, and metabolites in human urine, *Clin. Chem.*, **45**, 262–268 (1999).

77. G.P. Jackman, Differential assay for urinary catecholamines by use of liquid chromatography with fluorescence detection, *Clin. Chem.*, **27**, 1202–1204 (1981).

78. G.P. Jackman, A simple method for the assay of urinary metanephrines using high performance liquid chromatography with fluorescence detection, *Clin. Chim. Acta*, **120**, 137–142 (1982).

79. N.G. Abeling, A.H. van Gennip, H. Overmars and P.A. Voute, Simultaneous determination of catecholamines and metanephrines in urine by HPLC with fluorometric detection, *Clin. Chim. Acta*, **137**, 211–226 (1984).

80. G.M. Anderson, J.G. Young, P.I. Jatlow and D.J. Cohen, Urinary free catecholamines determined by liquid chromatography–fluorometry, *Clin. Chem.*, **27**, 2060–2063 (1981).

81. K. Mori and K. Imai, Sensitive high-performance liquid chromatography system with fluorometric detection of three urinary catecholamines in the same range, *Anal. Biochem.*, **146**, 283–286 (1985).

82. K. Mori, Automated measurement of catecholamines in urine, plasma and tissue homogenates by high-performance liquid chromatography with fluorometric detection, *J. Chromatogr.*, **218**, 631–637 (1981).

83. T. Seki, Y. Yamaguchi, K. Noguchi and Y. Yanagihara, Estimation of catecholamines by ion-exchange chromatography on Asahipak ES-502C, using glycylglycine as the post-derivatizing agent, *J. Chromatogr.*, **332**, 9–13 (1985).

84. S. Honda, M. Takahashi, Y. Araki and K. Kakehi, Postcolumn derivatization of catecholamines with 2-cyanoacetamide for fluorometric monitoring in high-performance liquid chromatography, *J. Chromatogr.*, **274**, 45–52 (1983).

85. H. Nohta, A. Mitsui and Y. Ohkura, High-performance liquid chromatographic determination of urinary catecholamines by direct pre-column fluorescence derivatization with 1,2-diphenylethylenediamine, *J. Chromatogr.*, **380**, 229–231 (1986).

86. H. Tsuchiya, M. Tatsumi, N. Takagi, T. Koike, H. Yamaguchi and T. Hayashi, High-performance chromatographic determination of urinary catecholamines by pre-column solid-phase dansylation on alumina, *Anal. Biochem.*, **155**, 28–33 (1986).

87. T.P. Davis, High-performance liquid chromatographic analysis of biogenic amines in biological materials as *o*-phthalaldehyde derivatives, *J. Chromatogr.*, **162**, 293–310 (1979).

88. A.A. Descombes and W. Haerdi, HPLC separation of catecholamines after derivatization with 9-fluorenylmethyl chloroformate, *Chromatographia*, **33**, 83–86 (1992).

89. H. Nohta, E. Yamaguchi, Y. Ohkura and H. Watanabe, Simultaneous high-performance liquid chromatographic determination of catecholamine-related compounds by post-column derivatization involving coulometric oxidation followed by fluorescence reaction, *J. Chromatogr.*, **467**, 237–247 (1989).

90. C. Bunyagidj and J.E. Girard, A comparison of coulometric detectors for catecholamine analysis by LC–EC, *Life Sci.*, **31**, 2627–2634 (1982).

91. J.W. Lenders, G. Eisenhofer, I. Armando, H.R. Keiser, D.S. Goldstein and I.J. Kopin, Determination of metanephrines in plasma by liquid chromatography with electrochemical detection, *Clin. Chem.*, **39**, 97–103 (1993).

92. J. Dutton, A.J. Hodgkinson, G. Hutchinson and B.R. Norman, Evaluation of a new method for the analysis of free catecholamines in plasma using automated sample trace enrichment with dialysis and HPLC, *Clin. Chem.*, **45**, 394–399 (1999).

93. W.A. Bartlett, Effects of mobile phase composition on the chromatographic and electrochemical behaviour of catecholamines and selected metabolites: reversed-phase ion-paired high-performance liquid chromatography using multiple-electrode detection, *J. Chromatogr.*, **493**, 1–14 (1989).

94. P. Hjemdahl, Catecholamine measurements by high-performance liquid chromatography, *Am. J. Physiol.*, **247**, 13–20 (1984).

95. R.E. Shoup and P.T. Kissinger, Determination of urinary normetanephrine, metanephrine, and 3-methoxytyramine by liquid chromatography with amperometric detection, *Clin. Chem.*, **23**, 1268–1274 (1977).

96. T.P. Moyer, N.S. Jiang, G.M. Tyce and S.G. Sheps, Analysis for urinary catecholamines by liquid chromatography with amperometric detection: methodology and clinical interpretation of results, *Clin. Chem.*, **25**, 256–263 (1979).

97. M. Goto, N. Nakamura and D. Ishii, Micro high-performance liquid chromatographic system with micro precolumn and dual electrochemical detector for direct injection analysis of catecholamines in body fluids, *J. Chromatogr.*, **226**, 33–42 (1981).

98. D.S. Goldstein, Modified sample preparation for high-performance liquid chromatographic–electrochemical assay of urinary catecholamines, *J. Chromatogr.*, **275**, 174–177 (1983).

99. J. Jouve, N. Mariotte, C. Sureau and H.P. Muh, High-performance liquid chromatography with electrochemical detection for the simultaneous determination of the methoxylated amines, normetanephrine, metanephrine and 3-methoxytyramine in urine, *J. Chromatogr.*, **274**, 53–62 (1983).

100. P.L. Orsulak, P. Kizuka, E. Grab and J.J. Schildkraut, Determination of urinary normetanephrine and metanephrine by radial-compression liquid chromatography and electrochemical detection, *Clin. Chem.*, **29**, 305–309 (1983).

101. E.D. Schleicher, F.K. Kees and O.H. Wieland, Analysis of total urinary catecholamines by liquid chromatography: methodology, routine experience and clinical interpretation of results, *Clin. Chim. Acta*, **129**, 295–302 (1983).

102. N.T. Buu, M. Angers, D. Chevalier and O. Kuchel, A new method for the simultaneous analysis of free and sulfoconjugated normetanephrine, metanephrine, and 3-methoxytyramine in human urine by HPLC with electrochemical detection, *J. Lab. Clin. Med.*, **104**, 425–432 (1984).

103. K. Kemper, E. Hagemeier, K.S. Boos and E. Schlimme, Direct clean-up and analysis of urinary catecholamines, *J. Chromatogr.*, **336**, 374–379 (1984).

104. A.H.B. Wu and T.G. Gornet, Preparation of urine samples for liquid-chromatographic determination of catecholamines: bonded-phase phenylboronic acid, cation-exchange resin, and alumina adsorbents compared, *Clin. Chem.*, **31**, 298–302 (1985).

105. N.C. Parker, C.B. Levtzow, P.W. Wright, L.L. Woodward and J.F. Chapman, Uniform chromatographic conditions for quantifying urinary catecholamines, metanephrines, vanillylmandelic acid, 5-hydroxyindoleacetic acid, by liquid chromatography with electrochemical detection, *Clin. Chem.*, **32**, 1473–1476 (1986).

106. A. Foti, S. Kimura, V. DeQuattro and D. Lee, Liquid-chromatographic measurement of catecholamines and metabolites in plasma and urine, *Clin. Chem.*, **33**, 2209–2213 (1987).

107. B.L. Lee, S.K. Chia and C.N. Ong, Measurement of urinary free catecholamines using high-performance liquid chromatography with electrochemical detection, *J. Chromatogr.*, **494**, 303–309 (1989).

108. P. Violin, Determination of free urinary catecholamines by high-performance liquid chromatography with electrochemical detection, *J. Chromatogr. B*, **655**, 121–126 (1994).

109. P. Kontour, R. Dawson and A. Monjan, Manipulation of mobile phase parameters for HPLC separation of endogenous monoamines in rat brain tissue, *J. Neurosci. Methods.*, **11**, 5–18 (1984).

110. B.M. Eriksson, S. Gustafsson and B.A. Persson, Determination of catecholamines in urine by ion-exchange liquid chromatography with electrochemical detection, *J. Chromatogr.*, **278**, 255–263 (1983).

111. S.R. Binder and M.E. Biaggi, Analysis of urinary catecholamines by high-performance liquid chromatography in the presence of labetalol metabolites, *J. Chromatogr.*, **385**, 241–247 (1987).

112. I.N. Mefford, Lauroyl sarcosine: a weak cation-exchange reagent for on-column modification of reversed-phase material, *J. Chromatogr.*, **393**, 441–446 (1987).

113. I.N. Mefford, M. Ota, M. Stipetic and W. Singleton, Application of a novel cation-exchange reagent, IGEPON T-77 (*N*-methyl oleoyl taurate), to microbore separations of alumina extracts of catecholamines from cerebrospinal fluid, plasma, urine and brain tissue with amperometric detection, *J. Chromatogr.*, **420**, 241–251 (1987).

114. T. Nieman, Detection based on solution-phase chemiluminescence systems, in *Chemiluminescence and Photochemical Reaction Detection in Chromatography*, J.W. Birks (Ed.), VCH, New York, 99–123 (1989).

115. G. Orosz, R.S.Givens and R.L. Schowen, A model for mechanism of peroxyoxalate chemiluminescence as applied to detection in liquid chromatography, *Crit. Rev. Anal. Chem.*, **26**, 1–27 (1996).

116. R.E. Milofsky and J.W. Birks, Laser photolysis study of the kinetics and mechanism of photoinitiated peroxyoxalate chemiluminescence, *J. Am. Chem. Soc.*, **113**, 9715–9723 (1991).

117. G.H. Ragab, H. Nohta, M. Kai, Y. Ohkura and K. Zaitsu, 1,2-Diarylethylenediamines as sensitive pre-column derivatizing reagents for chemiluminescence detection of catecholamines in HPLC, *J. Pharm. Biomed. Anal.*, **13**, 645–650 (1995).

118. M. Tsunoda, K. Takezawa, T. Santa and K. Imai, Simultaneous automatic determination of catecholamines and their 3-*O*-methylmetabolites in rat plasma by high-performance liquid chromatography using peroxyoxalate chemiluminescence, *Anal. Biochem.*, **269**, 386–392 (1999).

119. O. Nozaki, T. Iwaeda, H. Moriyama and Y. Kato, Chemiluminescent detection of catecholamines by generation of hydrogen peroxide with imidazole, *Luminescence*, **14**, 123–127 (1999).

120. O. Nozaki, H. Kawamoto and H. Moriyama, Total free catecholamines assay by identification of its two functional groups and micro-flow injection chemiluminescence, *Luminescence*, **14**, 369–374 (1999).

121. M. Israel and M. Tomasi, A chemiluminescent catecholamine assay: its application for monitoring adrenergic transmitter release, *J. Neurosci. Methods.*, **91**, 101–107 (1999).

122. D. Robertson, E.C. Heath, F.C. Falkner, R.E. Hill, G.M. Brillis and J.T. Watson, A selective and sensitive assay for urinary metanephrine and normetanephrine using gas chromatography–mass spectrometry with selected ion monitoring, *Biomed. Mass Spectrom.*, **5**, 704–708 (1978).

123. F.A. Muskiet, C.G. Thomasson, A.M. Gerding, D.C. Fremouw-Ottevangers, G.T. Nagel and B.G. Wolthers, Determination of catecholamines and their 3-*O*-methylated metabolites in urine by mass fragmentography with use of deuterated internal standards, *Clin. Chem.*, **25**, 453–460 (1979).

124. D.K. Crockett, E.L. Frank and W.L. Roberts, Rapid analysis of metanephrine and normetanephrine in urine by gas chromatography–mass spectrometry, *Clin. Chem.*, **48**, 332–337 (2002).

125. R.L. Taylor and R.J. Singh, Validation of liquid chromatography–tandem mass spectrometry method for analysis of urinary conjugated metanephrine and normetanephrine for screening of pheochromocytoma, *Clin. Chem.*, **48**, 533–539 (2002).

126. A. Tornkvist, P.J. Sjoberg, K.E. Markides and J. Bergquist, Analysis of catecholamines and related substances using porous graphitic carbon as separation media in liquid chromatography–tandem mass spectrometry, *J. Chromatogr. B*, **801**, 323–329 (2004).

127. S.A. Lagerstedt, D.J. O'Kane and R.J. Singh, Measurement of plasma free metanephrine and normetanephrine by liquid chromatography-tandem mass spectrometry for diagnosis of pheochromocytoma, *Clin. Chem.*, **50**, 603–611 (2004).

128. B.A. Callingham and M.A. Barrand, The catecholamines: adrenaline, noradrenaline, dopamine, in *Hormones in Blood*, C.H. Gary and V.H.T. James (Eds), Academic Press, London, 143–207 (1979).

129. P. Moleman, Preservation of urine samples for assay of catecholamines and their metabolites, *Clin. Chem.*, **31**, 653–654 (1985).

130. T.B. Weir, C.C.T. Smith, J.M. Round and D.J. Betteridge, Stability of catecholamines in whole blood, plasma, and platelets, *Clin. Chem.*, **32**, 882–883 (1986).

131. F. Boomsma, G. Alberts, L. Vaneijk, A.J. Manintreld, M.A.D.H. Schalekamp, Optimal collection and storage conditions for catecholamine measurements in human plasma and urine, *Clin. Chem.*, **39**, 2503–2508 (1993).

132. Z.S.K. Lee and J.A.J.H. Critchley, Simultaneous measurement of catecholamines and kallikrein in urine using boric acid preservative, *Clin. Chim. Acta*, **276**, 89–102 (1998).

133. K. Miki and A. Sudo, Effect of urine pH, storage time, and temperature on stability of catecholamines, cortisol, and creatinine, *Clin. Chem.*, **44**, 1759–1762 (1998).

134. R.T. Peaston and C. Weinkove, Measurement of catecholamines and their metabolites, *Ann. Clin. Biochem.*, **41**, 17–38 (2004).

135. P.M. Bouloux and D. Perrett, Interference of labetalol metabolites in the determination of plasma catecholamines by HPLC with electrochemical detection. *Clin. Chim. Acta*, **150**, 111–117 (1985).

136. Y. Yui, T. Fujita, T. Yamamoto, Y. Itokawa and C. Kawai, Liquid-chromatographic determination of norepinephrine and epinephrine in human plasma. *Clin. Chem.*, **26**, 194–196 (1980).

137. A.Yamatodani and H. Wada, Automated analysis for plasma epinephrine and norepinephrine by liquid chromatography, including a sample cleanup procedure. *Clin. Chem.*, **27**, 1983–1987 (1981).

138. Y. Yui and C. Kawai, Comparison of the sensitivity of various post-column methods for catecholamine analysis by high-performance liquid chromatography. *J. Chromatogr.*, **206**, 586–588 (1981).

139. M.E. Swartz, Ultra performance liquid chromatography (UPLC): an introduction, *LC–GC North Am. Suppl. (Ultra Performance LC Separation Science Redefined)*, 8–14 (2005).

8

Chromatographic Measurement of Volatile Organic Compounds (VOCs)

Larry A. Broussard

Department of Clinical Laboratory Sciences, LSU Health Sciences Center, 1900 Gravier, New Orleans, LA 70112, USA

8.1 INTRODUCTION

Volatile organic compounds (VOCs) measured in clinical and forensic laboratories may be grouped into several categories, including ethanol and other volatile alcohols, anesthetics, alkyl nitrites, industrial solvents and chemicals and multiple volatile compounds often abused and categorized as inhalants. There is no classification system based on clinical effect or chemical structure and these compounds may exist in solid, liquid and gaseous forms. The common feature of these compounds is volatility, the property of existing in or being able to be converted to a gaseous form. This volatility affects the necessity for proper collection, handling and analysis of specimens and profoundly affects the ability to determine accurately the concentrations of these compounds in body fluids and tissues following ingestion or exposure. Reasons for measuring these compounds include confirmation of ingestion (or inhalation), stratification of severity of effects, including intoxication, detection or monitoring of exposure, and verification of administration.

8.2 GENERAL CONSIDERATIONS

The volatility of these compounds facilitates the use of gas chromatography (GC) for separation and quantitation. There are many factors to consider when choosing or implementing a chromatographic method for the analysis of one or more VOCs, including the intended use [medical diagnosis, forensic applications, general screen, quantitation of specific

Chromatographic Methods in Clinical Chemistry and Toxicology   Edited by R. L. Bertholf and R. E. Winecker
© 2007 John Wiley & Sons, Ltd

compound(s) for monitoring of exposure, verification of use, post-mortem analysis, etc.], volatility of the compound, available resources and specimen(s) in addition to the chromatographic parameters. Many of these factors will be reviewed in order to provide information to be considered when choosing or implementing a method.

8.3 INTENDED USE

The choice of a chromatographic method may be dependent on the intended use. The appropriate method when the intent is to establish a general screen for inhalant abuse may be very different from the method chosen to monitor administration of a particular anesthetic or exposure to industrial chemicals in the workplace. Likewise, forensic applications may vary from a general screen to post-mortem quantitation of particular suspected VOCs in blood and tissues. The number of methods developed for specific purposes (or compounds) is beyond the scope of this chapter, but it is hoped that the information provided will assist the analyst in finding or adapting chromatographic methods suitable for their particular situation.

8.4 VOLATILITY OF COMPOUNDS

The partitioning of a volatile compound into phases (air, blood and tissue) influences pharmacological and toxicological effects of the compound and must be considered when developing chromatographic methods. This distribution between phases is expressed by the partition coefficient (K), which may be calculated by dividing the concentration of the analyte in the sample phase by that in the gaseous phase. The lower the partition coefficient, the more readily the compound partitions into the gas phase and the greater is the risk of loss due to evaporation. For example, loss of hexane with a K of 0.14 (air–water system at 40 °C) is more likely than loss of ethanol with a K of 1355 (air–water system at 40 °C).[1] Hence preventing loss due to evaporation during specimen collection and handling is more critical for samples to be analyzed for hexane than for those containing ethanol. Analytical sensitivity can be increased during extraction and/or analysis lowering the K using techniques such as increasing the temperature or adding salts to decrease solubility. Partition coefficients of VOCs are typically listed for air–water but analysts are concerned with coefficients involving the biological samples to be analyzed. Blood–air, blood–tissue and tissue–air partition coefficients for some solvents[2] and anesthetics[3] have been determined, but the analyst may have to determine these parameters during the development of a particular method.

In addition to the partition coefficient, the analyst must be aware of the phase ratio, defined as the relative volume of the space above the sample (headspace) compared with the volume of the sample in the container, and its effect on loss of analyte. Each time the container is opened, analyte in the headspace is lost and the equilibrium of the compound's concentration in the sample and headspace is re-established. The larger the headspace in relation to the sample size, the greater is the loss due to evaporation. Collecting and storing samples in sealed containers with minimal headspace may minimize the loss of VOCs. Using partition coefficients and measuring phase ratios, the percentage loss after equilibrium for five inhalation anesthetics in closed 5-mL tubes at 37 °C has been calculated.[3] Partition coefficients and phase ratios impact both method sensitivity and loss of analyte. In general, the concentration of volatile analytes in the headspace is higher when conditions are adjusted to give the lowest values for both the partition coefficient and the phase ratio. Practical applications to achieve these results include increasing temperature, agitation, addition of inorganic salts (sodium or ammonium chloride or sulfate, potassium carbonate, sodium citrate, etc.) to decrease the solubility of polar organic volatiles in aqueous matrices and using sample containers with minimal headspace.

8.5 SAMPLE COLLECTION, HANDLING AND STORAGE

Precautions taken in the collection and storage of the specimen affect the likelihood of detecting exposure to volatile substances. Recommended precautions include collection of blood using an anticoagulant (lithium heparin) into a glass container with a cap lined with metal foil if possible.[4] Soft rubber stoppers are permeable to toluene and other VOCs. Samples should be collected and stored in containers with minimal headspace and addition of an internal standard after collection will minimize errors due to evaporation during storage or tissue homogenization. Samples should be stored, transported and handled at temperatures between −5 and 4 °C.[5] For highly volatile compounds, additional precautions should be employed. The preparation of calibration (standard) solutions of gaseous analytes poses additional problems. Yang et al.[6] performed all procedures on ice and used vacuumed tubes and precooled (−20 °C) gas-tight syringes to prevent analyte loss during spiking of standard and control samples. Liquid and solid analytes can be weighed in directly. Streete et al.[7] prepared calibration standards of gaseous analytes by transferring a volume of vapor from a cylinder into a glass gas sampling bulb fitted with a septum port and then injecting appropriate volumes (using a gas-tight syringe) into calibrated septum bottles filled with nitrogen. Broussard et al.[8] described a process for the preparation of a methanolic stock standard of a VOC (difluoroethane) utilizing a cylinder of the VOC in the gaseous state. Other authors prepare calibration standards in water,[9] methanol[10] and aqueous solutions of ethanol or n-propanol.[9]

8.6 HEADSPACE GAS CHROMATOGRAPHIC METHODS

Although some direct injection methods for determining ethanol[11] and other VOCs have been developed, headspace (HS) techniques are most frequently used. These HS methods include static mode variations such as solid-phase microextraction (HS-SPME), cryogenic focusing (HS-CF), cryogenic oven trapping (HS-COT) and the dynamic mode method of purge and trap.[5] For HS methods, an aliquot of HS air is sampled from a sealed vial containing specimen plus internal standard. As previously discussed, VOCs in the specimen are partitioned into the HS above the sample and the sample is taken when equilibrium has been established. Evaporation may be accelerated using heating, addition of salt and/or agitation. For tissue samples, homogenization at low temperature in a closed container or treatment with a proteolytic enzyme such as subtilisin A may be used to enhance evaporation.[5]

Solid-phase microextraction (SPME) is a sampling and concentration technique used to increase the sensitivity of HS methods. This technique is utilized for arson analysis and environmental monitoring purposes and also for clinical and forensic procedures. Short, narrow diameter, fused-silica optical fibers coated with stationary phase polymers are either immersed in the sample or the HS and compounds are adsorbed or absorbed (depending on the type of fiber). Extraction can be enhanced by adjustment of conditions (pH, temperature, time) and employing techniques such as agitation and addition of salt. Fibers are chosen based on the desired selectivity; in general, like attracts like, e.g. polar fibers are used for polar analytes and non-polar fibers for non-polar analytes. A review by Mills and Walker[12] summarizes the general principles of SPME and includes a list of SPME devices with some of their properties.

One of the most commonly used fibers for VOC analysis is coated with a 100-μm polydimethylsiloxane phase (PDMS) and is available as a unit with a stainless-steel guide rod housed in a hollow septum-piercing needle (Supelco, Bellefonte, PA, USA). The fiber can be withdrawn into the needle for protection during handling and the depth of fiber exposure can be controlled using the adjustable holder. In HS techniques, the needle can be used to pierce the septum of the sample vial and to introduce the fiber into the injection port of the gas chromatograph. In the sample vial, the extraction onto the fiber reaches equilibrium fairly quickly for volatile compounds. Mills and Walker's review[12] includes tabular summaries of HS-SPME–GC methods for the detection of alcohols, drugs, solvents and chemicals in blood and urine.

Another way to increase sensitivity is to increase the volume injected into the column. Lowering the temperature of the entire column in cryogenic oven trapping (COT) methods or for the injection port at the inlet of the column in cryogenic focusing (CF) methods allows the injection of as much as 10 times the typical sample volume. Liquid nitrogen or liquid carbon dioxide is used to lower the temperatures to well below 0 °C (-180 °C for nitrogen and -90 °C for carbon dioxide). These techniques result in better peak shapes in addition to increased sensitivity.

In dynamic HS analysis, also called purge and trap, a carrier gas constantly passes above the sample and evaporated volatiles are trapped in a cryogenic and/or adsorbent trap.[5] Analytes are released on to the column by extensive heating. This technique results in increased sensitivity, particularly for water-soluble compounds. It is a fairly difficult technique and problems include artifacts/interfering peaks due to impurities in the purge gas or water and foaming from biological samples. Procedures using an antifoam agent for biological samples and a dehydration agent have been developed. Wille and Lambert[5] have tabulated analytical data for all of the HS techniques discussed.

8.7 COLUMNS AND DETECTORS

Initially packed columns were used, but capillary columns are most frequently used today. A variety of polar and non-polar column phases have been used to determine VOCs depending on the particular compound(s) analyzed. Numerous examples were summarized by Wille and Lambert.[5] Similarly, a range of detection devices have been utilized. Flame ionization detection (FID) and mass spectrometry (MS) in the selected-ion monitoring mode (SIM) are most frequently used, but other techniques include electron-capture detection (ECD) and Fourier transform infrared (FTIR) detection.

8.8 IDENTIFICATION, QUANTITATION AND CONFIRMATION

The identification and quantification of VOCs utilizing GC is accomplished with the same techniques as used for non-volatile compounds. Most methods do not require a derivatization step because of the volatility of the compounds of interest, but testing for some of the polar metabolites of VOCs does require derivatization. The partition coefficient and measurement of the headspace concentration could be used to determine the concentration of a VOC, but typically quantification is performed using calibration curves and an internal standard. Problems involved with the preparation of standard (calibration) and control samples and techniques used to minimize evaporation have been discussed previously (Section 8.5). Pure standards in solvents or as a gas in a cylinder may be purchased from chromatography/ reference material suppliers. Because many VOCs are used for industrial purposes, standards containing multiple analytes in methanol are available for US Environmental Protection Agency (EPA) methods. For example, approximately 60 VOCs in environmental samples are detected by EPA Methods 502.1, 502.2, 524.1 and 524.2 and reference materials containing these compounds in methanol are available.

The matrix effect may cause problems in some biological samples, particularly post-mortem samples. Techniques that can be used in these cases include standard addition, multiple HS extraction and full evaporation.[5] The standard addition method, involving comparison of the signal in a sample before and after addition of known amounts of the VOC, is a relatively routine procedure. The multiple HS technique is based on the fact that each time an aliquot is removed from the HS, re-equilibrium occurs and the concentration reduces logarithmically. After multiple injections, the original concentration can be determined by extrapolation. Heating a small amount of sample at a high temperature in a closed container causes evaporation of (almost) all of the analyte. This full evaporation technique is another technique for avoiding the matrix effect.

Acceptable forensic practice recommends confirmation of the detection of an analyte (drug) by a second or definitive method. For VOCs, two acceptable GC confirmation procedures are re-sampling and re-analysis using a different column or use of mass spectrometry for absolute identification.

8.9 ETHANOL AND OTHER VOLATILE ALCOHOLS

Ethanol, the most frequently measured VOC in clinical and forensic laboratories, may be measured using enzymatic or chromatographic methods. Laboratories measuring ethanol alone almost always utilize enzymatic (alcohol dehydrogenase) methods performed on automated analyzers, but laboratories offering a broader test menu employ chromatographic methods. The National Academy of Clinical Biochemistry (NACB) laboratory medicine practice guidelines (LMPG) recommend that laboratories have the stat capability, with a turnaround time of 1 h or less, of performing quantitative assays for certain analytes including ethanol, methanol and ethylene glycol.[13] In clinical and forensic laboratories, the quantitative analysis of ethanol and methanol is frequently part of a volatiles panel, including 2-propanol and acetone, commonly performed using HS-GC. Both packed and capillary columns have been used, although currently capillary columns are more frequently used. Many types of packing materials have been used for ethanol analysis, including Carbowax, Porapak, DB-1, DB-WAX, DB-624 (Agilent Technologies, Wilmington, DE, USA) and columns developed specifically for such analysis such as DB-ALC1, DB-ALC2 (J&W Scientific, Folson, CA, USA) or Rtx-BAC1 and Rtx-BAC2 (Restek, Bellefonte, PA, USA). Commonly used internal standards are *n*-propanol, 2-butanone (methyl ethyl ketone) and *tert*-butanol.[14,15]

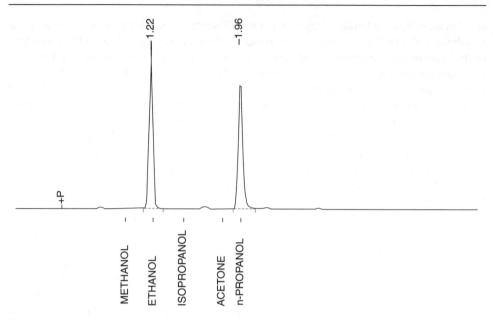

Figure 8.1 Chromatogram of ethyl chloride (retention time 1.22 min) and internal standard *n*-propanol (retention time 1.96 min.) on an Rtx-BAC1 column (30 m × 0.32 mm i.d., 1.8 μm). Static headspace sampling following a 5-min equilibration time at 70 °C and isothermal GC separation at 45 °C for 3.5-min. Note that ethyl chloride and ethanol have the same retention time.

These methods for ethanol and volatile alcohols may be modified to be used as a screening method for the detection of multiple VOCs[9,10] or to quantitate a particular volatile substance.[16] In some instances, the first evidence of inhalant abuse or solvent exposure may be the presence of an unknown peak during the analysis of ethanol and volatile alcohols. If inhalation or exposure to a particular VOC is suspected, a potential time-saving first step in method development would be HS injection of the analyte compound using the laboratory's procedure for ethanol. As mentioned previously, a second complementary analytical technique should be used for all forensic toxicological confirmation analyses,[15,17] including ethanol analysis. For example, Laferty[18] reported co-elution of ethanol and ethyl chloride on 0.3 and 0.2% Carbowax and 60/80 Carbopack C columns and Broussard *et al.*[19] reported similar results using an Rtx-BAC1 column. Figures 8.1 and 8.2 illustrate how ethyl chloride could be mistaken for ethanol in a one-column system but not in a two-column system.

8.10 INHALANTS AND SCREENING FOR MULTIPLE VOCs

The abuse of a wide range of volatile chemicals (often called inhalants) that may be inhaled accidentally or intentionally is a worldwide problem, particularly in the adolescent population. There are more than 1000 ordinary relatively inexpensive household products that may be abused via inhalation. VOCs such as toluene, chloroform, butane, propane and acetone are present in many commercial products (solvents, contact adhesives, correction fluid,

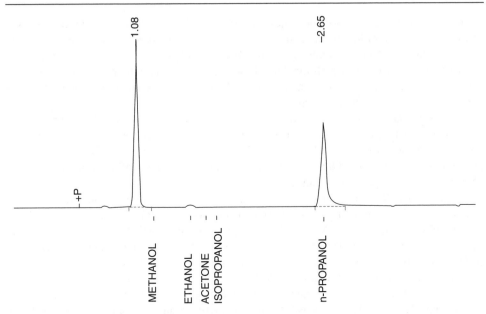

Figure 8.2 Chromatogram of ethyl chloride (retention time 1.08 min) and internal standard *n*-propanol (retention time 2.65 min.) on an Rtx-BAC2 column (30 m × 0.32 mm i.d., 1.2 μm). Static headspace sampling following a 5-min equilibration time at 70 °C and isothermal GC separation at 45 °C for 3.5 min. Note that ethyl chloride and ethanol are separated.

gasoline, lighter fluid, refrigerants, fire extinguishers, etc.) and halogenated hydrocarbons are the propellants for virtually all aerosol products. Three of the most frequently abused volatile substances or classes of substances are toluene, halogenated hydrocarbons (solvents such as chloroform, carbon tetrachloride, methylene chloride, 1,1,1-trichloroethane in correction fluid, trichloroethylene used for dry cleaning and ozone-depleting chlorofluorocarbons and their non-ozone-depleting replacements) and butane.

Inhalants in small doses produce euphoria characterized by loss of inhibition but also have the potential to cause cardiac arrhythmias ('sudden sniffing death'). Large doses produce acute problems, including convulsions, coma and death. Depending on the product, volatile substances may be inhaled directly from the container (snorting or sniffing), from a plastic bag (bagging) – particularly if the product is an aerosol or a viscous liquid such as glue – or from a saturated cloth (huffing).[16] Of these routes of administration, bagging usually results in the highest concentration, snorting the lowest and huffing an in-between concentration. In addition to screening for abuse, other reasons for detection, identification and quantitation of these compounds include monitoring of exposure in the industrial setting and post-mortem cause-of-death investigations.

For isolated cases, the first indication of the presence of a volatile substance may be the presence of an unknown peak when a sample is analyzed for ethanol and other volatile alcohols (see the previous section). In these instances, method development focuses on the compound encountered and frequently involves modification of the ethanol analysis procedure. Numerous procedures for single VOCs or classes of VOCs have been developed and summarized in review articles.[5,6,12,20,21] HS-GC remains the most frequently used method, although high-performance liquid chromatography (HPLC) is used for the detection of the more polar and less volatile urine metabolites.[5,22] Blood is the specimen of choice, although analysis of urine metabolites sometimes extends the time frame for detection of exposure. For fatalities in which inhalant abuse is suspected, tissues (brain, lung, fat, liver, heart, kidney) should be analyzed in addition to blood. Typically the highest concentrations are found in blood and brain tissue.

Several HS-GC procedures, listing the retention data of as many as hundreds of compounds, for the screening of multiple VOCs have been published.[4,7,9,10,23–25] Although limits of detection (LODs) and linear analytical ranges vary with particular methods, most procedures have typical LODs of $0.1 \, \text{mg L}^{-1}$ and upper limits of linearity (ULOL) to concentrations as high as $5000 \, \text{mg L}^{-1}$.

Table 8.1 lists the retention times of common volatiles for the procedure described by Sharp.[9] This HS procedure utilizes dual-column elution on non-polar DB-1 (30 m × 0.32 mm i.d., 5.0-μm film thickness) (Agilent Technologies) and polar DB-WAX (30 m × 0.32 mm i.d., 0.25-μm film thickness) (Agilent Technologies) columns with flame ionization detection (FID). HS vials (20 mL) containing 0.1 mL of sample and 1.5 mL of *tert*-butanol (internal standard) are incubated for 30 min at 70 °C and aliquots are injected using a 20:1 split injection through a Y connector. The injector and detector temperatures were 150 and 220 °C, respectively. The total run time was 20 min with a temperature program of 40 °C for 9 min, then ramped at $10 \, °\text{C min}^{-1}$ to 150 °C. Method validation included comparison on a single specialty column (DB-624; 30 m × 0.53 mm i.d., 3.0-μm film thickness) (Agilent Technologies) system and confirmation of peak identities by mass spectrometry (MS). The procedure provides detection and identification of more than 40 compounds. For the few compounds that do not display unique retention times on both columns, either the use of a DB-624 column or HS-GC–MS have been shown to be acceptable confirmation procedures. For confirmation on the DB-624 column, samples were incubated for 30 min at 60 °C. The injector and detector temperatures were 150 and 250 °C, respectively. The total run time was 15 min with a temperature program of 45 °C for 5 min, then ramped at $9 \, °\text{C min}^{-1}$ to 90 °C and held 5 min. Columns for HS-GC–MS confirmation were DB-1 (30 m × 0.25 mm i.d., 0.25-μm film thickness) and DB-WAX (15 m × 0.25 mm i.d., 0.25-μm film thickness). Although this method is used as a screen for forensic purposes, it can be adapted for use in the detection and/or quantitation of compounds for workplace monitoring purposes.

8.11 INTERPRETATION

Clinical chemists and toxicologists should be aware that the detection of volatile substances in blood does not always indicate inhalant abuse or occupational exposure. Acetone and other volatile compounds may be found in ketoacidotic patients and some inborn errors of metabolism result in the accumulation of volatile compounds. Even though many studies and case reports have included concentrations of volatile substances in blood, definitive correlations between these blood concentrations and the clinical features of toxicity have not been demonstrated for any of these compounds.

The detection of urinary metabolites has also been used to detect occupational exposure or to confirm inhalation of volatile substances. Urinary metabolites such as phenol (benzene metabolite), trichloroacetic acid (tetrachloroethylene), hippuric acid (toluene) and methylhippuric acid (xylene) have been detected and measured.[16] Results are often

Table 8.1 HS-GC retention times (min) of VOCs[a,b]

Compound	DB-1 run program	DB-WAX 40–150 °C	DB-624 40–150 °C
Acetaldehyde	2.33	1.55	2.20
Acetone	4.14	1.86	3.49
Acetonitrile	3.76	3.61	3.86
Amyl acetate[c]	19.96	7.17	>15
Amyl acetate, artifact	>20	10.01	>15
Benzene	13.15	2.71	7.77
Butanol, 1-	12.94	8.99	8.84
Butanol, 3-methyl-1-[d]	15.72	11.55	11.09
Butanol, 4-chloro-1-, artifact	>20	14.29	>15
Butanol, 4-chloro-1-[c]	11.31	2.10	6.76
Butanol, iso-(2-methyl-1-propanol)	11.27	6.29	7.70
Butanol, *terti*-	5.45	2.43	4.25
Butyl acetate, *n*-	18.16	5.36	13.29
Chloral hydrate[e]	>20	>20	>15
Chlorobenzene	19.49	11.06	>15
Chlorobutane, 1-	12.51	1.96	7.32
Chloroform	10.61	3.98	6.84
Cyclohexane	13.58	1.59	7.22
Diethyl ether	5.04	1.46	3.15
Ethanol	3.39	2.75	3.02
Ethchlorvynol	>20	>20	>15
Ethyl acetate	10.44	2.27	6.42
Ethylbenzene	19.93	6.92	>15
Formaldehyde[d]	15.42	2.38	10.48
Heptane, *n*-	15.09	1.54	8.26
Hexane, isomer 2	9.46	1.36	4.52
Hexane, isomer 1	12.01	1.51	6.07
Hexane, *n*-	10.50	1.42	4.99
Hexanone, 2-	17.40	5.63	12.93
Methanol	2.33	2.36	2.31
Methyl acetate	5.60	1.93	3.94
Methyl ethyl ketone	8.88	2.36	6.28
Methyl *terti*-butyl ether	8.12	1.53	4.49
Methyl-2-pentanone, 4-	15.84	3.69	10.87
Methylene chloride	5.73	2.63	4.10
Octane, *n*-	18.35	1.78	11.55
Octanol[e]	>20	18.25	>15
Pentane	5.07	1.38	2.92
Pentanol, 1-[c]	16.77	12.77	12.29
Pentanol, 3-	14.49	7.14	9.73
Pentanone, 4-hydroxy-4-methyl-(HMP)[e]	18.94	17.24	>15
Propanol, 2-	4.46	2.68	3.66
Propanol, *n*-	6.98	4.54	5.47
Toluene	17.14	4.29	11.22
Trichloroethane, 1,1,1-	12.40	2.20	7.13

(*continued*)

Table 8.1 (*Continued*)

Compound	DB-1 run program	DB-WAX 40–150 °C	DB-624 40–150 °C
Trichloroethanol, 2,2,2-	>20	>20	>15
Trichloroethylene	14.69	3.47	8.90
Trimethylpentane, 2,2,4-	14.76	1.49	7.92
Xylene, *m*-[c]	19.93	6.86	>15
Xylene, *o*-[c]	>20	7.19	>15
Xylene, *p*-	>20	7.20	>15

[a](Columns: DB-1, 30 m × 0.32 mm i.d., 5.0-μm film thickness; DB-WAX, 30 m × 0.32 mm i.d., 0.25-μm film thickness; DB-624, 30 m × 0.53 mm i.d., 3.0-μm film thickness (Agilent Technologies, Wilmington, DE, USA).
[b]Retention time entries denote compounds readily detected at 100 mg%.
[c]Artifact peaks detected.
[d]Split peak.
[e]Primary peak is an artifact or breakdown product.

Reproduced from the *Journal of Analytical Toxicology* by permission of Preston Publications, A Division of Preston Industries, Inc. Sharp M.E., A comprehensive screen for volatile organic compounds in biological fluids, *J. Anal. Toxicol.* 2001, **25**, 631–636.

expressed as a ratio to the urine creatinine concentration in order to normalize results in relation to urine volume and fluid intake. Caution must be used when interpreting results. For example, urinary hippuric acid may be due to ingestion of benzoate preservatives in foods and not to exposure to toluene.

8.12 CONCLUSION

Compounds classified as VOCs are typically measured using HS-GC. Many methods have been developed for the screening, detection, identification and quantitation of single or multiple VOCs. Due to the volatility of these compounds, sample collection, handling and storage are extremely important. In order to choose and/or develop methods that accurately detect and quantitate VOCs, the analyst must understand and utilize the principles of phase equilibrium, including partition coefficients and phase ratios. In clinical and forensic laboratories, the occurrence of an unknown peak during the analysis of ethanol and volatile alcohols by HS-GC may be the first indication of the presence of a VOC and the substance may be identified and quantitated by modification of the method. Several screening methods for multiple VOCs have been developed and details of one such method has been presented in this chapter.

REFERENCES

1. LabHut Education Centre. URL: http://www.labhut.com/education/headspace/introduction03.php (accessed 26 September, 2006).

2. C.H. Pierce, R.L. Dills, G.W. Silvey, D.A. Kalman, Partition coefficients between human blood or adipose tissue and air for aromatic solvents, *Scand. J. Work Environ. Health*, **22**, 112–118 (1996).

3. N.C. Yang, H.F. Wang, K.L. Hwang, W.M. Ho, A novel method for determining the blood/gas partition coefficients of inhalation anesthetics to calculate the percentage of loss at different temperatures, *J. Anal. Toxicol.*, **28**, 122–127 (2004).

4. R.J. Flanagan, P.J. Streete, J.D. Ramsey, *Volatile Substance Abuse: Practical Guidelines for Analytical Investigation of Suspected Cases and Interpretation of Results*. United Nations, Geneva, 1997. URL: http://www.unodc.org/pdf/technical_series_1997-01-01_1.pdf (accessed 26 September 2006).

5. S.M.R. Wille, W.E.E. Lambert, Volatile substance abuse – post-mortem diagnosis, *Forensic Sci. Int.*, **142**, 135–156 (2004).

6. N.C. Yang, K.L. Hwang, C.H. Shen, H.F. Wang, W.M. Ho, Simultaneous determination of fluorinated inhalation anesthetics in blood by gas chromatography–mass spectrometry combined with a headspace autosampler, *J. Chromatogr. B*, **759**, 307–318 (2001).

7. P.J. Streete, M. Rupah, J.D. Ramsey, R.J. Flanagan. Detection and identification of volatile substances by headspace capillary gas chromatography to aid the diagnosis of acute poisoning, *Analyst*, **117**, 1111–1127 (1992).

8. L. Broussard, T. Brustowicz, T. Pittman, K.D. Atkins, L. Presley, Two traffic fatalities related to the use of difluoroethane, *J. Forensic Sci.*, **42**, 1186–1187 (1997).

9. M.E. Sharp, A comprehensive screen for volatile organic compounds in biological fluids, *J. Anal. Toxicol.*, **25**, 631–635 (2001).

10. I.A. Wasfi, A.H. Al-Awadhi, Z.N. Al-Hatali, F.J. Al-Rayami, N.A.A. Katheeri, Rapid and sensitive static headspace gas chromatography–mass spectrometry method for the analysis of ethanol and abused inhalants in blood, *J. Chromatogr. B*, **799**, 331–336 (2004).

11. A. Tangerman, Highly sensitive gas chromatographic analysis of ethanol in whole blood, serum, urine, and fecal supernatants by the direct injection method, *Clin. Chem.*, **43**, 1003–1009 (1997).

12. G.A. Mills, V. Walker, Headspace solid-phase microextraction procedures for gas chromatographic analysis of biological fluids and materials, *J. Chromatogr. A*, **902**, 267–287 (2000).

13. A.H.B. Wu, C. McKay, L.A. Broussard, R.S. Hoffman, T.C. Kwong, T.P. Moyer, *et al.*, National Academy of Clinical Biochemistry Laboratory Medicine Practice Guidelines: recommendations for the use of laboratory tests to support poisoned patients who present to the emergency department, *Clin. Chem.*, **49**, 357–379 (2003).

14. B. Levine, Y.H. Caplan, Alcohol, in *Principles of Forensic Toxicology*, B. Levine (Ed.), AACC Press, Washington, DC, 2003, pp. 157–172.

15. A.W. Jones, M. Fransson, Blood analysis by headspace gas chromatography: does a deficient sample volume distort ethanol concentration?, *Med. Sci. Law*, **43**, 241–247 (2003).

16. L. Broussard, Inhalants, in *Principles of Forensic Toxicology*, B. Levine (Ed.), AACC Press, Washington, DC, 2003, pp. 341–348.

17. B.K. Logan, G.A. Case, E.L. Kiesel, Differentiation of diethyl ether/acetone and ethanol/acetonitrile solvent pairs, and other common volatiles by dual column headspace gas chromatography, *J. Forensic Sci.*, **39**, 1544–1551 (1994).

18. P.I. Laferty, Ethyl chloride: possible misidentification as ethanol, *J. Forensic Sci.*, **39**, 261–265 (1994).

19. L. Broussard, A.K. Broussard, T. Pittman, D.K. Lirette, Death due to inhalation of ethyl chloride, *J. Forensic Sci.*, **45**, 223–225 (2000).

20. A. Astier, Chromatographic determination of volatile solvents and their metabolites in urine for monitoring occupational exposure, *J. Chromatogr.*, **643**, 389–398 (1993).

21. A. Uyanik, Gas chromatography in anaesthesia I. A brief review of analytical methods and gas chromatographic detector and column systems, *J. Chromatogr. B*, **693**, 1–9 (1997).

22. L. Broussard, The role of the laboratory in detecting inhalant abuse, *Clin. Lab. Sci.*, **13**, 205–209 (2000).

23. D.J. Tranthim-Fryer, R.C. Hansson, K.W. Norman, Headspace/solid-phase microextraction/gas chromatography–mass spectrometry: a screening technique for the recovery and identification of volatile organic compounds (VOC's) in postmortem blood and viscera samples, *J. Forensic Sci.*, **46**, 934–946 (2001).

24. Restek Applications Note 59548, *GC Analysis of Commonly Abused Inhalants in Blood Using Rtx®-BAC1 and Rtx®-BAC2.* URL: http://www.restekcorp.com/Fantasia/pdfCache/59548a.pdf (accessed 26 September 2006).

25. J. Liu, H. Kenji, S. Kashimura, M. Kashiwagi, T. Hamanaka, A. Miyoshi, M. Kageura, Headspace solid-phase microextraction and gas chromatographic–mass spectrometric screening for volatile hydrocarbons in blood, *J. Chromatogr. B*, **748**, 1–9 (1997).

9

Chromatographic Techniques for Measuring Organophosphorus Pesticides

H. Wollersen and F. Musshoff

Department of Legal Medicine, University of Bonn, Stiftsplatz 12, 53111 Bonn, Germany

9.1 INTRODUCTION

The term 'pesticides' comprises a large number of substances belonging to many completely different chemical groups, the only common characteristic being that they are effective against pests. Pesticides may be classified based on their chemical structure, their pharmacokinetic or pharmacodynamic profile and also their formulation, application and target organism.[1–3] Some substances are not easily categorized by standard methods as they can be used against two or more targets or in different formulations, resulting in varying pharmacokinetic and/or pharmacodynamic properties. A major classification system links pesticides to the pest that is controlled, e.g. insecticides, herbicides and fungicides controlling insects, weeds and fungi, respectively (Table 9.1).

A second classification categorizes the pesticides according to their chemical nature, i.e. inorganic and organic compounds. Most pesticides used today are organic compounds, a small number of which are either derived from plant products or extracted directly from plants. However, most modern pesticides are synthetic compounds, this fact being primarily responsible for the rapid expansion of pesticides since the 1940s: in 1997, for example, 15 400 t of herbicides, 8400 t of fungicides, 3300 t of other pesticides and 900 t of insecticides were sold in Germany.[4] Although the amount of insecticides appears to be small in comparison with other substances, it has to be taken into consideration that 81 t of the total of 900 t of insecticides were used for private gardening. This means that 10% of the insecticides produced were utilized by persons without appropriate expertise.

Chromatographic Methods in Clinical Chemistry and Toxicology   Edited by R. L. Bertholf and R. E. Winecker

Table 9.1   Classes of pesticides

Class	Pest
Acaricide	Mites
Bactericide	Bacteria
Fungicide	Fungi
Herbicide	Weeds
Insecticide	Insects
Molluscicide	Snails
Nematicide	Nematodes
Rodenticide	Rodents (mouse and rat)

Today's commercially available pesticides are characterized by a vast variety of chemical structures and functional groups making their chemical classification quite complex. The major chemical groups are as follows:

- *Organophosphates*: these are esters of phosphoric, phosphonic, phosphorothionic and related acids.

- *Organochlorines*: these are hydrocarbons with one or more chlorine atoms.

- *Carbamates*: these substances are formed by salts or esters of carbamic acids.

- *Pyrethroids*: these substances are synthesized from pyrethrums, natural plant products in different kinds of chrysanthemum.

- *Triazines*: these substances are classical herbicides consisting of a number of substituted 1,3,5-triazines.

- *Substituted ureas*: this group comprises substances such as phenylureas, sulfonylureas and benzoylureas, which are mainly herbicides.

For humans, pesticides exclusively targeted against vermin hold a great toxicological risk. Others, e.g. herbicides, have only a small if any influence on the human organism, as they were developed to interfere with plant physiology. Insecticides can be subdivided into three groups based on their mechanism of action: pyrethroids, organochlorine compounds and organophosphorus pesticides together with carbamates are distinguished, the last type being acetylcholinesterase inhibitors.

To estimate the toxicological risk of a given pesticide, a knowledge of exposure levels is a crucial step in the risk-evaluation process and can be achieved by measuring the dose entering the body. However, the analytical detection of pesticides in food or different parts of the environment (e.g. air, soil, water) and detection in human biological tissues are difficult, for many reasons: First, current analytical methods are highly sophisticated and require specialized laboratories. Pure pesticide standards, and even more so standards for their metabolites, are rarely commercially available. Therefore, only a few validated methods exist that have been approved by reference organizations. In field studies on pesticide exposure, it is difficult to collect representative samples and to define a correct sampling time. Furthermore, permissible exposure limits and biological exposure indexes are available for only a limited group of compounds. For these and other reasons, a detailed discussion of which would go beyond the scope of this chapter, biological monitoring of pesticide exposure is a very complex issue.

Based on these considerations, the following sections will focus on organophosphorus insecticides and their toxicological properties and well review various analytical methods for the qualitative detection and quantitative determination of different substances of this class.

9.2 ORGANOPHOSPHORUS PESTICIDES (OPs)

Substances belonging to the group of OPs are esters of phosphoric, phosphonic, phosphorothionic and related acids. In Table 9.2 the chemical structures of the most commonly used OPs and their LD_{50} values (rat) are shown.

9.2.1 Mechanism of action

OPs act through inhibition of acetylcholinesterase activity in the central nervous system (CNS). This enzyme is essential for the hydrolysis of acetylcholine (ACh), which is a neurotransmitter within the human body. ACh is crucial for most inter-cell communication: upon stimulation of nerve endings, ACh molecules are released and diffuse through the synaptic gap. The interaction of ACh with its receptor evokes a sodium and calcium influx into the downstream cell, resulting in an electrochemical stimulus. This stimulus is then either passed on using intracellular signaling pathways (nerve cells) or used to implement a specific response in the recipient cell (other organs). After transferring their information, it is necessary to remove the ACh molecules from their receptor binding sites. This is realized by an enzymatic cleavage of ACh into choline and acetic acid, thereby restoring the receptors and enabling them to process new signals.

The enzyme cholinesterase, effecting the degradation of ACh, consists of two active sites, the anionic and the esteratic site (Figure 9.1). In the first step, ACh is attached to the anionic site and the acetyl group of the ACh molecule is transferred to the amino acid serine. The resulting serine ester is then cleaved hydrolytically and the enzyme is regenerated.[5]

The first step of the reaction of cholinesterase with the OPs is similar to its interaction with ACh as its catalytic serine residue is phosphorylated and the acid component is released from the educt (Figure 9.2). This phosphorylation is nearly irreversible and the cholinesterase is no longer able to cleave ACh, causing a continuous stimulus. This uncontrolled reaction results in an intoxication.[6]

Thiophosphate insecticides are protoxins, as they are unable to phosphorylate the serine until after the first liver passage, during which the thiono group is replaced by an oxo group. Therefore, they are called indirect acetylcholinesterase inhibitors.

9.2.2 Intoxication

OPs are very lipophilic and are readily absorbed both enterally and percutaneously. The adsorption rate is increased by simultaneous intake/application of organic solvents or plant oils.

Table 9.2   Chemical structures of the most popular organophosphorus insecticides

Name	Chemical structure	LD_{50}, rat, oral (mg kg^{-1})
Mevinphos	CH₃O, CH₃O–P(=O)–O–C(CH₃)=CH–COOCH₃	3.7
Parathion-methyl	CH₃O, CH₃O–P(=S)–O–C₆H₄–NO₂	6
Parathion-ethyl	C₂H₅O, C₂H₅O–P(=S)–O–C₆H₄–NO₂	7
Oxydemeton-methyl	H₃CO, H₃CO–P(=O)–S–(CH₂)₂–S(=O)–C₂H₅	50
Azinphos-methyl	CH₃O, CH₃O–P(=S)–S–CH₂–N (benzotriazinone ring)	80
Chlorpyrifos	trichloropyridinyl–O–P(=S)(OC₂H₅)(OC₂H₅)	245
Fenthion	CH₃O, CH₃O–P(=S)–O–C₆H₃(CH₃)–S–CH₃	250
Dimethoate	CH₃O, CH₃O–P(=O)–S–CH₂–C(=O)–NH–CH₃	300
Fenitrothion	CH₃O, CH₃O–P(=S)–O–C₆H₃(CH₃)–NO₂	500
Malathion	CH₃O, CH₃O–P(=S)–S–CH(COOC₂H₅)–CH₂–COOC₂H₅	1400
Temephos	CH₃O, CH₃O–P(=S)–O–C₆H₄–S–C₆H₄–O–P(=S)(OCH₃)(OCH₃)	2000
Bromophos	CH₃O, CH₃O–P(=S)–O–C₆H₂(Cl)(Br)(Cl)	3750

The onset of intoxication symptoms is dependent on the pathway of absorption: within a few minutes of inhalation, from 15 min to 1 h after swallowing, and 2–3 h after cutaneous resorption, a toxic concentration will be reached in the blood. With indirect acetylcholinesterase inhibitors, e.g. parathion, symptoms of poisoning occur later. No exact data exist about the extent of bioavailability in the human body.

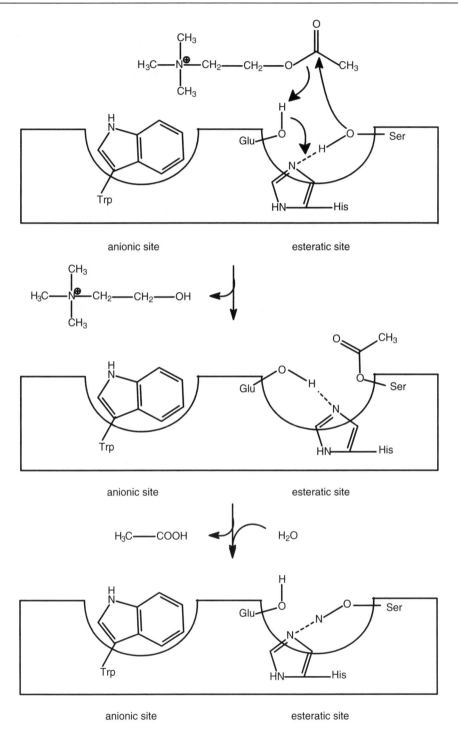

Figure 9.1 Chemical reaction of acetylcholine (ACh) with the enzyme acetylcholinesterase (Ser, serine; His, histidine; Glu, glutamate; Trp, tryptophan)

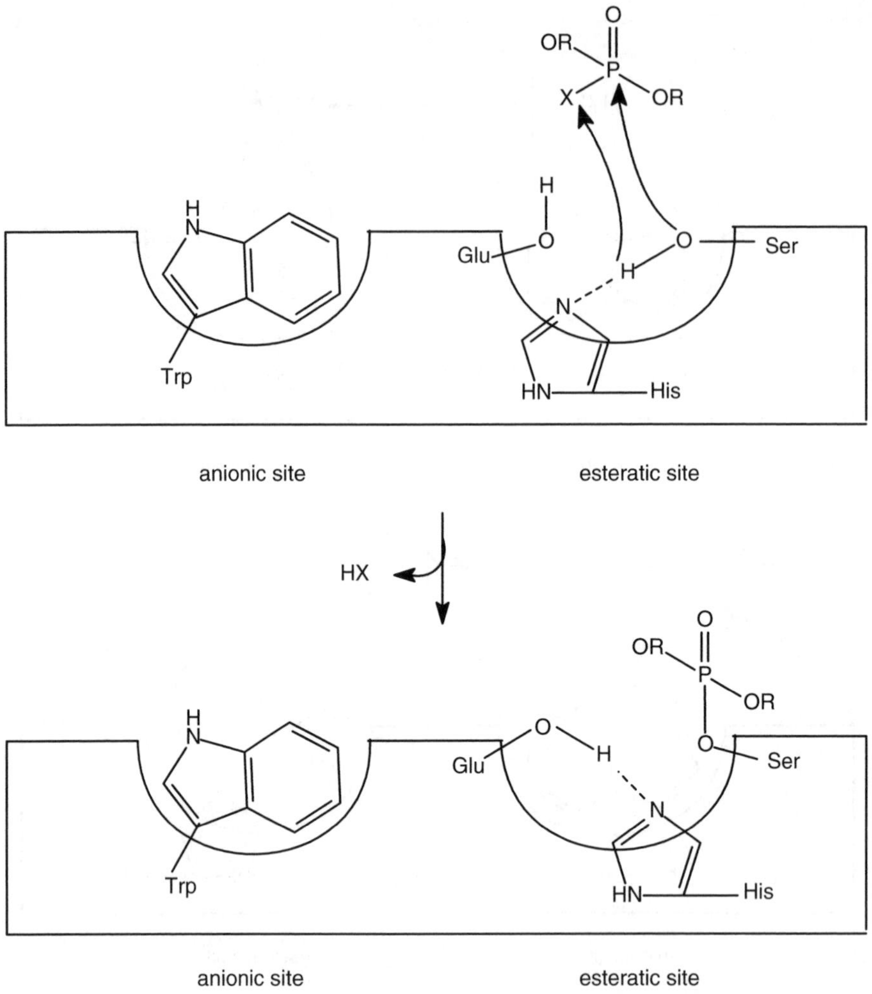

Figure 9.2 Chemical reaction of OPs with the enzyme acetylcholinesterase (Ser, serine; His, histidine; Glu, glutamate; Trp, tryptophan; X, acid group)

The most frequently observed symptoms of mild intoxication with OPs are miosis, lacrimation, increased secretion from mucous membranes, hypersalivation, nausea, emesis, gastrointestinal spasms, diarrhea and profuse sweating. Progressive and massive intoxications result in agitated melancholia, dyspnea, algospasms, coma, confusion, tremor, muscular cramps, muscle stiffening, amyasthenia and central and peripheral respiratory paralysis. Diagnostic clues for intoxication with OPs are blue or red coloration of the concoction, of the vomit or of residues in the mouth of the patient. Also, an intense onion-like smell is indicative of OP intoxication.[7]

The toxicity of a given OP is variable, depending on its chemical structure, lipophilic character, affinity to the esterase and the readiness with which it is hydrolyzed. With increasing lipophilicity, OPs are increasingly well absorbed through the skin and through the mucous membranes of the gastric and intestinal tract. Furthermore, high lipophilicity in

OP*s* facilitates their CNS penetration. In contrast, a polar character of a given OP increases its affinity to the cholinesterase molecule, i.e. to its anionic site. If the phosphoric ester bond in an OP is easily hydrolyzable, its toxicity will be low as the cleavage products are unable to interact with the enzyme. However, stable phosphate esters that contain a hydrolyzable functional group (e.g. alkyl groups) are extremely toxic. After dissociation of this cleavable group from the OP–esterase complex, the aggregate is stabilized (Figure 9.2) resulting in irreversible inhibition of the enzyme. This process is called 'aging' of the enzyme, with OP*s* optimized towards fast aging being used as chemical warfare agents.[8]

9.2.3 Progression of intoxication and longer term risks

In cases of attempted suicide, often fatal intoxications occur where, in spite of early hospitalization, patients are beyond remedy. Initially, signs of recovery may be observed, but after 2–3 days patients relapse with a measurable increase in OP plasma concentrations. This phenomenon is called intermediary syndrome (IMS), first described by Senananyake *et al.* in 1987,[9] and is possibly a consequence of a severe intoxication with permanent depolarization of the neuromuscular end-plates and constant excitation of the nicotinic acetylcholine receptors in the CNS of the patient. The IMS is clinically characterized by acute respiratory paresis, weakness of facial, palatal, external ocular, nuchal and proximal limb muscles and depressed tendon reflexes. Some authors[10,11] propose that an insufficient therapy with oximes or atropine (see Section 9.2.5) and inadequate artificial respiration in the early stages of intoxication may cause the occurrence of an IMS. It is further remarkable that only some distinct OP agents (e.g. fenthion, dimethoate, monocrotophos and methamidophos) seem capable of producing the IMS.

A third neurological manifestation, besides the acute toxicity which appears within 30 min after resorption and the IMS which develops about 2–3 days after resorption, is the so-called organophosphate-induced delayed neurotoxicity (OPIDN).[12,13] Symptoms of OPIDN appear about 2–3 weeks after acute exposure to OP compounds. OPIDN is caused by inhibition of another esterase enzyme, neuropathy target esterase (NTE), and establishes a second structure–activity relationship (SAR) in OP toxicology as a predominantly motoric, distal symmetrical polyneuropathy emerges. OPIDN generally follows acute cholinergic symptoms. Muscle cramps and pain in the legs are the usual initial complaint, sometimes followed by distal numbness and paresthesias. Progressive leg weakness is then accompanied by a depression of tendon reflexes. After several days, similar symptoms may appear in the hands and forearms. Sensory loss sometimes develops, initially in the legs and then in the arms, but is often mild or inconspicuous. The importance of this clinical picture is underlined by the fact that more than 40 000 cases of OPIDN have been documented from 1899 to 1989. Permanent damage remains mostly after a severe intoxication, and differentiation between such damage and OPIDN is often not possible.

In spite of the extensive knowledge about different kinds of consequences of an intoxication with OP*s*, data on toxic blood concentrations are non-existent for most of the substances. When known, the values vary strongly between the respective OP*s*. For example, the toxic and comatose-fatal (see case report) blood concentrations of parathion are in the ranges 10–50 and 50–80 ng mL^{-1}, respectively. In contrast, higher concentrations of malathion are required for toxicity: doses of 500 ng mL^{-1} and 175 µg mL^{-1} were described as being toxic and fatal, respectively.[14,15]

9.2.4 Therapy

In case of intoxication, the first measure is to stabilize respiration and circulation. This can be achieved by administration of atropine, which is a non-depolarizing inhibitor of acetylcholine receptors. In the early stages of intoxication, treatment with oximes, for example toxogonin, provides the opportunity to reactivate the enzyme. In the first step of the reactivation process, the bound OP molecule undergoes a nucleophilic attack by the oxime that causes cleavage of the ester bond, thereby unblocking the catalytic serine residue of the esterase. The NO–phosphate ester bond in the oxime–OP complex is very unstable and degrades easily, releasing the hydrolyzed OP. The mechanism of the reactivation reaction is shown in Figure 9.3.[16]

9.2.5 Analytical procedures

For detection of intoxication with OPs, many different analytical methods have been reported. One alternative is to measure the cholinesterase activity, because an acute intoxication is characterized by a 20% decrease in acetylcholinesterase activity. Another approach includes the determination of the unmetabolized OPs in blood or other tissues of

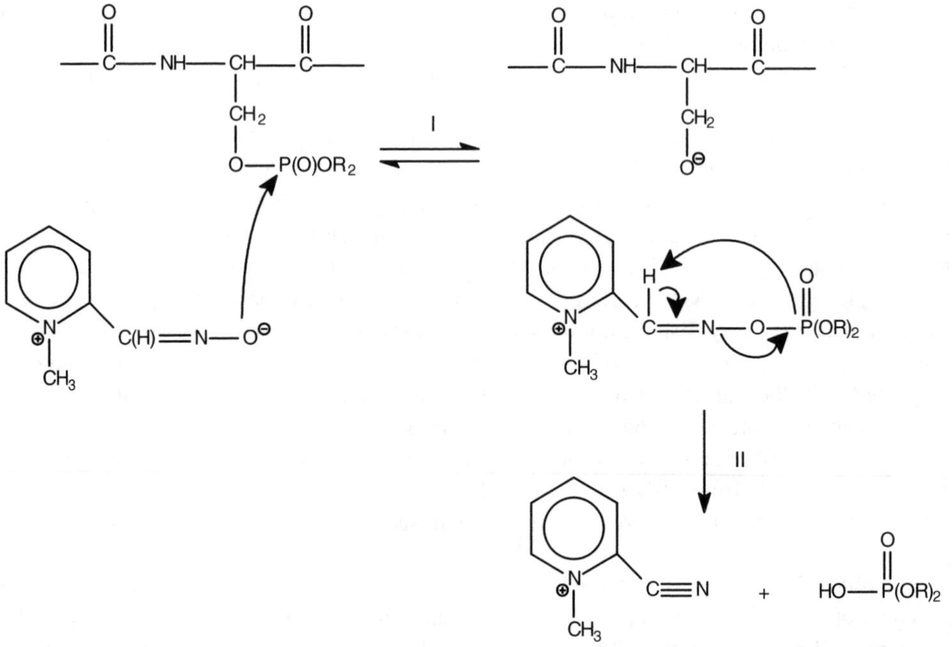

Figure 9.3 Nucleophilic attack of an oxime on the enzyme acetylcholinesterase with cleavage of nitrile and dialkyl phosphate

the contaminated person. Furthermore, the analytical detection of different metabolites of OPs in urine has often been described. In subsequent sections, these different detection methods for biological samples (blood, urine and tissues) with their advantages and disadvantages are illustrated.

Analytical methods concerning OPs do not focus exclusively on the determination or quantitation of the OPs themselves, the latter often being an integral part of toxicological screenings for drugs, pharmaceuticals and their metabolites: Pelander et al.[17] developed a toxicological screening method using 1 mL of urine sample and Turbolon spray liquid chromatography–time-of-flight mass spectrometry (LC-TOFMS) in the positive ionization mode with continuous mass measurement. The substance database consisted of exact monoisotopic masses for 637 compounds, for 392 of which an LC retention time was available. Lacassie et al.[3] presented sensitive and specific multiresidue methods for the determination of pesticides of various classes in human biological matrices. These methods involved rapid solid-phase extraction using polymeric material. For volatile pesticides such as OPs and organochlorines, GC-MS was employed, and for thermolabile and polar pesticides such as carbamates, LC-MS was used. A GC-MS method for the determination of 15 pesticides in whole blood was developed by Frenzel et al.[18] For analysis, whole blood was hemolyzed and subsequently deproteinized. After extraction of the supernatant, the pesticide concentration was determined using GC-MS. In addition to the above methods, a multitude of additional GC-MS and LC-MS methods for the determination of pesticides, especially OPs, in different matrices such as water, vegetables and fruits have been described. A discussion of these methods, however, would go beyond the scope of this chapter.

9.2.5.1 Serum and erythrocyte cholinesterase activity

In addition to the specific acetylcholinesterase that is situated in the erythrocytes, an unspecific cholinesterase called 'pseudo-cholinesterase' exists in plasma. This unspecific cholinesterase hydrolyzes the molecule ACh significantly slower than other cholinesterases, which is why the activity of this cholinesterase is of less importance for an intoxication with OPs. Nonetheless, the measurement of the esterase activity comprises both esterases, which results in misleading data.

Methods for the detection of cholinesterase activity can generally be divided into four groups, electrometric,[19–23] colorimetric,[24–28] titrimetric[29,30] and tintometric[31,32] methods, with the last being the one mainly used in the field. The colorimetric method developed by Ellman et al.[24] is the most frequently used procedure. Titrimetry is extremely accurate and precise, but rarely used because of its high cost and complexity. Electrometric methods are less sensitive than colorimetric methods and less accurate than titrimetric methods.

The colorimetric method is based on the hydrolysis of the substrate acetylthiocholine to acetate and thiocholine as performed by the cholinesterase. Thiocholine is then reacted with 5,5′-dithiobis(2-nitrobenzoic acid) (DTNB) to form a yellow anion (5-thio-2-nitrobenzoate). The latter is quantitated by spectrometric analysis at 405 nm, with the concentration being proportional to the cholinesterase activity in the given sample. Also for a few days post-mortem the cholinesterase activity in different tissues is measurable.[33]

Despite the advantages of biologically monitoring the exposure to pesticides via the degree of cholinesterase inhibition, the enormous intra- and inter-individual variations in cholinesterase activity pose a serious problem. For example, plasma cholinesterase activity may be depressed by cirrhosis, chronic hepatitis or other liver diseases and also by drug use and abuse.[34] There are no differences in cholinesterase activity associated with race in general, but plasma cholinesterase activity in North American black races has been reported to be lower than in caucasians of the same sex.[24] Any results obtained using enzyme inhibition monitoring should therefore preferably be compared with values for the innate cholinesterase activity of each subject, if possible the median of three samples obtained in a pre-exposure period.[35]

9.2.5.2 Unmetabolized OPs

In cases of poisoning, blood and/or urine or gastric contents can be tested for unmetabolized OPs to confirm exposure. In fatalities, the CNS and other tissues such as liver or kidney should be screened for the same unchanged substances. In the literature, different screening methods for OPs and analytical methods for the detection of single components have been described. A review of available analytical methods and their main characteristics is given in Table 9.3.

Screening methods

In cases where an acute intoxication with OPs is suspected, it is necessary to rapidly obtain information about the absorbed substance, which is why the application of a screening method is very suitable. So far a small number of different screening techniques using different principles of sample preparation and/or detection have been developed. Liu *et al.*[36] presented a simple and rapid method for the isolation of 11 OPs from human urine and plasma with Sep-Pak C_{18} cartridges. Detection of the pesticides was achieved by wide-bore capillary gas chromatography with flame ionization detection (GC-FID). Pesticide-containing samples were initially diluted with water before being applied to the cartridges. Elution of analytes was achieved using chloroform/2-propanol. Separation of most compounds from each other and from impurities was generally satisfactory as judged from the gas chromatograms. The recoveries were close to 100% for many compounds and none was less than 60%.

This method is a good example of the evolution in pesticide sample preparation: today, liquid–liquid extraction of aqueous samples is often replaced by solid-phase extraction (SPE), which has its commercial roots in the late 1970s. Since then, it has become a common and effective technique for extracting analytes from complex matrices. In contrast to liquid–liquid extraction, the solvent consumption in SPE is low, long evaporation times for solvent removal therefore cease to apply, interfering impurities are removed to a larger extent and, in addition, the sample is concentrated. Another advantage of SPE is the flexible application: SPE can be performed 'off-line' i.e. manually, semi-automated (e.g. using the ASPEC from Gilson), or 'on-line' (using Prospect from Spark Holland, OSP-2 from Merck, etc.), which is especially useful for large-scale operations.[1]

Despite the advantages of SPE, it has to be considered that the parallel extraction of different analytes, such as in screening procedures, can be problematic: due to the distinct physico-chemical properties of various analytes, the universal use of a single SPE matrix for all substances can result in low recoveries for individual analytes. This issue will be addressed in the context of the following examples.

Table 9.3 Analytical procedures for the determination of organophosphorous pesticides in biological samples

Analyte	Matrix	Sample preparation	Apparatus	LOD	Recovery (%)	RSD (%)	Reference
Azinphos-ethyl	Blood, urine, gastric lavage liquid	Extraction with benzene	GLC-FPD, TLC	$1 \mu g\ L^{-1}$	92–102	—	44
Chlorpyriphos	Serum, gastric content	Extraction with diethyl ether	GC-MS, SIM m/z 314	$25\ ng\ mL^{-1}$	102–104	2–3	46
	Blood	Extraction with acetone	GC-FPD	—	—	—	47
Chlorpyrifos-methyl, dichlorvos	Blood, tissues	Extraction with diethyl ether	GC-FID	—	—	—	45
Diazinon	Blood, tissues	Extraction with acetonitrile and hexane	GC-FID	—	—	—	93
Dichlorvos	Blood, tissues	Extraction with ethyl acetate	GC-FID and GC-MS	—	—	—	51
Dichlorvos	Blood	SPMEM (solid-phase microextraction membrane)	GC-MS (SIM)	—	—	—	52
Dichlorvos and malathion	Blood, tissues	Extraction with methanol and SPE purification	GC-NPD	$12\ ng\ mL^{-1}$ (malathion); $14\ ng\ mL^{-1}$ (dichlorvos)	>75	0.1–8.5	50
Dichlorvos and metrifonate	Blood	Extraction with toluene	GC-MS and GC-NPD	$0.8\ \mu mol\ L^{-1}$ (metrifonate); $50\ nmol\ L^{-1}$ (dichlorvos)	87–99	2–9	48
Disulfoton	Plasma	Extraction with hexane[a]	GC-FID	—	—	—	94
	Blood, tissues	SPE Extrelut[b]	GC-FPD and GC-MS	0.2 ng	104 ± 2.7	—	95

Table 9.3 (*Continued*)

Analyte	Matrix	Sample preparation	Apparatus	LOD	Recovery (%)	RSD (%)	Reference
Fenitrothion[c]	Blood, gastric lavage liquid	SPE Extrelut	GC-FID, GC-MS and GC-FPD	—	—	—	58
Fenitrothion	Blood, tissues	—	GC-FPD	$0.005\ \mu g\ mL^{-1}$	—	5	57
Fenthion	Blood, urine, tissues	SPE C_8	HPLC-DAD and GC-MS (EI) scan[d]	$0.25\ \mu g\ mL^{-1}$ (HPLC-DAD), $0.10\ \mu g\ mL^{-1}$ (GC-MS)	71 ± 12.5 (blood), 46 ± 8 (liver)	—	59
Malathion	Blood	HS-SPME	GC-MS (EI)	$1\ \mu g\ g^{-1}$	86.4 ± 6.3	4	53
	Blood, tissues	Extraction with diethyl ether	GC-MS (EI)	$1\ \mu g\ g^{-1}$	—	—	55
Methamidophos	Blood, tissues	Extraction with solvent	GC-NPD	$82\ ng\ mL^{-1}$	—	—	96
	Blood, brain of exposed animals	Treatment with acetonitrile and ethyl acetate	GC-MS (EI), SIM m/z 94	—	75–80	—	97
Metrifonate	Serum	SPE C_{18}	GC-flame thermionic detection	$2.5\ ng\ mL^{-1}$	99.4 ± 2.9	—	98
Oxydemeton-metyl	Blood, tissues	SPE C_{18}	LC-MS (ion trap)	$1\ ng\ mL^{-1}$	79–88	1–6	60
Parathion, paraoxon	Plasma, tissues	Extraction with isooctane (plasma), hexane (tissues)	GC-EC	—	79.4–110.3	1–9	56

Parathion-methyl	Blood, tissues	HS-SPME	GC-NPD	1 ng mL^{-1}	46–54	0.9–5.1	54
Screening							
(23 substances)	Blood, urine, serum	Extraction with 1 mL toluene	GC-NPD	—	50–133	—	39, 40
(22 substances)	Blood	HS-SPME	GC-MS (EI)	0.01–0.3 µg g^{-1}	70–95	3–31	42
(11 substances)	Plasma, urine	SPE C$_{18}$	GC-MS (EI) / GC-FID	—	>60	—	36
(11 substances)	Serum, urine	Deproteinization with acetonitrile	HPLC–DAD (230 nm)	0.05–6.8 µg mL^{-1}	97.4–101.4	0.5–6.5 (serum), 1.4–3.1 (urine)	37
(25 substances)	Serum	SPE C$_{18}$ or extraction with n-hexane–ethyl acetate	HPTLC	0.12–1.1 µg mL^{-1}	27.6–101.8	1.3–14.4	38
(10 substances)	Blood	SPE C$_{18}$	GC-NPD	—	71.8–97.1	—	41

[a]After oxidation with potassium permanganate.

[b]After extraction oxidation with potassium permanganate.

[c]Urine samples were analyzed with the same procedure for determination of 3-methyl-4-nitrophenol, aminofenitrothion, aminofenitroxon, acetylaminofenitroxon and S-methylfenitrothion.

[d]After derivatization with diazomethane.

In contrast to the above mentioned GC methods, Cho et al.[37] developed a simple and rapid method for measuring 11 OPs and one metabolite of fenitrothion in serum and urine using high-performance liquid chromatography (HPLC) with diode-array detection (230 nm). Sample preparation consisted of protein precipitation with acetonitrile. Without further cleaning, the deproteinated sample was directly injected into the HPLC system with a C_{18} column and acetonitrile/water as the mobile phase. Detection and identification of the compounds were achieved by UV spectral data analysis. Using this approach, two cases of fatal poisoning were examined: starting with the sample delivery, a period of only 2 h was needed for data acquisition and interpretation in both cases, thereby proving this method to be simple and rapid enough to be applied to biological samples in an emergency.

Futagami et al.[38] published a method for the identification of 25 commonly used OPs in human serum using high-performance thin-layer chromatography (HPTLC). The sample preparation method includes both liquid–liquid and solid-phase extraction. The OPs were separated on plates with three different developing systems within 6–18 min and detected by means of UV radiation and chromogenic reactions with 4-(4-nitrobenzyl)pyridinetetraethy-lenepentamine reagent or palladium chloride reagent. The limits of detection (LODs) of this method are within the range of those of other GC and HPLC methods (see Table 9.3). The recoveries of dichlorvos and trichlorfon using SPE were markedly lower than those using liquid–liquid extraction. As both substances are highly water soluble, these differences can be attributed to unwanted elution of the analytes during a water washing step in the SPE method.

Tarbah et al.[39] also used a single-step liquid–liquid extraction method with toluene for the determination of 23 OPs in urine, blood, serum and food samples using GC with nitrogen–phosphorus selective detection (GC-NPD) and electron ionization mass spectrometry (GC-EIMS). The recoveries for spiked human plasma ranged between 50% (demethoate) and 133% (dialifos). Akgur et al.[40] successfully measured OP concentrations in 28 cases of acute OP poisoning using the above application.

In contrast, Garcia-Repetto et al.[41] presented a GC-NPD method for the identification and quantification of 10 pesticides in human blood using SPE on C_{18} cartridges. This method, with mean recoveries between 71.8 and 97.1%, replaced an earlier method involving liquid–liquid extraction of a mixture of n-hexane and benzene.

Musshoff et al.[42] used headspace (HS) solid-phase microextraction (SPME) in combination with GC-MS for the determination of 22 OPs in human blood. The recoveries for spiked blood samples ranged between 70 and 95%. With this rapid HS-SPME method, no elaborate sample preparation is necessary. A carry-over of matrix compounds on to the GC column is effectively excluded in HS-SPME, hence the chromatograms did not show any interference (Figure 9.4). Advantages of this matrix-free chromatographic technique are a prolonged lifespan of the GC columns and liners and a low cleaning frequency of the ion source, as in HS-SPME these components are not burdened with analytically irrelevant contaminants. With the above procedure, four cases of intoxication were detected within 12 months, the OPs involved being parathion-ethyl on two occasions, malathion and bromophos-ethyl.

Recently, LC-MS techniques have become increasingly popular as an approach to analytical problems. In keeping with this development, Saint-Marcoux et al.[43] described a general unknown screening procedure for serum samples, using SPE and subsequent liquid chromatography–electrospray mass spectrometry (LC-ESIMS). This method allows for the detection and identification of pesticides and also other drugs and toxic compounds.

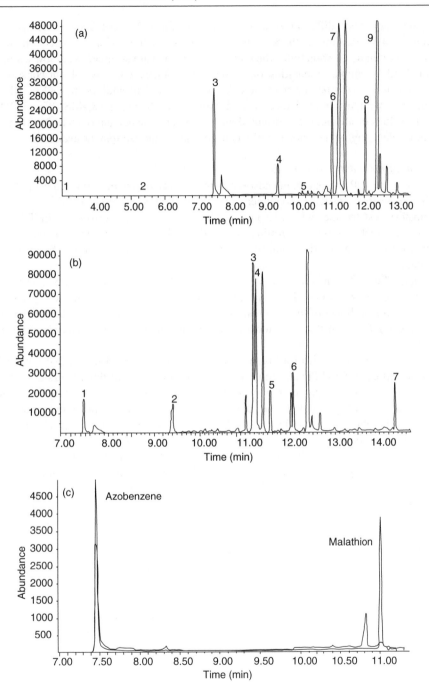

Figure 9.4   SIM chromatograms with SPME-GC-MS.[42] (a) and (b) blood samples spiked with two different mixtures of OPs. (a) 1-Dichlorvos; 2-mevinphos; 3-azobenzene; 4-diazinon; 5-parathion-methyl; 6-malathion; 7-chlorpyriphos; 8-chlorfenvinphos; 9-bromophos-ethyl. (b) 1-azobenzene; 2-disulfoton; 3-fenthion; 4-parathion-ethyl; 5-bromophos-methyl; 6-quinalphos; 7-edifenphos. (c) Blood sample from a person intoxicated with malathion

In summary, very different screening methods for OPs have been described in the literature. Although some authors prefer a conventional liquid–liquid extraction method, SPE is also often used. Sometimes the recoveries obtained are comparatively low using SPE, but this kind of sample preparation offers other advantages, e.g. low solvent consumption. Separation of analytes has been carried out with HPTLC in addition to GC or HPLC in combination with different detectors, and more recently using LC-MS. The LOD and recovery values of all methods presented are comparable, hence analytical laboratories can therefore select any of the above methods according to the equipment available.

Detection of individual compounds

In cases of acute OP poisoning, the adsorbed substance can often be identified using the label on the respective pesticide container. On these occasions, no screening but only a confirmatory test for the individual target substance has to be performed. In Table 9.3 a selection of such specific determination methods is listed. In the following section, exemplary extraction and detection methods for individual substances will be discussed in more detail.

Marques[44] detected the OP azinphos-ethyl in blood, urine and liquid from gastric lavage in 35 cases of intoxications. The concentrations determined were subsequently correlated with the degree of illness of the patients. The samples were extracted with benzene and then analyzed using GC with flame photometric detection (FPD) with a recovery of between 92 and 102%.

Chlorpyrifos is an OP with a higher LD_{50} than azinphos-ethyl (see Table 9.2) and has been detected using GC-FPD and more recently GC-MS.[45–47] All three methods (see Table 9.3) employed a liquid–liquid extraction step using different solvents (diethylether, acetone). Nolan *et al.*[47] investigated with the above approach the kinetics of chlorpyrifos and its principal metabolite in six healthy male volunteers. In contrast, Martinez *et al.*[46] reported a mild case of self-poisoning with a chlorpyrifos formulation following oral ingestion.

The determination of dichlorvos (DDVP), another OP, is often accomplished in combination with other OPs. For example, Villen *et al.*[48] developed an analytical method for the simultaneous quantification of metrifonate, an OP used in the treatment of *Schistosoma haematobium*, and dichlorvos in whole blood. Metrifonate is chemically unstable as in neutral and alkaline aqueous solutions it is transformed to dichlorvos and hydrolysis products.[49] The method employs a liquid–liquid extraction with toluene and subsequent GC-MS for detection. The within-assay coefficients of variation were 2 and 5% at 225 and 50 nmol L^{-1} (LOD), respectively. Garcia-Repetto *et al.*[50] analyzed DDVP and malathion in blood and tissues of rats using GC-NPD. Samples were homogenized, extracted with cold methanol and purified by SPE on a C_{18} column. The detection limit reached with this method was comparatively low (14 ng mL^{-1}). Using the above methodology, the authors were able to propose a novel approach to toxicokinetic modeling: rather than using only data from blood or plasma, as was usually done, this work categorized several different tissues into compartments. Shimizu *et al.*[51] used ethyl acetate as extraction solvent for the determination of DDVP in serum and tissues samples by GC-MS and GC-FID. No validation data were given. Yang and Xie[52] presented a new, simple and solventless preparation technique, with a solid-phase micro-extraction membrane (SPMEM), for the extraction of DVPP from blood. SPMEM merges sampling, extraction and concentration into a single step and combines the advantages of both the solid-phase micro-extraction (SPME) and membrane separation. For extraction the SPMEM strips were incubated in the blood sample for 1 h, and were then taken out, washed and dried with filter-paper. In the next step, the adhering compounds were extracted with ethanol and analyzed using GC-MS. The authors were able to show that the proposed method is suitable for the detection of DDVP, but they did not present any validation data.

In addition to this novel SPMEM method, two SPME methods for the detection of the OPs malathion[53] and methyl-parathion[54] have also been described. As already discussed above, Musshoff et al.[42] used this technique for the screening of 22 OPs. Tsoukali et al.[54] examined post-mortem biological samples (i.e. tissues and blood) from a woman fatally poisoned by intravenous injection of methyl-parathion. Methyl-parathion was extracted on polyacrylate fibers and analyzed using GC-NPD. The recoveries were found to be low, 46% in whole blood and 53–54% in tissues. Namera et al.[53] also applied the SPME technique when extracting malathion from blood. The SPME method implied exposing the extraction fiber in the headspace of a vial heated at 90°C for 5 min. The low background noise levels in the selected-ion monitoring (SIM) (GC-MS) chromatograms were due to the high absorption potential of the SPME fiber for the vaporized drug. The authors reported recoveries of around 86%. In contrast, Thompson et al.[55] used conventional liquid–liquid extraction for the determination of malathion in a case of fatal OP poisoning. Whole blood and homogenized tissues samples were extracted three times with 1 mL aliquots of methyl butyl ether. The analyses were performed using a GC-MS system. The LOD of 1 µg g^{-1} is similar to the value obtained using the SPME technique.

The OP parathion and its active metabolite paraoxon were simultaneously determined in plasma and tissues by Abbas and Hayton.[56] Their method involved a simple liquid–liquid extraction with isooctane with subsequent GC-electron capture detection and yielded recoveries from 79–110% for tissues and 91–100% for plasma. Fenitrothion, another OP, was detected in blood samples and tissues by Yoshida et al.[57] using a GC-FPD method. Kojima et al.[58] reported a case of attempted suicide by ingestion of a fenitrothion emulsion: fenitrothion and its metabolites were extracted from body fluids by an Extrelut column extraction method and subsequently detected by GC with either FID or FPD. Data were confirmed using GC-MS. No validation data were given.

In addition to the often used GC coupled with different detection systems for the determination of OPs, the use of HPLC has also been described. For example, Meyer et al.[59] presented an analytical method for the detection of fenthion in post-mortem samples by HPLC with diode-array detection (DAD) preceded by SPE on C$_{18}$ columns. Peak identification is accomplished via comparison with the UV spectrum of the standard substance. Comparison of this SPE method with traditional liquid–liquid extraction shows similar recoveries for both procedures (71 ± 12.5%). The LOD is lower using GC-MS (0.10 µg mL^{-1}) than HPLC–DAD (0.25 µg mL^{-1}). A completely validated LC-MS method for the determination of oxymeton-methyl (ODM) and its main metabolite demeton-S-methylsulfon (DSMS) in human blood and various tissue samples was established by Beike et al.[60] After SPE using C$_{18}$ cartridges, the extracts were analyzed by HPLC-MS (ESI ion trap, positive mode). The LOD was with 1 ng g^{-1} blood for the parent substance ODM and 2 ng g^{-1} blood for the metabolite DSMS, comparatively low. The recoveries were in the 82–88% range for ODM and 79–84% for DSMS.

In summary, for the determination of single OP compounds in blood or organic tissues, different GC methods in combination with various detection methods (MS, NPD, FPD) are used. For sample preparation, some authors prefer the traditional liquid–liquid extraction, others use SPE on conventional C$_{18}$ materials. In addition to these methods, novel techniques such as SPMEM or SPME have also been described. The latter offer the advantage of extremely reduced background noise in the acquired chromatograms. In addition to GC, analytical methods involving HPLC and LC-MS have also been described. Judging by the validation data, the LC-MS approach outperforms the other methods, which is why it is to be expected that LC-MS will gain popularity in the field of OP analysis.

9.2.5.3 Alkylphosphate metabolites

OPs are unstable in aqueous solution and even more so in blood owing to the presence of esterase. After ingestion, OPs decompose in the human body to yield alkyl phosphates. Typical degradation products are dimethyl phosphate (DMP), dimethyl phosphorothionate (DMTP), dimethyl phosphorodithioate (DMDTP), diethyl phosphate (DEP), diethyl phosphorothionate (DETP) and diethylphosphorodithioate (DEDTP), which are formed by the hydrolysis of the ester bond in the OP molecule. In Figure 9.5 the chemical structures of the respective alkylphosphates are shown.[61]

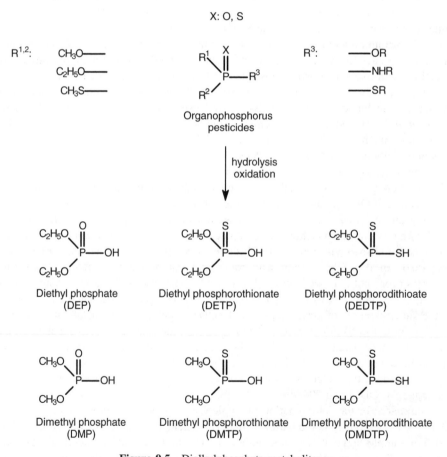

Figure 9.5 Dialkylphosphate metabolites

The most commonly used analytical methods for the quantification of the six alkyl phosphates are based on their purification from acidified samples via liquid–liquid extraction. This is followed by their conversion to volatile derivatives which can be detected using GC-MS or GC-FPD. In Table 9.4 some analytical procedures for the determination of the above substances in biological samples are listed.

In contrast to the determination of the unmetabolized OPs, the determination of their alkyl phosphate metabolites is more complex, as it includes several sample preparation steps for which mostly liquid–liquid extraction is used. Tarbah et al.[62] presented a GC-MS method following toluene extraction of the phosphamidon metabolite DMP from blood, serum, urine and gastric fluid after intoxication with the mother substance. The average recovery of DMP in plasma samples was 60% with an LOD below $0.06 \, mg \, L^{-1}$. Whyatt and Barr[63] used a GC-MS-MS method for the determination of the six alkyl phosphate metabolites in post-partum meconium as a potential biomarker of prenatal OP exposure. Following lyophilization, the meconium was extracted with methanol and after the subsequent derivatization it was analyzed using GC-MS-MS. The recoveries varied between 18 and 66% depending on the compound. The LODs with this method are in a similar range to those for other methods. Hardt and Angerer[64] also used liquid–liquid extraction to enrich the six metabolites from acidified urine into a mixture of diethyl ether and acetonitrile. After derivatization, the extracts were analyzed using GC-MS, giving higher LODs than reported by Tarbah et al.[62] and Whyatt and Barr[63] (see also Table 9.4).

Barr et al.[65] and Bravo et al.[66] used a different technique for sample preparation and determination of alkyl phosphates in urine: the urine was spiked with internal standard and then concentrated to dryness via azeotrope codistillation with acetonitrile. The residue was dissolved in acetonitrile, derivatized and then analyzed using GC-MS. Both studies reported satisfactory validation data with coefficients of variation of <20%.

Moate et al.[67] and Kupfermann et al.[68] presented analytical methods involving SPE for the extraction of alkyl phosphates from urine. Moate et al.[67] used SPE for sample cleanup followed by azeotrope distillation and then derivatization of the analytes. Determination of the analytes was achieved using GC-FPD. In contrast, Kupfermann et al.[68] purified the derivatized extract of urine samples from a case of OP poisoning on silica SPE columns followed by quantitative analysis by GC-MS. The LODs ranged from 3 to $6 \, ng \, mL^{-1}$.

For the detection of the alkyl phosphate metabolites using GC-MS, derivatization of the compounds is necessary. Most often the derivatization agents are diazoalkanes,[69,70] triazenes,[71] anilinium hydroxide[72,73] and pentafluorobenzyl bromide.[64,74–78] The last compound has the advantage of yielding a single reaction product for both DMTP and DETP, whereas the others form two isomers for each metabolite. Diazomethane derivatization is popular because of its rapid and efficient reaction with dialkyl phosphates. However, it is not a suitable reagent to measure DMP accurately: on derivatization with diazomethane, DMP and inorganic phosphate, which is endogenous in human urine, both form the same product, i.e. trimethyl phosphate.[73]

9.2.5.4 Other metabolites

In addition to the above-described alkyl phosphates, a few less common metabolites can be mentioned, e.g. 3,5,6-trichloro-2-pyridinol (TCP). TCP is a major product of esterase cleavage of the OPs chlorpyrifos and chlopyrifos-methyl.[47,79,80] Urinary TCP, like other

Table 9.4 Analytical procedures for the determination of alkyl phosphates in biological samples

Analyte	Matrix	Sample preparation	Apparatus	LOD	Recovery (%)	RSD (%)	Reference
DMP, DEP, DMTP, DETP, DEDTP, DMDTP	Urine	Extraction with solvents, evaporations, SPE	GC-MS	3–6 ng L^{-1}	—	3.8–11.2	68
DMP	Blood, urine, gastric fluid	Extraction with toluene	GC-MS	<60 µg L^{-1}	60	1.9–9.3	62
DMP, DEP, DMTP, DETP, DEDTP, DMDTP	Urine	Azeotropic codistillation with acetonitrile	GC-MS	0.58 µg L^{-1} (DMP), 0.18 µg L^{-1} (DMTP), 0.08 µg L^{-1} (DMDTP), 0.2 µg L^{-1} (DEP), 0.09 µg L^{-1} (DETP), 0.05 µg L^{-1} (DEDTP)	—	—	65
DMP, DEP, DMTP, DETP, DEDTP, DMDTP	Meconium	Extraction with methanol, derivatization	GC-MS-MS	0.51 µg L^{-1} (DMP), 0.18 µg L^{-1} (DMTP), 0.08 µg L^{-1} (DMDTP), 0.2 µg L^{-1} (DEP), 0.09 µg L^{-1} (DETP), 0.05 µg L^{-1} (DEDTP)	18–66	—	63
DMP, DEP, DMTP, DETP, DEDTP, DMDTP	Urine	Azeotropic codistillation with acetonitrile, derivatization	GC-MS-MS	1.0 mg L^{-1} (DMP), 1.6 mg L^{-1} (DMTP), 0.78 mg L^{-1} (DMDTP), 0.72 mg L^{-1} (DEP), 0.38 mg L^{-1} (DETP), 0.25 mg L^{-1} (DEDTP)	17–65	11.2–19.7	66
DMTP, DETP, DEDTP, DMDTP	Urine	Anion-exchange disk extraction	GC-FPD	9 µg L^{-1} (DMTP), 12 µg L^{-1} (DMDTP), 12 µg L^{-1} (DETP), 13 µg L^{-1} (DEDTP)	17.9–77.2	3.2–13.2	99

Analytes	Matrix	Extraction method	Detection	Detection limit	Recovery (%)	RSD (%)	Reference
DMP, DEP, DMTP, DETP, DEDTP, DMDTP	Urine	Freeze-drying and derivatization	GC-FPD	50 µg L^{-1} (DMP), 10 µg L^{-1} (DMTP), 10 µg L^{-1} (DMDTP), 10 µg L^{-1} (DEP), 5 µg L^{-1} (DETP), 5 µg L^{-1} (DEDTP)	—	6.1–23.0	100
DMP, DEP, DMTP, DETP, DEDTP, DMDTP	Urine	Extraction with acetonitrile	GC-FPD	2 µg L^{-1} (DMTP, DMDTP, DETP, DEDTP), 3 µg L^{-1} (DMP, DEP)	85.8–101.0	1.9–11.4	74, 101
DMP, DEP, DMTP, DETP, DEDTP, DMDTP	Urine	Extraction with diethyl ether and acetonitrile	GC-MS	1 µg L^{-1} (DMTP, DMDTP, DETP, DEDTP, DEP), 5 µg L^{-1} (DMP)	68–114	9.7–15.5	64
DMP, DEP, DMTP, DETP, DMDTP	Urine	SPE	GC-FPD	10 µg L^{-1} (DMP, DEP) 2 µg L^{-1} (DMTP, DMDTP, DETP)	57–118	4.4–46.4	67

OP metabolites, may be used as an indicator of exposure to chlorpyrifos and chlorpyrifos-methyl, although the available data are still insufficient to define biological exposure limits. For the detection of TCP in urine, predominantly reversed-phase HPLC with UV detection and GC are used. In Table 9.5, exemplary analytical methods are listed. For example, Bartels and Kastl[81] presented a sensitive GC-negative ion chemical ionization mass spectrometric (GC-NCI-MS) method for measuring levels of TCP in human urine. The metabolite was isolated from urine by acid hydrolysis of urine aliquots followed by diethyl ether extraction. The LOD was very low ($0.5 \, \mu g \, mL^{-1}$) and the recoveries varied between 80.6 and 89.9%. Subsequently, Hill et al.[82] presented a method for measuring 12 analytes, pesticides or their metabolites, in urine. The sample preparation involved enzyme hydrolysis and solvent extraction through the use of laboratory robotics, followed by phase-transfer catalysis derivatization and silica clean-up. Samples were analyzed by GC-MS and the LODs were $1 \, \mu g \, mL^{-1}$ for most analytes. Aprea et al.[83] also used GC with different detection methods (MS or ECD) for the determination of urinary TCP in the general Italian population. The samples were hydrolyzed using hot acid, extracted with hexane and then derivatized. In contrast, Chang et al.[84] used a much simplified sample preparation scheme with subsequent HPLC analysis for the determination of TCP in urine. The detection limit was not as low as that obtained with GC-MS[81] and GC-ECD.[47] It was, however, about 5–20 times more sensitive than the HPLC study reported by Sultatos et al.[85], who reported an HPLC method with a sample preparation involving an ethyl acetate extraction and evaporation step.

Another product of esterase cleavage is *p*-nitrophenol (PNP), which is generated in the human body by the degradation of parathion, parathion-methyl and parathion-ethyl.[86] Some methods for the determination of PNP in urine are summarized in Table 9.5. PNP can, for example,[87] be determined in the urine of occupationally exposed subjects by means of GC-ECD. Sample preparation involves acid hydrolysis, extraction with diethyl ether, derivatization with diazoethane and purification on silica gel columns. The LOD was $20 \, \mu g \, L^{-1}$ with a recovery of 85–98%. This method has also been used in a study on selected pesticide residues and metabolites in urine from a survey of the US general population.[88] In the context of another survey study, a more complicated technique to detect PNP in urine has recently been used by Hill et al.[89] involving GC-MS-MS with positive chemical ionization after derivatization with 1-chloro-3-iodopropane. The authors also used this method for the determination of the metabolite TCP and 10 other analytes in urine.[82] Sample preparation involved hydrolysis with β-glucuronidase, several extraction steps using different solvents and purification by SPE (silica gel column). The LOD was $1 \, \mu g \, L^{-1}$ with an inter-assay RSD of 24%.[90] Barr et al.[86] developed a rapid, high-throughput, selective method for quantifying PNP using LC-MS-MS. Sample preparation is less complicated than in the other methods: after hydrolysis with β-glucuronidase/sulfatase the hydrolyzate was acidified, extracted with dichloromethane and then analyzed. The method LOD was calculated to be $25 \, \mu g \, L^{-1}$. Although this LOD is higher than that of previously published methods (see above), the method was suitable to categorize samples into the various priority levels required for public health intervention.

The ester cleavage of fenitrothion leads to the formation of the metabolite 3-methyl-4-nitrophenol (MNP). Shafik et al.[87] used GC-ECD for the detection of MNP in occupationally exposed subjects. Sample preparation involved extraction with diethyl ether, derivatization with diazoethane and purification on silica gel columns. The LOD was determined as $50 \, \mu g \, L^{-1}$ with a recovery of 88–98%.

Table 9.5 Analytical procedures for the determination of different metabolites of organophosphorus pesticides in urine

Analyte	Hydolysis/sample preparation	Apparatus	LOD	Recovery (%)	RSD (%)	Reference
TCP	Hot HCl and extraction with diethyl ether	GC-NCI-MS	0.5 µg L⁻¹	80.6–89.9	0.4–2.6	81
	Hot HCl and extraction with toluene	GC-MS, GC-ECD	MS: 1.5 µg L⁻¹ ECD: 1.2 µg L⁻¹	— ECD: 91.7	MS: 8.2 ECD: 8.7	83
	Hot HCl and extraction with toluene	GC-ECD	5.0 µg L⁻¹	71.8	7	102
	Hot H$_2$SO$_4$ and SPE C$_{18}$	GC-ECD	10 µg L⁻¹	—	—	47
	β-Glucoronidase and extraction with 1-chlorobutane–diethyl ether and 1-chloro-3-iodopropane	GC-MS-MS	1 µg L⁻¹	—	24	82
	Hot acid hydrolysis and perchloric acid	HPLC-UV	2.2 ng per 20-µL injection	102 ± 15	<10	84
TCP in plasma and tissues	Extraction with ethyl acetate	HPLC-UV (290 nm)	—	85 ± 8	—	85

(continued)

Table 9.5 *(Continued)*

Analyte	Hydolysis/sample preparation	Apparatus	LOD	Recovery (%)	RSD (%)	Reference
PNP	Hot acid hydrolysis and extraction with diethyl ether	GC-ECD	20 µg L^{-1}	85–98	—	87
	β-Glucoronidase and extraction with 1-chlorobutane–diethyl ether and 1-chloro-3-iodopropane	GC-MS-MS	1 µg L^{-1}	—	—	89
	β-Glucoronidase/sulfatase, addition of HCl and extraction with dichloromethane	LC-MS-MS	25 µg L^{-1}	85 ± 8	—	86
MNP	Hot acid hydrolysis and extraction with diethyl ether	GC-ECD	50 µg L^{-1}	88–98	—	87
MCA and DCA	Extraction with diethyl ether, SPE	GC-FPD	MCA: 5 µg L^{-1} DCA: 20 µg L^{-1}	MCA: 98–104 DCA: 98–101	—	91
	Alkaline hydrolysis	GC-FPD	—	—	—	103

The respective mono- and dicarboxylic acid [malathion α-monocarboxylic acid (MCA) and malathion dicarboxylic acid (DCA)] are the main metabolites of the OP malathion generated by the hydrolysis of one or both of the side-chain diethylsuccinic esters.[35,91] Bradway et al.[91] presented a GC-FPD method for the determination of both substances in spiked urine and also in urine from rats exposed to malathion at five levels (see also Table 9.5). The clean-up procedure was modified based on a method developed by Shafik et al.[92] and involves extraction of acidified urine with diethyl ether, derivatization with diazomethane and purification on a silica gel column. Another method used by some authors involves the conversion of MCA and DCA into the respective alkyl phosphate metabolites DMTP and DMDTP (see Section 9.2.5.3) by alkaline hydrolysis. These compounds can then be derivatized with pentafluorobenzyl bromide and analyzed using GC-FPD.

9.3 CONCLUSION

In this chapter a short introduction to organophosphorus pesticides (OPs) and their mechanism of action has been given. Symptoms of intoxication in humans following exposure have been discussed along with the signs of progressing intoxication, long-term risks of OP poisoning and possible therapy regimes. Furthermore, a wide range of studies concerned with the analytical aspects of biological monitoring of OP exposure were reviewed.

Pesticides can be subdivided into three groups based on their mechanism of action. In addition to pyrethroid and organochlorine compounds, the OPs represent a large group within this substance class. OPs act through inhibition of the enzyme acetylcholinesterase in the CNS that is essential for the hydrolysis of the neurotransmitter acetylcholine. Symptoms of intoxication appear a few minutes after inhalation or swallowing and 2–3 h after cutaneous resorption of the respective OP. The more severe symptoms of OP intoxication include coma, tremor, muscle cramps and others. In cases of intoxication, the first measure is to stabilize respiration and circulation. In the next step, treatment with oximes provides the opportunity to reactivate the acetylcholinesterase, thereby alleviating the cholinergic symptoms.

For the detection of OPs, e.g. in the context of a medical diagnosis, many different analytical methods have been reported. One possibility is to measure the cholinesterase activity as an acute intoxication characterized by a 20% decrease in acetylcholinesterase activity. Furthermore, analytical methods involving OPs do not exclusively focus on the determination of the OPs themselves; the latter are rather an integral part of several toxicological screening procedures.

In cases of poisoning, blood and/or urine or tissue samples can be tested for unmetabolized OPs. A review of available analytical methods and their main characteristics is given in Table 9.3. For the screening methods and for the quantification of single compounds, often GC techniques in combination with various detection methods are used. Furthermore, several HPLC methods have been described. For sample preparation, some authors prefer the traditional liquid–liquid solvent extraction, whereas others use SPE on conventional materials. In addition to these, novel techniques such as SPME or SPMEM are also used.

In addition to the determination of the unmetabolized compounds to confirm exposure, the quantification of their metabolites, e.g. alkyl phosphate metabolites or other characteristic metabolites, is a possibility. The most commonly used analytical methods for the quantification of the six alkyl phosphates are based on their purification from acidified samples via liquid–liquid extraction. This is followed by their conversion to volatile derivatives which can be detected using GC-MS or GC-FPD (see Table 9.4). A review of analytical methods for the determination of other metabolites, e.g. 3,5,6-trichloro-2-pyridinol (TCP), the major product of ester cleavage of chlorpyrifos, is given in Table 9.5. These metabolites can be quantified with GC but also using HPLC or HPLC-MS   methods.

REFERENCES

1. Pico Y, Blasco C, Font G. Environmental and food applications of LC-tandem mass spectrometry in pesticide-residue analysis: an overview. *Mass Spectrom Rev* 2004;**23**:45–85.
2. Schmoldt A. Schädlingsbekämpfungsmittel gegen tierische Schädlinge. In Madea B, Brinkmann B, eds. *Handbuch Gerichtliche Medizin*. Berlin: Springer, 2003, pp. 357–68.
3. Lacassie E, Marquet P, Gaulier JM, Dreyfuss MF, Lachatre G. Sensitive and specific multiresidue methods for the determination of pesticides of various classes in clinical and forensic toxicology. *Forensic Sci Int* 2001;**121**:116–25.
4. *Wirkstoffe in Pflanzenschutz- und Schädlingsbekämpfungsmitteln: Physikalische-chemische und Toxikologische Daten.* BLV Verlags-Gesellschaft/Industrieverband Agrar, 1990.
5. Mutschler E. *Arzneimittelwirkungen, Lehrbuch der Pharmakologie und Toxikologie*, 7th edn. Stuttgart: Wissenschaftliche Verlagsgesellschaft, 1996, pp. 297–9.
6. Marquardt H, Schäfer SG (Eds). *Lehrbuch der Toxikologie*. Heidelberg: Spektrum Akademischer Verlag, 1997, pp. 460–5.
7. Forth W, Henschler D, Rummel W, Starke W (Eds). *Allgemeine und Spezielle Pharmakologie und Toxikologie*. Heidelberg: Spektrum Akademischer Verlag, 1996, pp. 858–61.
8. Kupfermann N. Instrumentelle Metabolitenanalytik von insektiziden Organophosphaten in biologischem Material. *Thesis*, Universität Hamburg, Fachbereich Chemie, 2003.
9. Senanayake N, Karalliedde L. Neurotoxic effects of organophosphorus insecticides. An intermediate syndrome. *N Engl J Med* 1987;**316**:761–3.
10. De BJ, Willems J, Van Den NK, De RJ, Vogelaers D. Prolonged toxicity with intermediate syndrome after combined parathion and methyl parathion poisoning. *J Toxicol Clin Toxicol* 1992;**30**:333–45.
11. De BJ, Van Den NK, Willems J. The intermediate syndrome in organophosphate poisoning: presentation of a case and review of the literature. *J Toxicol Clin Toxicol* 1992;**30**:321–9.
12. Abou-Donia MB, Lapadula DM. Mechanisms of organophosphorus ester-induced delayed neurotoxicity: type I and type II. *Annu Rev Pharmacol Toxicol* 1990;**30**:405–40.
13. Lotti M, Becker CE, Aminoff MJ. Organophosphate polyneuropathy: pathogenesis and prevention. *Neurology* 1984;**34**:658–62.
14. Schulz M, Schmoldt A. Therapeutic and toxic blood concentrations of more than 800 drugs and other xenobiotics. *Pharmazie* 2003;**58**:447–74.
15. Flanagan RJ. Guidelines for the interpretation of analytical toxicology results and unit of measurement conversion factors. *Ann Clin Biochem* 1998;**35**:261–7.
16. Stark I. Insektizide und Nervengase: Vergiftung und Therapie. *Chem Unsere Zeit* 1984; **18**(3): 96–106.
17. Pelander A, Ojanpera I, Laks S, Rasanen I, Vuori E. Toxicological screening with formula-based metabolite identification by liquid chromatography/time-of-flight mass spectrometry. *Anal Chem* 2003;**75**:5710–8.
18. Frenzel T, Sochor H, Speer K, Uihlein M. Rapid multimethod for verification and determination of toxic pesticides in whole blood by means of capillary GC-MS. *J Anal Toxicol* 2000;**24**:365–71.

19. Tammelin LE. An electrometric method for the determination of cholinesterase activity. I. Apparatus and cholinesterase in human blood. *Scand J Clin Lab Invest* 1953;**5**:267–70.

20. Diamant H, Tammelin LE. An electrometric method for the determination of cholinesterase activity. II. Cholinesterase in brain tissue. *Scand J Clin Lab Invest* 1953;**5**:271–2.

21. Heilbronn E. An electrometric method for the determination of cholinesterase activity. III. Cholinesterase activity in blood spots on filter paper. *Scand J Clin Lab Invest* 1953;**5**:308–11.

22. Mohammad FK, St Omer VE. Modifications of Michel's electrometric method for rapid measurement of blood cholinesterase activity in animals: a minireview. *Vet Hum Toxicol* 1982;**24**:119–21.

23. Mohammad FK, Faris GA, al-Kassim NA. A modified electrometric method for measurement of erythrocyte acetylcholinesterase activity in sheep. *Vet Hum Toxicol* 1997;**39**:337–9.

24. Ellman GL, Courtney KD, Andres V, Jr, Featherstone RM. A new and rapid colorimetric determination of acetylcholinesterase activity. *Biochem Pharmacol* 1961;**7**:88–95.

25. Worek F, Mast U, Kiderlen D, Diepold C, Eyer P. Improved determination of acetylcholinesterase activity in human whole blood. *Clin Chim Acta* 1999;**288**:73–90.

26. Lewis PJ, Lowing RK, Gompertz D. Automated discrete kinetic method for erythrocyte acetylcholinesterase and plasma cholinesterase. *Clin Chem* 1981;**27**:926–9.

27. Mason HJ, Lewis PJ. Intra-individual variation in plasma and erythrocyte cholinesterase activities and the monitoring of uptake of organo-phosphate pesticides. *J Soc Occup Med* 1989;**39**:121–4.

28. Kolf-Clauw M, Jez S, Ponsart C, Delamanche IS. Acetyl- and pseudo-cholinesterase activities of plasma, erythrocytes, and whole blood in male beagle dogs using Ellman's assay. *Vet Hum Toxicol* 2000;**42**:216–9.

29. Augustinsson KB. A titrimetric method for the determination of plasma and red blood cell cholinesterase activity using the thiocholine esters as substrates. *Scand J Clin Lab Invest* 1955;**7**:284–90.

30. Coye MJ, Lowe JA, Maddy KT. Biological monitoring of agricultural workers exposed to pesticides: I. Cholinesterase activity determinations. *J Occup Med* 1986;**28**:619–27.

31. Jeyaratnam J, Lun KC, Phoon WO. Blood cholinesterase levels among agricultural workers in four Asian countries. *Toxicol Lett* 1986;**33**:195–201.

32. McConnell R, Magnotti R. Screening for insecticide overexposure under field conditions: a reevaluation of the tintometric cholinesterase kit. *Am J Public Health* 1994;**84**:479–81.

33. Oehmichen M, Besserer K. Forensic significance of acetylcholine esterase histochemistry in organophosphate intoxication. Original investigations and review of the literature. *Z Rechtsmed* 1982;**89**:149–65.

34. Carmona GN, Baum I, Schindler CW, Goldberg SR, Jufer R, Cone E, *et al.* Plasma butyrylcholinesterase activity and cocaine half-life differ significantly in rhesus and squirrel monkeys. *Life Sci* 1996;**59**:939–43.

35. Aprea C, Colosio C, Mammone T, Minoia C, Maroni M. Biological monitoring of pesticide exposure: a review of analytical methods. *J Chromatogr B* 2002;**769**:191–219.

36. Liu J, Suzuki O, Kumazawa T, Seno H. Rapid isolation with Sep-Pak C18 cartridges and wide-bore capillary gas chromatography of organophosphate pesticides. *Forensic Sci Int* 1989;**41**:67–72.

37. Cho Y, Matsuoka N, Kamiya A. Determination of organophosphorus pesticides in biological samples of acute poisoning by HPLC with diode-array detector. *Chem Pharm Bull (Tokyo)* 1997;**45**:737–40.

38. Futagami K, Narazaki C, Kataoka Y, Shuto H, Oishi R. Application of high-performance thin-layer chromatography for the detection of organophosphorus insecticides in human serum after acute poisoning. *J Chromatogr B* 1997;**704**:369–73.

39. Tarbah FA, Mahler H, Temme O, Daldrup T. An analytical method for the rapid screening of organophosphate pesticides in human biological samples and foodstuffs. *Forensic Sci Int* 2001;**121**:126–33.

40. Akgur SA, Ozturk P, Solak I, Moral AR, Ege B. Human serum paraoxonase (PON1) activity in acute organophosphorus insecticide poisoning. *Forensic Sci Int* 2003;**133**:136–40.

41. Garcia-Repetto R, Gimenez MP, Repetto M. New method for determination of ten pesticides in human blood. *J AOAC Int* 2001;**84**:342–9.

42. Musshoff F, Junker H, Madea B. Simple determination of 22 organophosphorus pesticides in human blood using headspace solid-phase microextraction and gas chromatography with mass spectrometric detection. *J Chromatogr Sci* 2002;**40**:29–34.

43. Saint-Marcoux F, Lachatre G, Marquet P. Evaluation of an improved general unknown screening procedure using liquid chromatography–electrospray-mass spectrometry by comparison with gas chromatography and high-performance liquid chromatography–diode array detection. *J Am Soc Mass Spectrom* 2003;**14**:14–22.

44. Marques EG. Acute intoxication by azinphos-ethyl. *J Anal Toxicol* 1990;**14**:243–6.

45. Moriya F, Hashimoto Y, Kuo TL. Pitfalls when determining tissue distributions of organophosphorus chemicals: sodium fluoride accelerates chemical degradation. *J Anal Toxicol* 1999;**23**: 210–5.

46. Martinez MA, Ballesteros S, Sanchez de la Torre C, Sanchiz A, Almarza E, Garcia-Aguilera A. Attempted suicide by ingestion of chlorpyrifos: identification in serum and gastric content by GC-FID/GC-MS. *J Anal Toxicol* 2004;**28**:609–15.

47. Nolan RJ, Rick DL, Freshour NL, Saunders JH. Chlorpyrifos: pharmacokinetics in human volunteers. *Toxicol Appl Pharmacol* 1984;**73**:8–15.

48. Villen T, Abdi YA, Ericsson O, Gustafsson LL, Sjoqvist F. Determination of metrifonate and dichlorvos in whole blood using gas chromatography and gas chromatography–mass spectrometry. *J Chromatogr* 1990;**529**:309–17.

49. Reiner E, Krauthacker B, Simeon V, Skrinjaric-Spoljar M. Mechanism of inhibition *in vitro* of mammalian acetylcholinesterase and cholinesterase in solutions of *O,O*-dimethyl 2,2,2-trichloro-1-hydroxyethyl phosphonate (Trichlorphon). *Biochem Pharmacol* 1975;**24**:717–22.

50. Garcia-Repetto R, Martinez D, Repetto M. Malathion and dichlorvos toxicokinetics after the oral administration of malathion and trichlorfon. *Vet Hum Toxicol* 1995;**37**:306–9.

51. Shimizu K, Shiono H, Fukushima T, Sasaki M, Akutsu H, Sakata M. Tissue distribution of DDVP after fatal ingestion. *Forensic Sci Int* 1996;**83**:61–6.

52. Yang R, Xie W. Preparation and usage of a new solid phase micro-extraction membrane. *Forensic Sci Int* 2004;**139**:177–81.

53. Namera A, Yashiki M, Nagasawa N, Iwasaki Y, Kojima T. Rapid analysis of malathion in blood using head space–solid phase microextraction and selected ion monitoring. *Forensic Sci Int* 1997;**88**:125–31.

54. Tsoukali H, Raikos N, Theodoridis G, Psaroulis D. Headspace solid phase microextraction for the gas chromatographic analysis of methyl-parathion in post-mortem human samples. Application in a suicide case by intravenous injection. *Forensic Sci Int* 2004;**143**:127–32.

55. Thompson TS, Treble RG, Magliocco A, Roettger JR, Eichhorst JC. Case study: fatal poisoning by malathion. *Forensic Sci Int* 1998;**95**:89–98.

56. Abbas R, Hayton WL. Gas chromatographic determination of parathion and paraoxon in fish plasma and tissues. *J Anal Toxicol* 1996;**20**:151–4.

57. Yoshida M, Shimada E, Yamanaka S, Aoyama H, Yamamura Y, Owada S. A case of acute poisoning with fenitrothion (Sumithion). *Hum Toxicol* 1987;**6**:403–6.

58. Kojima T, Yashiki M, Miyazaki T, Chikasue F, Ohtani M. Detection of *S*-methylfenitrothion, aminofenitrothion, aminofenitroxon and acetylaminofenitroxon in the urine of a fenitrothion intoxication case. *Forensic Sci Int* 1989;**41**:245–53.

59. Meyer E, Borrey D, Lambert W, Van PC, Piette M, De LA. Analysis of fenthion in postmortem samples by HPLC with diode-array detection and GC-MS using solid-phase extraction. *J Anal Toxicol* 1998;**22**:248–52.

60. Beike J, Ortmann C, Meiners T, Brinkmann B, Kohler H. LC-MS determination of oxydemeton-methyl and its main metabolite demeton-S-methylsulfon in biological specimens – application to a forensic case. *J Anal Toxicol* 2002;**26**:308–12.

61. Kupfermann, N. Instrumentelle Metabolitenanalytik von insektiziden Organophosphaten in biologischem Material. *Thesis, Universität Hamburg,* 2003.

62. Tarbah FA, Kardel B, Pier S, Temme O, Daldrup T. Acute poisoning with phosphamidon: determination of dimethyl phosphate (DMP) as a stable metabolite in a case of organophosphate insecticide intoxication. *J Anal Toxicol* 2004;**28**:198–203.

63. Whyatt RM, Barr DB. Measurement of organophosphate metabolites in postpartum meconium as a potential biomarker of prenatal exposure: a validation study. *Environ Health Perspect* 2001;**109**: 417–20.

64. Hardt J, Angerer J. Determination of dialkyl phosphates in human urine using gas chromatography–mass spectrometry. *J Anal Toxicol* 2000;**24**:678–84.

65. Barr DB, Bravo R, Weerasekera G, Caltabiano LM, Whitehead RD, Jr, Olsson AO, *et al.* Concentrations of dialkyl phosphate metabolites of organophosphorus pesticides in the U.S. population. *Environ Health Perspect* 2004;**112**:186–200.

66. Bravo R, Driskell WJ, Whitehead RD, Jr, Needham LL, Barr DB. Quantitation of dialkyl phosphate metabolites of organophosphate pesticides in human urine using GC-MS-MS with isotopic internal standards. *J Anal Toxicol* 2002;**26**:245–52.

67. Moate TF, Lu C, Fenske RA, Hahne RM, Kalman DA. Improved cleanup and determination of dialkyl phosphates in the urine of children exposed to organophosphorus insecticides. *J Anal Toxicol* 1999;**23**:230–6.

68. Kupfermann N, Schmoldt A, Steinhart H. Rapid and sensitive quantitative analysis of alkyl phosphates in urine after organophosphate poisoning. *J Anal Toxicol* 2004;**28**:242–8.

69. Shafik T, Bradway DE, Enos HF, Yobs AR. Gas–liquid chromatographic analysis of alkyl phosphate metabolites in urine. *J Agric Food Chem* 1973;**21**:625–9.

70. Lores EM, Bradway DE. Extraction and recovery of organophosphorus metabolites from urine using an anion exchange resin. *J Agric Food Chem* 1976;**25**:75–9.

71. Takade DY, Reynolds JM, Nelson JH. 1-(4-Nitrobenzyl)-3-(4-tolyl)triazene as a derivatizing reagent for the analysis of urinary dialkylphosphate metabolites of organophosphorus pesticides by gas chromatography. *J Agric Food Chem* 2005;**27**:746–53.

72. Moody RP, Franklin CA, Riedel D, Muir NI, Greenhalgh R, Hladka A. A new GC-on-column methylation procedure for analysis of DTMP (*O,O*-dimethyl phosphorothioate) in urine of workers exposed to fenitrothion. *J Agric Food Chem* 1985;**33**:464–7.

73. Weisskopf CP, Seiber JN. New approaches to analysis of organophosphate metabolites in the urine of fields workers. *ACS Symp Ser* 1989;**382**:206–14.

74. Aprea C, Sciarra G, Lunghini L. Analytical method for the determination of urinary alkylphosphates in subjects occupationally exposed to organophosphorus pesticides and in the general population. *J Anal Toxicol* 1996;**20**:559–63.

75. Reid SJ, Watts RR. A method for the determination of dialkyl phosphate residues in urine. *J Anal Toxicol* 1981;**5**:126–32.

76. Bradway DE, Moseman R, May R. Analysis of alkyl phosphates by extractive alkylation. *Bull Environ Contam Toxicol* 1981;**26**:520–3.

77. Fenske RA, Leffingwell JT. Method for determination of dialkyl phosphate metabolites in urine for studies of human exposure to malathion. *J Agric Food Chem* 1985;**37**:995–8.

78. Loewenherz C, Fenske RA, Simcox NJ, Bellamy G, Kalman D. Biological monitoring of organophosphorus pesticide exposure among children of agricultural workers in central Washington State. *Environ Health Perspect* 1997;**105**:1344–53.

79. Richardson RJ. Assessment of the neurotoxic potential of chlorpyrifos relative to other organophosphorus compounds: a critical review of the literature. *J Toxicol Environ Health* 1995;**44**:135–65.

80. Sultatos LG, Shao M, Murphy SD. The role of hepatic biotransformation in mediating the acute toxicity of the phosphorothionate insecticide chlorpyrifos. *Toxicol Appl Pharmacol* 1984;**73**:60–8.

81. Bartels MJ, Kastl PE. Analysis of 3,5,6-trichloropyridinol in human urine using negative-ion chemical ionization gas chromatography–mass spectrometry. *J Chromatogr* 1992;**575**:69–74.

82. Hill RH, Jr, Shealy DB, Head SL, Williams CC, Bailey SL, Gregg M, *et al.* Determination of pesticide metabolites in human urine using an isotope dilution technique and tandem mass spectrometry. *J Anal Toxicol* 1995;**19**:323–9.

83. Aprea C, Betta A, Catenacci G, Lotti A, Magnaghi S, Barisano A, *et al.* Reference values of urinary 3,5,6-trichloro-2-pyridinol in the Italian population – validation of analytical method and preliminary results (multicentric study). *J AOAC Int* 1999;**82**:305–12.

84. Chang MJ, Lin CY, Lo LW, Lin RS. Biological monitoring of exposure to chlorpyrifos by high performance liquid chromatography. *Bull Environ Contam Toxicol* 1996;**56**:367–74.

85. Sultatos LG, Costa LG, Murphy SD. Determination of organophosphorus insecticides, their oxygen analogs and metabolites by high pressure liquid chromatography. *Chromatographia* 1982;**15**:669–71.

86. Barr DB, Turner WE, DiPietro E, McClure PC, Baker SE, Barr JR, *et al.* Measurement of *p*-nitrophenol in the urine of residents whose homes were contaminated with methyl parathion. *Environ Health Perspect* 2002;**110**(Suppl 6):1085–91.

87. Shafik TM, Sullivan HC, Enos HR. Multiresidue procedure for halo- and nitrophenols. Measurement of exposure to biodegradable pesticides yielding these compounds as metabolites. *J Agric Food Chem* 1973;**21**:295–8.

88. Kutz FW, Cook BT, Carter-Pokras OD, Brody D, Murphy RS. Selected pesticide residues and metabolites in urine from a survey of the U.S. general population. *J Toxicol Environ Health* 1992;**37**:277–91.

89. Hill RH, Jr, Head SL, Baker S, Gregg M, Shealy DB, Bailey SL, *et al.* Pesticide residues in urine of adults living in the United States: reference range concentrations. *Environ Res* 1995;**71**: 99–108.

90. Hill RH, Jr, Shealy DB, Head SL, Williams CC, Bailey SL, Gregg M, *et al.* Determination of pesticide metabolites in human urine using an isotope dilution technique and tandem mass spectrometry. *J Anal Toxicol* 1995;**19**:323–9.

91. Bradway DE, Shafik TM. Malathion exposure studies. Determination of mono- and dicarboxylic acids and alkyl phosphates in urine. *J Agric Food Chem* 1977;**25**:1342–4.

92. Shafik MT, Bradway D, Enos HF. A cleanup procedure for the determination of low levels of alkyl phosphates, thiophosphates, and dithiophosphates in rat and human urine. *J Agric Food Chem* 1971;**19**:885–9.

93. Poklis A, Kutz FW, Sperling JF, Morgan DP. A fatal diazinon poisoning. *Forensic Sci Int* 1980;**15**:135–40.

94. Futagami K, Otsubo K, Nakao Y, Aoyama T, Iimori E, Urakami S, *et al.* Acute organophosphate poisoning after disulfoton ingestion. *J Toxicol Clin Toxicol* 1995;**33**:151–5.

95. Yashiki M, Kojima T, Ohtani M, Chikasue F, Miyazaki T. Determination of disulfoton and its metabolites in the body fluids of a Di-Syston intoxication case. *Forensic Sci Int* 1990;**48**: 145–54.

96. Akgur S, Ozturk P, Yemiscigil A, Ege B. Rapid communication: postmortem distribution of organophosphate insecticides in human autopsy tissues following suicide. *J Toxicol Environ Health A* 2003;**66**:2187–91.

97. Singh AK. Improved analysis of acephate and methamidophos in biological samples by selective ion monitoring gas chromatography–mass spectrometry. *J Chromatogr* 1984;**301**:465–9.

98. Ameno K, Fuke C, Ameno S, Kiriu T, Ijiri I. A rapid and sensitive quantitation of Dipterex in serum by solid-phase extraction and gas chromatography with flame thermionic detection. *J Anal Toxicol* 1989;**13**:150–1.

99. Lin WC, Kuei CH, Wu HC, Yang CC, Chang HY. Method for the determination of dialkyl phosphates in urine by strong anion exchange disk extraction and in-vial derivatization. *J Anal Toxicol* 2002;**26**:176–80.

100. Oglobline AN, O'Donnell GE, Geyer R, Holder GM, Tattam B. Routine gas chromatographic determination of dialkylphosphate metabolites in the urine of workers occupationally exposed to organophosphorus insecticides. *J Anal Toxicol* 2001;**25**:153–7.

101. Mills PK, Zahm SH. Organophosphate pesticide residues in urine of farmworkers and their children in Fresno County, California. *Am J Ind Med* 2001;**40**:571–7.

102. Fenske RA, Elkner KP. Multi-route exposure assessment and biological monitoring of urban pesticide applicators during structural control treatments with chlorpyrifos. *Toxicol Ind Health* 1990;**6**:349–71.

103. Fenske RA. Correlation of fluorescent tracer measurements of dermal exposure and urinary metabolite excretion during occupational exposure to malathion. *Am Ind Hyg Assoc J* 1988;**49**:438–44.

10

Chromatographic Analysis of Nerve Agents

Jeri D. Ropero-Miller

Research Triangle Institute, Center for Forensic Sciences, 3040 Cornwallis Road, P.O. Box 12194, Research Triangle Park, NC 27709-2194, USA

10.1 INTRODUCTION

The word 'paralytic' is medically defined as 'relating to paralysis' or 'a person affected with paralysis'.[1] The analyses of paralytics in the clinical laboratory include therapeutic drug monitoring (TDM), diagnosis of overdose and treatment of medical emergencies following exposure. Various chromatographic techniques are employed by the clinical laboratory for the routine analysis of paralytics whereas other techniques such as tandem mass spectrometry are reserved for clinical research. The different chromatographic methods utilized are dependent on the agent and include gas chromatography (GC), high-performance liquid chromatography (HPLC) and liquid chromatography–mass spectrometry (LC-MS and LC-MS-MS). This chapter focuses on the chromatography and detection of therapeutic and natural paralytic agents including neuromuscular blocking agents and saxitoxin. While botulinum neurotoxins have recently been analyzed for clinical purposes, chromatographic techniques are exclusively used to separate and detect sequence variations in genes and, consequently, relevance to this chapter is limited and will not be discussed further.[2]

10.2 NEUROMUSCULAR BLOCKERS

10.2.1 Background and uses

The use of neuromuscular blocking agents (NMBAs) dates back as early as the 1500s, when South American Indians created deadly arrows, by dipping them into various curare poisons,

Chromatographic Methods in Clinical Chemistry and Toxicology   Edited by R. L. Bertholf and R. E. Winecker
© 2007 John Wiley & Sons, Ltd

to paralyze their prey. Sir Walter Raleigh and other explorers brought many forms of curare back to Europe upon returning from the Americas. It was not until 1932 that highly purified curare was first introduced to patients with tetanus and spastic disorders. It is still utilized today for this clinical purpose. Ten years later, Griffith and Johnson first reported curare for purposes of muscle relaxation during anesthesia.[3]

Today, there are five main therapeutic uses of neuromuscular blockers. The primary use of NMBAs is as an adjuvant to promote muscle relaxation during surgical anesthesia to reduce the amount of anesthesia needed. Additionally, neuromuscular blockades can effectively prevent asynchronous breathing and barotraumas, particularly in neonates and small infants.[4] Orthopedists employ NMBAs to facilitate correction of dislocations and alignments of fractures. NMBAs are also given to prevent trauma in electroconvulsive shock therapy and to treat severe cases of tetanus. Finally, neuromuscular blockers can be used in assisting extended mechanical ventilation (i.e. endotracheal intubation). Several US surveys, spanning more than a decade, assessing the use of sedating drugs and NMBAs in patients requiring mechanical ventilation demonstrated, however, that agents such as opiates and benzodiazepines were used more frequently (36 vs 20%; 68 vs 13%)[5-7] and in different situations than NMBAs. One survey concluded that the use of NMBAs was associated with longer duration of mechanical ventilation, weaning time and stay in the intensive care unit (ICU) and higher mortality.[6]

10.2.2 Classification, mechanism and duration of action

Mechanistically, neuromuscular blockers combine with nicotinic receptors on the postsynaptic membrane to block competitively acetylcholine (ACh) binding. This prevents conformational changes of, and sodium passage through, the ion channels in the membrane. Once applied to the neuromuscular end-plate, NMBAs desensitize the muscle cells to motor-nerve impulses and ACh.[3] Chemical structures of acetylcholine and representative NMBAs are depicted in Figure 10.1.

Neuromuscular blocking agents are divided into two classes according to their actions: non-depolarizing, for which the prototype is curare, and depolarizing agents, such as succinylcholine. Curare, also known as d-tubocurarine, is a plant alkaloid, which acts in the opposite manner to nicotine. The systematic name of the active principle is 7′,12′-dihydroxy-6,6′-dimethoxy-2,2′,2′-trimethyltubocuraranium hydrochloride. This structure and variants of this structure comprise the non-depolarizing neuromuscular blocking agents. Non-depolarizing neuromuscular blocking agents create a competitive blockade of acetylcholine receptors, thereby preventing depolarization of motor end-plate. Non-depolarizing agents do not cause fasiculations and their effects can be reversed by anticholinesterase agents.

Succinylcholine (suxamethonium) and decamethonium (C-10) are depolarizing neuromuscular blockers that act as acetylcholine analogs by binding directly to nicotinic receptors causing persistent depolarization of the motor end-plate. Because these drugs are slowly hydrolyzed, repolarization of the muscle is delayed. Depolarization by these drugs is divided into Phase I and II responses. For a Phase I response, brief initial twitches known as 'fasiculations' are followed by flaccid paralysis. Flaccid paralysis is induced by a refractory state of the nerve awaiting a rapid increase of membrane potential resulting in an action potential. Acetylcholinesterase (AChE) potentiates the Phase I blockade and competitive

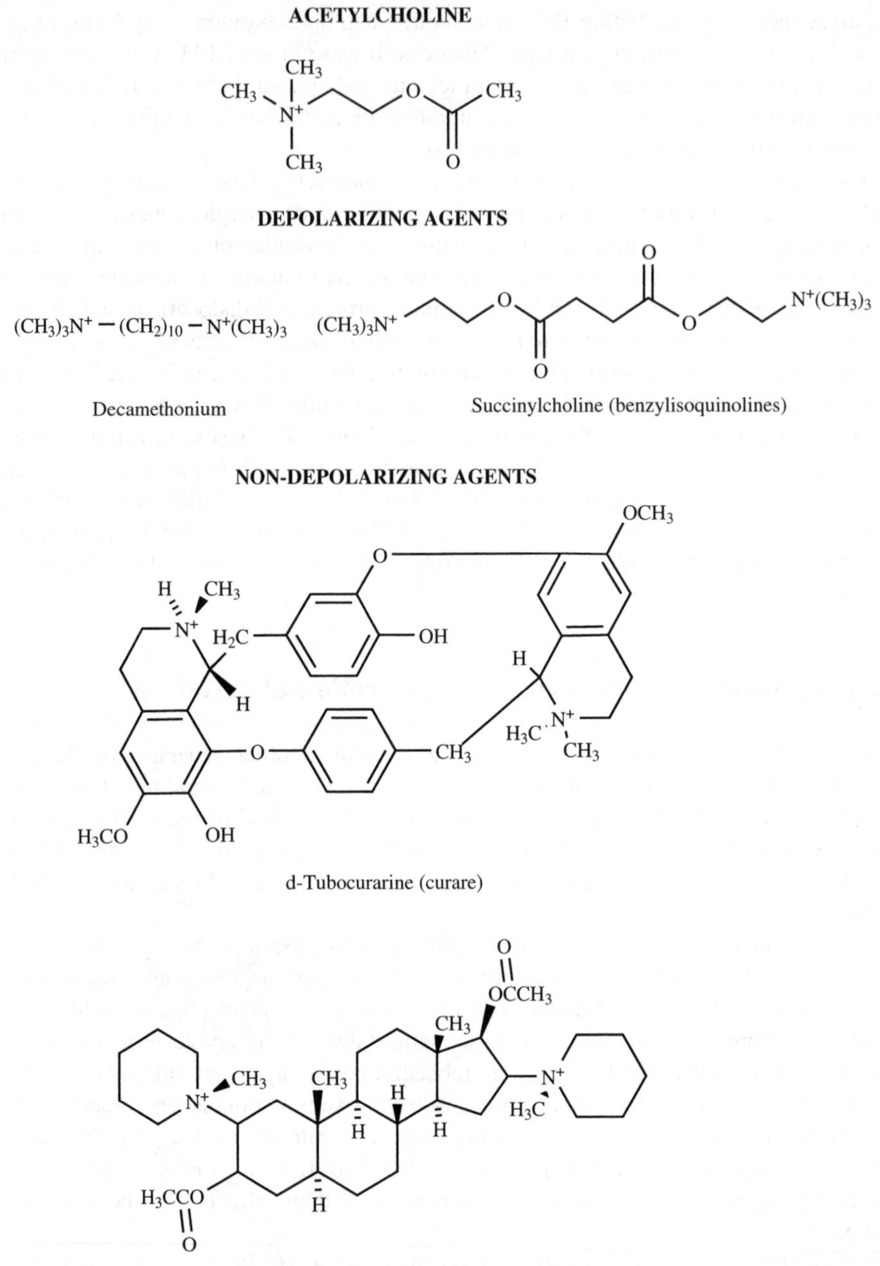

ACETYLCHOLINE

DEPOLARIZING AGENTS

Decamethonium   Succinylcholine (benzylisoquinolines)

NON-DEPOLARIZING AGENTS

d-Tubocurarine (curare)

Pancuronium (aminosteroid prototype)

Figure 10.1   Structures of acetylcholine and representative neuromuscular blocking agents

blockers antagonize it. A Phase II response maintains a flaccid paralysis by desensitizing nicotinic receptors with a long blockade or a high concentration of NMBA. Desensitized receptors result in a smaller effect for a given degree of receptor occupancy, essentially competitive inhibition. Therefore, acetylcholinesterases antagonize the Phase II blockade whereas competitive blockers potentiate it.[3] The muscle fasiculations of depolarizing agents are followed by relaxation. Paralysis caused by depolarizing agents is terminated by metabolism of the agent by pseudocholinesterase. Succinylcholine is primarily used for short-acting paralysis such as intubation. Decamethonium has a higher risk of fatality and is no longer used therapeutically in the USA.

NMBAs are further differentiated by their duration of action during anesthesia. Succinylcholine and mivacurium are common ultra-short-acting competitive NMBAs (5–10 min). An intermediate duration of action (30–45 min) is maintained with the use of atracurium, cisatracurium, rocuronium and vecuronium. A long-lasting duration of action (90–100 min) is observed with *d*-tubocurarine, doxacurium, metocurine, pancuronium and pipecuronium.

Less commonly, NMBAs are classified based on their chemical nature to include natural alkaloids and their congeners (e.g. *d*-tubocurarine), aminosteroids (e.g. cisatracurium, vecuronium) and benzylisoquinolines (e.g. succinylcholine). This classification scheme also relates to mechanism of actions and associated effects.

10.2.3 Effects and toxicity

Neuromuscular blockers are ionized (i.e. quaternary nitrogen), prohibiting oral ingestion and CNS effects since they do not cross the blood–brain barrier. As such, NMBAs are injected intravenously, causing rapid onset of their effects. Following an NMBA injection, skeletal muscles experience motor weakness followed by flaccid paralysis. Muscle paralysis occurs first for fine muscles located in eyes and fingers, followed by appendages, neck, trunk and intercostals muscles, and finally the diaphragm and muscles of respiration. Recovery occurs in reverse order. Death is attributed to paralysis of respiratory muscles. In addition to skeletal muscle paralysis, competitive NMBAs can block nicotinic receptors at the autonomic ganglia and adrenal medulla, resulting in hypotension and tachycardia, decreased gastrointestinal muscle tone and motility. Finally, competitive NMBAs block histamine release, potentially leading to hypotension, bronchospasm and excessive bronchial and salivary secretions. Toxicity of NMBAs is summarized in Table 10.1. The blockade of competitive neuromuscular agents can be overcome with the use of cholinesterase inhibitors (e.g. physostigmine), which increase ACh survival in the neuromuscular junction, thereby increasing the probability that the receptor will become occupied by ACh and not an NMBA.[3]

10.2.4 Analysis

Neuromuscular blockers can be analyzed using HPLC with fluorescence, ultraviolet (UV), or electrochemical detection (ECD). As an alternative to HPLC, GC with nitrogen–phosphorus detection (NPD) and LC coupled with electrospray ionization tandem mass spectrometry (ESI-MS-MS) have also been investigated as viable techniques to measure neuromuscular blockers (Table 10.2).

Regardless of the chromatographic technique, NMBAs are purified from biological tissues and fluids by both liquid–liquid extractions (LLE) and solid-phase extractions (SPE). Because many NMBAs have a short half-life (e.g. suxamethonium, $t_{1/2} = 0.7$ min),

Table 10.1 Toxicity of neuromuscular blockers

Mechanism of toxicity	Non-depolarizing NMBAs	Depolarizing NMBAs
Skeletal muscle blockade	Prolonged apnea (>20 s)	Prolonged apnea
Ganglionic blockade		
	Hypotension	Hypotension
	Tachycardia	Tachycardia
	Decreased GI motility/tone	
Histamine release	Hypotension	
	Bronchospasm	
	Increased secretions	
Calcium release from		Malignant hyperthermia
sarcoplasmic reticulum		• increased muscle tone (rigidity)
		• increased cellular metabolic
		rate (hyperthermia)
Loss of potassium in		Hyperkalemia
muscle cells		
Abnormal cholinesterase		Prolonged paralysis and
activity in individual		delayed recovery

urine may need to be collected and analyzed in addition to blood.[8] Furthermore, metabolites of a drug may be the only analyte identified after administration of an NMBA since the parent drug may be metabolized to undetectable amounts in a matter of minutes (e.g. atracurium and mivacurium). LLEs require ion-pair formation between the NMBA and a salt such as potassium iodide. SPE methods are performed with silica-based hydrocarbon sorbents, typically C_{18} or C_8. HPLC methods to detect NMBAs are usually optimized for one or two agents. A majority of HPLC techniques rely on either UV or fluorescence detection. As such, HPLC coupled with UV detection is limited to NMBAs that absorb UV light, namely mivacurium, cisatracurium, atracurium and its degradation products laudanosine and quaternary monoacrylate. Similarly, HPLC with fluorescent detection requires agents that fluoresce or that are labeled with a fluorescent marker, and electro-chemical detectors require analytes that are either oxidizible or reducible. For example, Reich et al.[4] successfully detected cisatracurium and vecuronium by fluorescent and electrochemical detection, respectively, in 19 neonates and small infants infused with these drugs during congenital heart surgery.

Pitts et al.[9] successfully determined succinylcholine in plasma by HPLC with ECD. Owing to rapid butyrylcholinesterase (BuChE) hydrolysis of succinylcholine to succinyl-monocholine and choline in blood, the parent drug is difficult to determine. Since succinylmonocholine is further hydrolyzed to choline and inactive succinic acid at physiological pH, these investigators measured succinylcholine concentrations indirectly by determining the concentration of its breakdown product, choline. Complete hydrolysis of succinylcholine to choline and succinic acid is a two-step reaction, yielding choline at each step. Under the assayed conditions, none of the measured choline was derived from succinylmonocholine. In this method, choline was measured in two blood or plasma samples, in one of which succinylcholine hydrolysis was inhibited immediately (e.g. endogenous choline or hydrolyzed aliquot) by the addition of physostigmine (10^{-5} M),

Table 10.2 Chromatographic Detection of Neuromuscular Blockers

Author (Year)	Neuromuscular Blocker(LOQ)	Technique & Specimen Type	Method Summary
Pitts (2000)[9]	Succinylcholine (SUC; 16 ng ml^{-1})	HPLC-ECD Plasma (2 mL)	*Pretreatment:* • Total choline: addition of physostigmine (10^{-5} M, 0.01%) • Endogenous choline: addition of BuChE (200 mU at 37°C) in 0.5 mL of 0.05 MPB (pH 7.4) • Reaction stopped: 1 mL ice cold 0.1 M perchloric acid. • Centrifuge: 5000xg at 4°C for 5 min to obtain plasma • Storage: −70°C *Sample Preparation:* • Choline Standards: prepared in 52 mM acetic acid (pH 5.5) with 0.01% antimicrobial agent, Kathon (BAS- Bioanalytical Systems, Lafayette, IN) *Reverse Phase Column:* • Physostigmine trapping column: pre-column 20 μL injection loop (MF-6262, BAS) • ACh analytical column (BAS)- 96 mm × 5 mm i.d. • Choline retained 11.1 min then passed through 300 mm long immobilized enzyme column containing AChE (60U) and choline oxidase (60U) from BAS analytical kit • Platinum working electrode: measured peroxide reaction product • Electrode potential: 0.5 mV relative to Ag/AgCl reference electrode • Resulting current measured with LC-4B amperometric detector (BAS) • Output: 5–20 nA full scale depending on concentration and electrode sensitivity *Mobile Phase:* • Flow rate: 0.7 mL min^{-1} • 0.05 M Tris/NaClO$_4$ (pH 8.5) with 0.5% antimicrobial agent, Kathon (BAS)

(continued)

Table 10.2 (Continued)

Author (Year)	Neuromuscular Blocker(LOQ)	Technique & Specimen Type	Method Summary
			Detection: • Electrochemical Signal measured: analyte peak height relative to concentration *Run Time:* Not Reported
Reich (2004)[4]	Cisatracurium (CST; 20 ng ml^{-1})	HPLC-FL Plasma	*Pretreatment:* • Acidified and frozen *Standard Preparation:* • SPE on C18 Bond Elut column (Varian, Palo Alto, CA) *Reverse Phase Column:* Regis C18 Spherisorb Little Champ (Regis Technologies Inc., Morton Grove, IL) 5 cm × 4.6 mm, 5 μm *Mobile Phase:* • 0.03 M NPB (pH 3.0) with 0.14% (CH3)$_4$NH$_4$OH in ACN (60:40, v/v) *Detection:* • Fluorescence • Wavelength not reported *Run Time : Not Reported:*
Reich (2004)[4]	1. Vecuronium (VE; 15 ng ml^{-1}) 2. 3-OH Vecuronium (3OH-VE; 15 ng ml^{-1})	HPLC-ECD Plasma	*Pretreatment:* Acidified and frozen *Sample Preparation:* • SPE on C1 Bond Elut column (Varian, Palo Alto, CA) • Precondition: MeOH & H$_2$O • Wash: H$_2$O, ACN, MeOH • Elute: 0.01 M NaClO$_4$ in MeOH • Evaporate in solvent Dissolve: 0.1 mL mobile phase

Reverse Phase Column:

Waters Nova-Pak CN (Waters Corp., Milford, MA)

3.9 (units NR) × 150 mm, 4 μm

Mobile Phase:

- 0.033 M NPB (pH 5.5) in ACN (60:40, *v/v*)

Detection

- Electrochemical

Run Time: Not Reported:

Pretreatment:

- Heparinized blood centrifuged 5 min (2500 U/min-1)
- To prevent hydrolysis of ROC, decanted plasma acidified by diluting equally with 0.8 M PB (pH 5.4)
- Stability study to confirm samples stable for 1 year. 600 μL of pretreated sample (−70°C) at different concentrations were evaluated.

Sample Preparation:

- IS: 3OH-VE; LLE: 500 μL sample (250 μL specimen/250 μL PB) treated with saturated KI solution (500 μL) and MeCL₂ (5 mL); Rotary mix: 30 min and decant MeCL₂ (Breda Scientific); Evaporate: 60°C in SpeedVac Plus SC 110A (Savant; Bierbeek, Belgium) CAN Dissolve: 200 μL Derivatize: MTBSTFA at 70°C overnight Evaporate: 60°C in SpeedVac; Dissolve: 100 μL Acetone

Alternate Sample Preparation Method:

- IS: 3OH-VE; SPE: 500 μL sample (250 μL specimen/250 μL PB) & saturated KI solution (500 μL) added to luer locked Extrelut cartridge (Merck; Darmstadt, Germany); Add: MeCL₂ (5 mL) and equilibrate for 5 min; Evaporate: 60°C in SpeedVac; Dissolve: 200 μL CAN; Derivatize: MTBSTFA at 70°C overnight; Dissolve: 100 μL Acetone

GC Column:

DB1 (J &W Scientific; Koln, Germany)

- 10m × 0.32 mm i.d. × 0.25 μm

(*continued*)

Probst (1997)[11]

GC-NPD

Plasma

1. Rocuronium (ROC; 10 ng ml⁻¹)

2. 17-OH-ROC (17OH-ROC; 50 ng ml⁻¹)

Table 10.2 (*Continued*)

Author (Year)	Neuromuscular Blocker(LOQ)	Technique & Specimen Type	Method Summary
			GC System: • injector: 300°C, split (1:25); injection volume: 2 µL • oven preramp: 5 min at 120°C • oven ramp: 30°C min-1 to 300°C hold 6.5 min • carrier gas (helium, 5.0): 3 ml min-1; make up gas (nitrogen): 25 ml min^{-1}; hydrogen: 3.5 ml min^{-1}; synthetic air: 175 ml min^{-1} *Detector:* Nitrogen specific detector- TSD (Varian, Darmstadt, Germany) TSD bead probe modified by applying rubidium dotted glass bead Detector temp: 320°C Detector current: 2.9-3.0 A *Run Time:* 13 min:
Gutteck-Amsler (2000)[12]	a) Pancuronium (IS)- PE-10 ng ml^{-1} b) ROC- 5 ng ml^{-1} c) VE-15 ng ml^{-1} d) 3OH-VE-15 ng ml^{-1}	LC-ESI-MS Plasma	*Pretreatment:* preservative- 1 mol L^{-1} sodium hydrogen phosphate solution *Sample Preparation:* • Add: IS (50 µL) & 6 mol/L KI (1 mL) • LLE: toluene (5 mL); shaken for 20 min • Centrifuge, separate, & evaporate organic layer • Dissolve: 0.1 mL mobile phase *Reverse Phase Column:* Nucleosil C18 12.5 cm × 4mm, 5µm Macherey-Nagel *Guard Column* • Nucleosil C 18; 8 × 4 mm, 5 µm Macherey-Nagel *Mobile Phase:* 50 µM formate buffer (pH 3) in MeOH (73:27, *v/v*)

Detection

LCQ ion trap MS (Thermoquest, San Jose, CA)
Positive ESI
Voltage: 3.8 kV
Capillary: 230°C
Product ions (*m/z*)-a) PA: 286.4 → 236.7; b) ROC: 529.4 →
487.7; c) VE: 279.2 → 249.4; d) 3OH-VE: 258.5 → 100.2
Retention Time: a) PA: 8.2 min; b) RO: 6.6 min; c) VE: 15.4 min; d)
OHV: 9.7 min

Sample Preparation:

- Extract: Blood (1 mL) mixed with saturated KI (1 mL)
  and 0.8 M PB (NaH$_2$PO$_4$, 1 mL, pH 5.4) and MeCl$_2$ (5 mL)
- Shake: 15 min at 100 cycles min-1
- Centrifuge 10 min (4000 rpm)
  Decant: Organic phase and evaporate
  Reconstitute: 30 µL mobile phase
Reverse Phase Column:
- Nucleosil C18
- 150 cm × 1.5 mm i.d., 5 µm
- Protected by 0.5 µm frit
- Injection volume 2 µL

Mobile Phase:

50 mM NH$_4$COOH (pH 3) in MeOH using formic acid to adjust pH
Linear gradient: 30 to 90% MeOH in 6 min
Flow: continuous 50 µL min-1
Detection:

- LCQ ion trap MS (PE Sciex API-100; Positive ESI; Voltage: 4.2 kV;
  Capillary: NR
- Nebulizing gas: Nitrogen (99.95%)
- Ion source orifice: 25 µm at +40V
- Full scan mode: 200 to 720 *m/z*
Fragment ions (*m/z*) d-TBC: **609**, 552,521; PA: **430**, 472, 617; ALC:
333, 312, 711; ROC: **487**, 530; OHV: **515**, 258, 356; VE: **557**,

(*continued*)

Cirimele
(2003)[14]

LC-ESI-MS
Blood
Urine

a) d-TBC
b) PA
c) ALC
d) ROC
e) OHV
f) VE
g) ATC
h) MVC
i) MBZ

Table 10.2 (*Continued*)

Author (Year)	Neuromuscular Blocker(LOQ)	Technique & Specimen Type	Method Summary
			356, 398; ATC: **464**, 516, 570; MVC: **514**, 600, 671; MBZ: **294**, 276, 208; (Quant ion bolded)
			Retention Time (min): d-TBC: 11.7; PE: 12.3; ALC: 12.4; ROC: 12.4; OHV: 12.4; VE: 12.7; ATC: 13.2; MVC: 13.4; MBZ: 15.1;
Kerskes (2004)[13]	a) ROC b) VE c) Suxamethonium (SUX, succinylcholine) d) Mivacurium (MVC) e) Gallamine (GAL) f) Pancuronium (IS) (PE-10 ng ml^{-1}) g) Atracurium (ATR) h) Laudanosine (LAU) i) Quaternary monoacrylate (QMA) LOQ-NR	LC-ESI-TMS Blood Serum Urine Bile Liver Muscle Brain	*Pretreatment:* • Incubate urine, bile, liver at 36°C, 3 h; 5 µL β-glucuronadase/aryl sulfatase (*helix pomatia*, Merck); Homogenate tissues (4:1 in $CH_8N_2O_3$); Vortex/Centrifuge; Use supernatant • All other matrices:1 mL in $CH_8N_2O_3$ (2 mL); Vortex/Centrifuge; Use supernatant *Sample Preparation:* • SPE: BondElute C18 HF (Varian, Harbor City, CA) • Precondition: MeOH & $CH_8N_2O_3$ (2 mL each) • Wash: Ammonium carbonate buffer (1.5 mL × 2) • Dry 5 min, add hexane (50 µL), & dry another 5 min • Elute: 0.1 M acetic acid in MeOH (1 mL) • Evaporate in heated, vacuumed centrifuge (Savant SpeedVac AE2000) for 25 min • Dissolve: 0.1 mL $NH_4C_2H_3O_2$ and ACN (7:3, *v/v*) Centrifuge 5 min (4000 rpm) *Reverse Phase Column:* Intertsil ODS2 (Chromopack) *Mobile Phase:* 50 µM $NH_4C_2H_3O_2$ (pH 5) in water; Flow: continuous; 0.4 mL min-1;

(*continued*)

Percent Gradient

Mins:	Buffer/ACN:
0→5	95/5
5→15	70/30
15→25	70/30
25→30	65/35
30→33	30/70
33→33.5	30/70
33.5→35	95/5

Detection:

- LCQ ion trap MS (Thermoquest, San Jose, CA); Positive ESI; Voltage: 3–46 kV; Capillary: 125–250°C; Sheath Gas: 45–85 psi; Collision Energy: 32–37%

Product ions (*m/z*)- PE: 286.4 → 236.6; 472.3; 430.2; 206.7; 100.3; ROC: 529.4 → 487.4 → 427.4; 376.3; 358.2; 418.2; 487.3; 400.3; 445.4 (MS/MS/MS); VE: 557.5 → 497.4; 398.2; 338.23; 416.2; 458.2; 356.3; SUX: 145.2 → 115.6; MVC: 514.4;

342.3 → 671.3; 357.1; 342.2; 600.3; 428.2; 325.2; GAL: 284.7; ATC: 464.3; 358.2; 570.2 → 327.0; 307.2; 370.2; 358.1; 296.1; LAU: 358.2 → 206.2; 327.0; QMA: 570.2 → 370.2; 327.0; 412.1; 256.1

- *Retention Time (min):* PE: 8.6; ROC: 7.9; VE: 9.5; SUX: 1.6; MVC: 18.5, 20.5, 22.2; GAL: 4.5; ATC: 17.1, 17.8, 18.4; LAU: 7.9; QMA: 23.9, 25.6

and the other which was completely hydrolyzed (e.g. total choline or inhibited aliquot) in 20 min by addition of BuChE (200 mU at 37 °C). These investigators used a commercially available acetylcholine/choline analytical kit [Bioanalytical Systems (BAS), West Layfayette, IN, USA] that measures these analytes by HPLC-ECD following reversed-phase column separation and pass through an immobilized enzyme reactor containing AChE and choline oxidase. In this process, only the oxidation of choline relates to the succinylcholine concentration. A 95% recovery of choline was achieved and the difference in the two choline measures represented the succinylcholine concentration. Choline calibration curves were linear from 16.3 to 20 900 ng mL^{-1} (156 pmol mL^{-1} to 200 nmol mL^{-1}) with correlation coefficients ≥ 0.999. The within-day ($n = 6$) and between-day precisions ($n = 10$) of HPLC choline detection were assessed by repeatedly injecting choline calibration standards multiple times within a day and over a 3-week period. Mean coefficients of variation (CVs) with standard deviations for this procedure were 3.7 ± 1.2 and $3.8 \pm 1.6\%$, respectively. The chosen chromatographic conditions (Table 10.2) did not yield any succinylcholine or interfering peaks.

GC methods are less common than LC for the analysis of NMBAs. Their mono- or bisquaternary ammonium structure, however, makes them ideal for nitrogen–phosphorus detection (GC-NPD). Most GC techniques for NMBAs follow similar parameters exemplified in a paper published by Furuta et al. in 1988.[10] They reported the use of GC-NPD to detect pancuronium (PA), pipecuronium (PIP), vecuronium (VE) and their 3-desacetyl metabolites (3-Des-) with detection limits of 2 ng mL^{-1} for the parent drugs and 4 ng mL^{-1} for their respective metabolites. Plasma samples from 106 surgical patients were acidified, derivatized at the 3-hydroxy steroidal position with an *O-tert*-butyldimethylsilyl derivatizing agent, washed with diethyl ether and finally extracted with an aliquot of dichloromethane. Lower limits of detection and upper limits of linearity were as follows: PA, VE, PIP, 2–5000; 3-Des-VE, 4–5000; 3-Des-PA, 8–500; and 3-Des-PIP, 25–1000 ng mL^{-1}. Adequate precision was demonstrated with coefficients of variation ranging from 2 to 20%. Reproducibility for clinically relevant concentrations of these drugs in human serum was 5–16% PA, 2–4% VE and 6–16% PIP. No interferences were observed in the analyses of over 100 clinical specimens.

In 1997, Probst et al.[11] modified Furata et al.'s method to quantify rocuronium and its 17-desacetyl metabolite in heparinized plasma. This group performed parallel studies of the previously described LLE extraction and an on-column ion pairing using Extrelut solid-phase extraction cartridges. The intra-assay precision ($n = 5$) at two concentrations (400 and 1052 ng mL^{-1}) for both parent drug and metabolite showed mean concentrations within 4–17% (CV = 9.7–10.5%) of the target concentration of rocuronium and within 7–8% (CV = 13.2–15.3%) of the target concentration of its metabolite. The inter-assay precision at three concentrations (400 ng mL^{-1}, $n = 15$; 800 ng mL^{-1}, $n = 13$; and 1480 ng mL^{-1}, $n = 15$) demonstrated similar results with mean concentrations within 0.5–11.7% (CV = 8.2–17.7%) and within 7–21% (CV = 16.4–21.3%), respectively. The mean accuracy ranged from 88 to 103% for rocuronium (ROC) from 108 to 121% for 17-desacetylrocuronium (17OH-ROC). The linearity of the ROC calibration curve was 50–6400 ng mL^{-1} and the linearity of 17OH-ROC was 80–6400 ng mL^{-1}. The lower limit of linearity was set as the limit of quantification determined as a reliable repeated signal-to-noise ratio between 5 and 10 of the peak area ratio of analyte to internal standard. Analytical interferences were not noted. These analytical modifications simplified the extraction procedure (e.g. separation of aqueous and organic phases not required) and improved sample preparation time (5–10 vs

30 min) in large-scale determinations as compared with LLE. In addition, further attempts to improve sensitivity such as use of alternative SPE cartridges (RP-8, RP-18, cyanopropyl or cation-exchange) or organic solvents (1,1,1-trichloroethane, trichloromethane, diethyl ketone, and ethyl methyl ketone) with and without potassium iodide were not successful. ROC and 17OH-ROC recoveries were between 30% and 48% for heparinized plasma and 52% and 73% for phosphate-buffered heparinized plasma, respectively. Despite these ROC recoveries, concentrations less than $100 \, ng \, mL^{-1}$ were successfully detected in patients' samples 4 h after a $22 \, mg \, h^{-1}$ ROC infusion. 17OH-ROC was not detected in any of the patients' samples.

Since 2000, a myriad of LC-MS techniques have been reported to detect an array of NMBAs. These methods are valuable because they are generally more sensitive and more practical because they can measure a number of NMBAs simultaneously. Likewise, pharmacokinetic studies of neuromuscular blockers require low detection limits ($<10 \, \mu g \, L^{-1}$) and acceptable reproducibility (CV $<10\%$). Hence more sensitive chromatographic techniques such as tandem mass spectrometry with electrospray ionization combined with liquid chromatographic separation (LC-ESI-MS-MS) are better suited for these types of clinical applications.[12] Brief descriptions of representative LC-MS methods follow.

In 2002, Kerskes et al.[13] reported an LC-ESI-MS method detecting seven neuromuscular blocking agents and two additional metabolites. ESI into an ion trap was the optimal interface for the charged structures of the NMBAs. A 35-min HPLC gradient using ammonium acetate buffer in acetonitrile effectively eluted all nine analytes on an Intertsil ODS2 Chromopack column within 26 min. Autotuning to optimize the tube lens offsets, octapole offsets and capillary voltage were carried out for each NMBA. Further tuning was done manually to determine the best sheath gas flow-rate and capillary temperature for each NMBA. For an unknown compound, the samples were screened using a method containing a suxamethonium (SUX) tune file from 0 to 6 min and an atracuronium (ATC) tune file from 6 to 35 min. For cases of known analyte, the specimens were analyzed with the specific tuning parameters for the given NMBA. The optimal collision energy for MS-MS measurements ranged from 32 to 37%. Performance studies for this procedure were not discussed by the authors.

NMBAs were analyzed by the MS configuration that gave the most intense ions in which the relative abundance was more than 10%. In the MS mode, the molecular ions were observed for gallamine ($m/z = 284.7$), pancuronium ($m/z = 286.4$), vecuronium ($m/z = 557.5$) and rocuronium ($m/z = 529.3$). In the MS-MS mode, the product ions of the last three molecular ions were unambiguous and easily identified. Gallimine did not have an appreciable signal in the MS-MS or MS-MS-MS modes. ROC and SUX were the only NMBAs requiring triple quadrapole identification, as only one peak (ROC, $m/z = 487.4$; SUX, $m/z = 115.6$) was observed in the MS-MS mode, hence a unique fragmentation pattern was not evident without further fragmentation. In addition to a small peak, mivacurium has a base peak molecular ion ($m/z = 514.4$) that can be fragmented into six product ions. Molecular ions were observed for atracurium ($m/z = 570.2$) and its metabolites, quaternary monoacrylate ($m/z = 570.2$) and laudanosine (protonated at eluent pH; m/z M + 1 = 358.2) and these precursor ions are additionally fragmented to four, four and two product ions, respectively. Mivacurium and atracurium both had several isomeric peaks between 16 and 22 min. Although there was no baseline separation of these two analytes, they could be uniquely identified based on their different mass spectra.

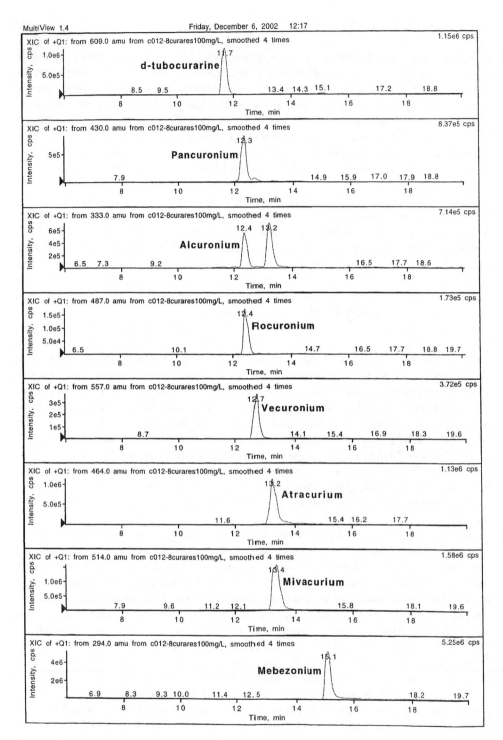

Figure 10.2 Typical chromatographic profiles for the eight quaternary muscle relaxants. Reprinted from *Journal of Chromatography B*, **789**, V. Cirimele, M. Villain, E. Pépin, B. Ludes and P. Kintz, Screening procedure for eight quaternary nitrogen muscle relaxants in blood by high-performance liquid chromatography–electrospray ionization mass spectrometry, 107–113, with permission from Elsevier

This multi-analyte LC-MS method for NMBAs investigated both clinical and post-mortem applications. First, suxamethonium and rocuronium were detected in serum and urine administered to a woman during a Caesarian section operation. Suxamethonium was detected in urine 35 min post-administration, but it was not present in the serum. About 25 min following the administration of rocuronium, it was detectable in both matrices. Another case investigated by this method involved a possible assisted suicide by pancuronium (Pavulon) injection. Biological specimens investigated by this method included blood, urine, bile, brain, muscle and liver collected during autopsy. Pancuronium was detected in all matrices with the highest concentration found in the liver ($2.4\,mg\,kg^{-1}$) and the lowest in the brain ($0.1\,mg\,kg^{-1}$). Two additional peaks were consistently present and thought to correspond to 3- and 17-hydroxypancuronium.

Another clinical and forensic method that identified and quantified eight quaternary nitrogen NMBAs in blood was reported by Cirmele *et al.* in 2003.[14] The muscle relaxants detected included *d*-tubocurarine (*d*-TBC), alcuronium (ALC), PA, VE, atracuronium (ATR), ROC and mebezonium (MBZ), with mivacurium (MVC) as the internal standard. The procedure used ion-pair extraction with methylene chloride (pH 5.4), reversed-phase HPLC separation with a binary mobile phase of 50 mM ammonium formate buffer (pH 3.0) in methanol and ESI-MS detection. Linearity of the procedure was established to be 0.1–$10\,mg\,L^{-1}$ with a correlation coefficient $r^2 > 0.929$. All analytes demonstrated a limit of detection of $0.1\,mg\,L^{-1}$ at a signal-to-noise >5.0. With eight replicates each, accuracy was demonstrated at a $1.0\,mg\,L^{-1}$ concentration with a CV between 6.9 and 17.8% and the relative extraction recovery was between 46.0 and 91.1% at $1.0\,mg\,L^{-1}$. Typical chromatographic profiles for the eight quaternary muscle relaxants are depicted in Figure 10.2. This screening procedure performed satisfactorily in the investigation of two fatalities. One successfully detected vecuronium (1.2 and $0.6\,mg\,L^{-1}$) and 3-hydroxyvecuroinium (4.4 and $0.7\,mg\,L^{-1}$) in blood and urine, respectively. Blood obtained at autopsy in a second case revealed the presence of MBZ at $6.5\,mg\,L^{-1}$. The full-scan chromatogram and specific mass spectrum of mebezonium in post-mortem blood are depicted in Figure 10.3. While the first case showed slightly elevated therapeutic concentrations of vecuronium in blood (0.4–$1.0\,mg\,L^{-1}$), the absence of medical control or assistance may have contributed to the death. The post-mortem mebezonium concentration in the second case was consistent with three other reported deaths.

A once exclusive instrumental technique for research, LC-MS is becoming ever more utilized in special applications. Since hospitals are requiring more robust, sensitive and specific assays for routine TDM analysis, it is inevitable, as demonstrated with the aforementioned NMBAs, that LC-MS methods will eventually become routine procedures in clinical settings.

10.3 PARALYTIC SHELLFISH POISONING: SAXITOXIN

10.3.1 Background

Shellfish poisoning occurs when shellfish feed on planktonic algae (e.g. dinoflagellates, especially *Gonyaulax* and *Alexandrium* species) that accumulate and metabolize toxins, which are subsequently ingested by humans. A particularly serious form of shellfish poisoning is paralytic shellfish poisoning (PSP) owing to its extreme toxicity caused by

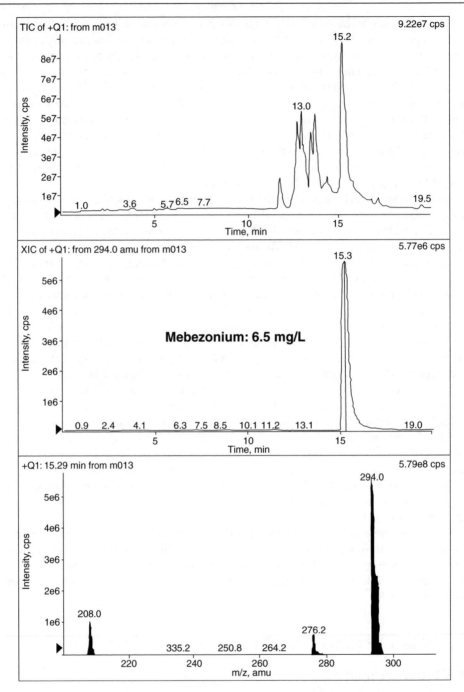

Figure 10.3 Full-scran chromatogram and specific mass spectrum of mebezonium in post-mortem blood. Reprinted from *Journal of Chromatography B*, **789**, V. Cirimele, M. Villain, E. Pépin, B. Ludes and P. Kintz, Screening procedure for eight quaternary nitrogen muscle relaxants in blood by high-performance liquid chromatography–electrospray ionization mass spectrometry, 107–113, with permission from Elsevier

over 21 structurally related toxins derived from saxitoxin (STX), a neurotoxin to mammals. STX toxins block the sodium channels in the excitable membranes of nerve cells and associated muscles; this depolarization blockade inhibits the action potentials and nerve transmission impulses, ultimately resulting in respiratory failure. Saxitoxin analogs fall into three subgroups based on their structure, with the least potent being carbamate toxin followed by sulfamate toxins and decarbamoyl gonyautoxins.

10.3.2 Toxicity

Saxitoxin manifests toxicoses, which may ultimately be fatal. Deadly outbreaks, such as that experienced in 1987 in Guatemala, have reported as many as 187 cases and 26 deaths occurring after the ingestion of a single exposure to a clam soup. The high mortality rate associated with PSP has led many countries to mandate long-term closure of water areas in which molluscan shellfish are harvested during particularly unsafe seasons (e.g. summer months) when planktonic blooms make PSP a natural phenomenon.[15] Today, PSP has relatively few outbreaks in the USA owing to strong control programs that prevent human exposure to toxic shellfish such as mussels, clams, cockles and scallops. During a 5-year period (1988–1992), there were five outbreaks of PSP (65 cases, two deaths) reported by the Center for Disease Control and Prevention (CDC) through its surveillance program for foodborne-disease outbreaks established in 1973.[16]

PSP has a rapid onset of symptoms occurring within 0.5–2 h after ingestion of the shellfish, depending on the amount of toxin consumed. Symptoms include

- tingling

- burning

- numbness

- drowsiness

- incoherent speech

- respiratory paralysis.

Without clinical support, death may occur most commonly by respiratory paralysis or occasionally by cardiovascular collapse, despite respiratory support, caused by the weak hypotensive action of the toxin. With less serious cases (e.g. lower concentration of saxitoxin in shellfish or smaller amount of shellfish consumed), complete recovery with no lasting side-effects is expected if respiratory support transpires within 12 h of exposure. In humans, 120–180 µg of STX can produce moderate symptoms such as tingling and numbness, 400–1060 µg of STX can cause deaths, and concentrations >2000 µg constitute a fatal dose.[17]

Although saxitoxin is best known as the agent producing PSP after ingestion of contaminated marine life, this neurotoxin is also deadly if inhaled; an individual may die within minutes after inhaling a lethal dose. When inhaled, saxitoxin acts by blocking nerve conduction directly and its lethality results from paralyzing muscles of respiration. Because of its threat as a chemical warfare agent and reported use as a biological weapon, saxitoxin has been placed on the Schedule of Chemical Warfare Agents. In comparison with other potential chemical weapons, saxitoxin is of the same magnitude of lethality as ricin and 50 times more potent than sarin gas based on LD_{50} measurements in mice.[18]

10.3.3 Analysis

Analysis of PSP toxins can be achieved by bioassays and chemical assays. Historically, *in vivo* and *in vitro* bioassays were introduced first, subsequently followed by biochemical assays such as receptor binding assays and immunoassays. Chemical assays employing various analytical techniques including fluorimetry and liquid chromatography (e.g. HPLC, LC-MS) are currently used for clinical applications. Although bioassays are less sensitive and specific than newer techniques, they are still widely utilized owing to their low cost, simplicity and regulatory acceptance by governmental agencies such as the US Food and Drug Administration (FDA).

The mouse biosassay first introduced for detection of PSP toxins in 1937 by Sommer and Meyer was adopted by the Association of Official Analytical Chemists in 1984 as a non-selective bioassay with animal symptoms and time to death utilized as the measure of PSP toxicity. Its LOD of 40 µg per 100 g of tissue is adequate to determine if a tolerance level of 80 µg for humans has been reached during a PSP episode and whether actions should be taken to maintain public safety.[19]

In the early 1990s, chemical assays began to replace bioassays as both a screening and confirmatory test to detect toxins within human fluids and tissues. *In vitro* receptor binding assays, such as the hippocampal slice assay[20] and the sodium channel blocking assays and saxitoxin-specific ligand binding assays, are utilized to detect saxitoxin in addition to immunoassay techniques such as enzyme-linked immunosorbent assay (ELISA).[21,22] These chemical assays can detect saxitoxin at nanogram to picogram levels with equivalency calculated at 2 µg STX eq. per 100 g of shellfish.[23–25]

HPLC is the most common chromatographic assay, predominantly utilizing fluorescence or mass spectrometric detection to measure presence of individual PSP toxins with a detection limit for saxitoxin of 20 fg per 100 g of tissue (0.2 ppm).[26] For clinical purposes, serum and urine are the preferred matrices for analysis of saxitoxin. For post-mortem analysis other organ tissues have been analyzed (e.g. heart, brain, liver, gastric, spleen).[27]

Two commonly used HPLC techniques employ reversed-phase liquid chromatography with fluorescence detection (HPLC-FLD). Sullivan *et al.*[28] used a polymer reversed-phase cyano column and gradient elution to separate the 10 most common PSP toxins in a 20-min run, and Oshima[29] used a C_8 bonded silica gel column and isocratic elution to separate all 21 PSP toxin fractions in three separate chromatographic runs for sulfamate toxin, gonyautox-ins and saxitoxins. Most HPLC methods reported in the literature are modifications of these procedures. Better resolution and chromatographic peaks by these methods are achieved through oxidation of the saxitoxin analogs. The oxidation can be performed before chromatographic separation using peroxide and a base such as sodium hydroxide, as depicted in Figure 10.4.[28] Alternatively, the use of periodic acid in sodium phosphate buffer in a heated post-column reaction coil to oxidize PSP toxins to fluorescent derivatives has been reported.[29,30] The latter method can give false identification of imposter peaks. To avoid misidentification of toxins, samples should be reanalyzed following purification by SPE (C_{18} Sep Pak, Waters, Milford, MA, USA) followed by ultrafiltration (5000 Da;

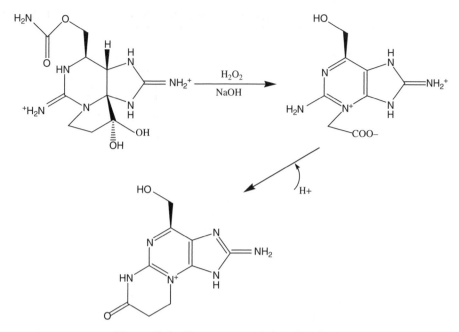

Figure 10.4 Fluorescent oxidation of saxitoxin

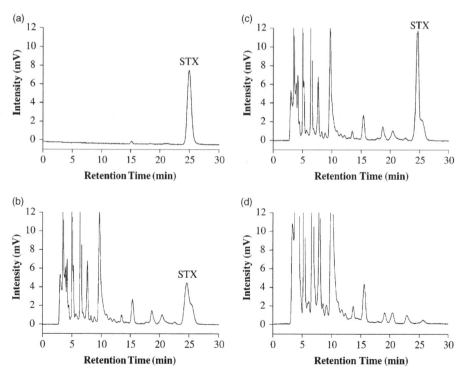

Figure 10.5 HPLC–FLD chromatogram of calibrated standard STX injection (a) compared with a toxic extract of *Octopus* (*Abdopus*) sp. 5 showing peale retention (b), spiking of the extract with an authentic STX standard (c) and with no post-column oxidation (d). Reprinted from *Toxicon*, **44**, A. Robertson, D. Stirling, C. Robillot, L. Llewellyn and A. Negri, First report of saxitoxin in octupi, 765–771, 2004, with permission from Elsevier

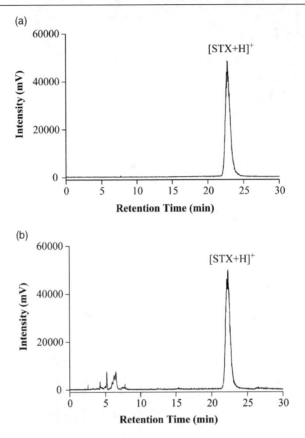

Figure 10.6 LC–MS chromatograms of an authenic STX standard (a) compared with extract of *Octopus* (*Abdopus*) sp. 5 (b) Reprinted from *Toxicon*, **44**, A. Robertson, D. Stirling, C. Robillot, L. Llewellyn and A. Negri, First report of saxitoxin in octupi, 765–771, 2004, with permission from Elsevier

Millipore, Bedford, MA, USA) and the oxidizing reagent should be replaced with distilled water to minimize oxidation. The fluorescence intensities of the peaks are compared to determine which peaks are true saxitoxins. Similarly, suspect toxin peaks such as tetrodotoxin, another marine toxin found in shellfish and octopi, can be identified using LC-MS performed in the single ion monitoring (SIM) mode.[31] Robertson *et al.*[31] presented representative chromatograms obtained using both HPLC-FLD and LC-MS, as depicted in Figures 10.5–10.7. Today, analysis of shellfish with simple toxin profiles will most likely employ the first method and, if the source of the PSP is unknown, then analysis by the second method is most appropriate. Representative LC techniques for the detection of saxitoxin are detailed in Table 10.3.

Advantages of using HPLC assays to measure saxitoxin include automated and continuous operation, increased sensitivity and increased precision. HPLC analysis of saxitoxin is limited in that it requires a skilled technician to operate the instrumentation, the analytical instrumentation is expensive, it requires extensive calibration and it is still not legally accepted for regulatory purposes. Advantages and limitations of LC-MS are similar to those

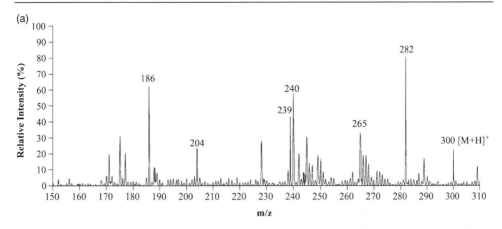

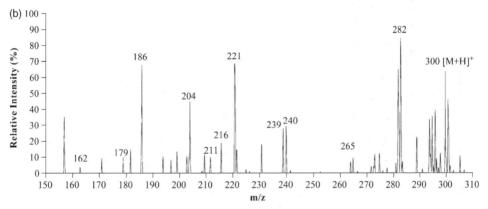

Figure 10.7 Fragment ion mass spectra for authenic STX (a) and an extract of *Octopus* (*Abdopus*) sp. 5 (b) Reprinted from *Toxicon*, **44**, A. Robertson, D. Stirling, C. Robillot, L. Llewellyn and A. Negri, First report of saxitoxin in octupi, 765–771, 2004, with permission from Elsevier

of HPLC; however, LC-MS still remains largely a research tool for the detection of saxitoxin owing its prohibitive cost and skill level.

10.4 SUMMARY

Chromatographic analysis of nerve agents has specialized purposes in many fields of analytical science, including clinical chemistry, food science, therapeutic drug monitoring and death investigation. As general analytical technologies have advanced, methodologies to improve the detection of nerve agents such as neuromuscular blocking agents and natural toxins such as saxitoxin have been introduced. Although chromatographic analyses of nerve agents are not heavily represented in the literature, techniques have been adequately presented and reproduced or modified by subsequent researchers. This chapter reviews both summarized and specific chromatographic techniques to analyze nerve agents presented in the past 20 years, in an effort to organize these technologies for the reader to evaluate carefully and understand.

Table 10.3 Liquid Chromatographic Detection of Saxitoxin

Author (Year)	STX Analogues (LOD equivalent)	Method Summary

Oshima (1995)	*Carbamate toxins*
	• STX (4 μg STX/100 g)
	• NEO (6 μg STX/100 g)
	• GTX 2 and 3 together
	  (<1 μg STX/100 g)
	  GTX 1 and 4 together
	  (2 & 3 μg STX/100 g, respectively)
		Sample Preparation:
		• Extract STX with 0.5 M acetic acid and 50% methanolic chloroform (2:1, v/v) in
		• Sonicate for 5 min
		• Stir for 30 min at RT and centrifuge
		• Reextract organic layer with 2 mL 0.5 M acetic acid
		• Aqueous phases combined and concentrated in a rotary evaporator and reconstituted in 0.5 M acetic acid
		• SPE (C18 Sep Psk, Waters Corp., Milford, MA) followed by ultrafiltration (5000 dalton; Millipore Co., MA)
		• Immediate analysis or store at −20°C
		
		Reverse Phase Column:
		• Prodigy 5μ C8
		• 150 × 4.6 mm² i.d.
		• 5 μm particles
		• 0.7 to 0.8 ml min⁻¹
		
		Mobile Phase:
		• 1 mM tetrabutylammonium phosphate (pH 5.8) for C toxin group
		• 2 mM 1-heptanesulfonic acid in 10 mM APB (pH 7.1) for gonyautoxin (GTXs)
		• 2 mM 1-heptanesulfonic acid in 30 mM APB (pH 7.1) & 3% CAN for saxitoxin (STXs)
		
		Post-column oxidation:
		7 mM periodic acid in 10 mM PPB (pH 9.0) in a Teflon tubing coil (0.5 mm × 10 m) at 65°C. 0.5 M acetic acid to halt rxn.
		
		Detection:
		• Fluorescence
		• $\lambda_{ex} = 330$ nm
		• $\lambda_{em} = 390$ nm
		Run Time <20 min

Asp (2004)	*Carbamate toxins*	*Sample Preparation:*
	1. STX (4 μg STX/100 g)	• 1. Extract STX with 0.1 N HCL at 90°C
	2. NEO (6 μg STX/100 g)	• 2. Protein Ppt. wit 5% tungstic acid at (9:1) with extract
	3. GTX 2 and 3 together	• 3. SPE Extraction (divinylbenzene/N-vinylpyrrolidone copolymer, Waters Corp., Milford, MA)
	(<1 μg STX/100 g) GTX	• 4. Microcentrifugation (0.2 μm filter)
	1 and 4 together (2 & 3 μg STX/100 g, respectively	*Reverse Phase Column*
		• Chrompack C8
		• 150 × 4.6 mm^2 i.d.
		• 5 μm particles
		Mobile Phase:
		• 2 mM sodium 1-heptanesulfonate in 30 mM APB (pH 7.1) in ACN (96:4)
		• 2 mM sodium 1-heptanesulfonate in 10 mM APB (pH 7.1) in ACN (96:4)
		Post-column oxidation
		• 7 mM periodic acid in 50 mM NPB (pH 9.0) in a reaction coil (0.5 mm × 10 m) at 85°C. 0.5 M acetic acid to halt rxn.
		• *Detection:*
		• Fluorescence
		• $\lambda_{ex} = 330$ nm
		• $\lambda_{em} = 390$ nm
		Run Time <20 min
Roberts on (2004)	1. STX	*Sample Preparation:*
		• Extract STX with 80% ethanol (pH2, HCl) at 5 mL/g tissue
		• Homogenize on ice
		• Sonicate for 10 min in ice bath. Repeat twice
		• Ultracentrifuge for 20 min at 4°C.
		• Repeat steps 1 thru 4 twice, pooling all supernatants
		• Ultrafiltration (0.2 μm nylon; Millipore Co, MA) and lyophilizing.
		• Dried extracts reconstituted in 0.05 M acetic acid (1 mL) and passed through 10,000 MWCO centrifugal ultra-filter (Microcon, Millipore)

(continued)

Table 10.3 (Continued)

Column:
- TSK-Gel Amide 80 column Toso Haas
- 250 × 4.6 mm 2 i.d.
- 5 μm particles

Mobile Phase:
- 40°C isocratic elution
- 2 mM ammonium formate, 3.6 mM formic acid in 70% ACN:water (at 0.2 ml min^{-1})

Special Treatment:
Ion spray interface introduction (9:1) waste to MS split
ESI interface

Detection:
MS/MS
- Positive ion
- HV capillary+4 kV
- Skimmer: 40 V
- Precursor ions: *m/z* 300 Da
- Isolation width: *m/z* 6
- Daughter ions scanned:
- *m/z* 100–400 Da
  - *Product ions (m/z)*
    STX: 300 → 282, 265, 240, 239, 204, 186

Run Time
<25 min

REFERENCES

1. *The American Heritage Stedman's Medical Dictionary*, 27th edn. Lippincott Williams and Wilkins. http://www.stedmans.com/. Accessed 12 July, 2005.
2. Franciosa, G., Pourshaban, M., De Luca, A., Buccino, A., Dallapiccola, B. and Aureli, P. Identification and Type A, B, E, and F botulinum neurotoxin genes and of botulinum neurotoxigenic clostridia by denaturing high-performance liquid chromatography. *Appl. Environ. Microbiol.*, **70**, 4170–4176 (2004).
3. Hardman, J. G., Limbard, L. E. and Goodman Gilman, A. (Eds), *Goodman & Gilman's The Pharmacological Basis of Therapeutics*, 10th edn. McGraw-Hill, New York, 2001.
4. Reich, D. L., Hollinger, I., Harrington, D. J., Seiden, H. S., Chakravorti, S. and Cook, D. R. Comparison of cisatracurium and vecuronium by infusion in neonates and small infants after congenital heart surgery. *Anesthesiology*, **101**, 1122–1127 (2004).
5. Rhoney, D. H. and Murry, K. R., National survey of the use of sedating drugs, neuromuscular blocking agents, and reversal agents in the intensive care unit. *J. Intensive Care Med.*, **18**, 139–145 (2003).
6. Arroliga, A., Frutos-Vivar, F., Hall, J., Esteban, A., Apezteguia, C., Soto, L. and Anzueto, A. Use of sedatives and neuromuscular blockers in a cohort of patients receiving mechanical ventilation. *Chest*, **128**, 496–506 (2005).
7. Hansen-Flaschen, J. H., Brazinsky, S., Basile, C. and Lanken, P. N. Use of sedating drugs and neuromuscular blocking agents in patients requiring mechanical ventilation for respiratory failure. A national survey. *JAMA*, **266**, 2870–2875 (1991).
8. Baselt, R. C. *Disposition of Toxic Drugs and Chemicals in Man,* 7th edn. Chemical Toxicology Institute, Foster City, CA, 2004, pp. 1043–1045.
9. Pitts, N. I., Deftereos, D. and Mitchell, G. Determination of succinylcholine in plasma by high pressure liquid chromatography with electrochemical detection. *Br. J. Anaesth.* **85**, 592–598 (2000).
10. Furata, T., Canfell, P. C., Castagnoli, K. P., Sharma, M. L. and Miller, R. D. Quantitation of pancuronium, 3-desacetylpancuronium, vecuronium, 3-desacetylvecuronium, pipecuronium and 3-desacetylpipecuronium in biological fluids by capillary gas chromatography using nitrogen-sensitive detection, *J. Chromatogr.*, **427**, 41–53 (1988).
11. Probst, R., Blobner, M., Luppa, P. and Neumeier, D. Quantification of the neuromuscular blocking agent rocuronium and its putative metabolite 17-desacetylrocuronium in heparinized plasma by capillary gas chromatography using nitrogen sensitive detector. *J. Chromatogr. B*, **702**, 111–117 (1997).
12. Gutteck-Amsler, U. and Rentsch, K. M. Quantification of the aminosteroidal non-depolarizing neuromuscular blocking agents rocuronium and vecuronium in plasma with liquid chromatography–tandem mass spectroscopy. *Clin. Chem.*, **46**, 1413–1414 (2000).
13. Kerskes, C. H. M., Lusthof, K. J., Zweipfenning, P. G. M. and Franke, J. P. The detection and identification of quaternary nitrogen muscle relaxants in biological fluids and tissues by ion-trap LC-ESI-MS, *J. Anal. Toxicol.*, **26**, 29–34 (2002).
14. Cirimele, V., Villain, M., Pépin, G., Ludes, B. and Kintz, P. Screening procedure for eight quaternary nitrogen muscle relaxants in blood by high-performance liquid chromatography–electrospray ionization mass spectrometry, *J. Chromatogr. B*, **789**, 107–113 (2003).
15. Various shellfish-associated toxins, in *Foodborne Pathogenic Microrganisms and Natural Toxins Handbook: The Bad Bug Book*. US Food and Drug Administration Center for Food Safety and Applied Nutrition, Washington, DC. Last update 28 January, 2004. http://www.cfsan.fda.gov/~mow/chap37.html. Accessed 8 July, 2005.
16. Bean, N. H., Goulding, J. S., Lao, C. and Angulo, F. J. *CDC. Surveillance for Foodborne-Disease Outbreaks – United States, 1988–1992*. MMWR 1991; 45(SS-5), pp 1–71, 25 October, 1996. http://www.cdc.gov/mmwr/PDF/SS/SS4505.pdf. Accessed 8 July, 2005.

17. Food and Drug Administration. *Action Levels for Poisonous or Deleterious Substances in Human Food and Animal feed.* US Department of Health and Human Services, Public Health Service, Washington, DC, 1985, pp. 1–13.

18. Franz, D. R. Defense against toxin weapons, in *Textbook of Military Medicine: Medical Aspects of Chemical and Biological Warfare,* 1994. http://www.usamriid.army.mil/education/defensetox/ toxdefbook.pdf. Accessed 2 June, 2005.

19. AOAC, Paralytic shellfish poisoning – biological method, final action. In *Official Methods of Analysis of the Association of Official Analytical Chemists,* 14th edn. AOAC, Arlington, VA, 1984, Sections 18.086–18.092.

20. Kerr, D. S., Briggs, D. M. and Saba, H. I. A neurophysiological method of rapid detection and analysis of marine algal toxins. *Toxicon,* **37,** 1803–1825 (1999).

21. Micheli, L., Di Stefano, S. Moscone, D., Palleschi, G., Marini, S., Coletta, M., Draisci, R. and delli Quadri, F. Production of antibodies and development of highly sensitive formats of enzyme immunoassay for saxitoxin analysis. *Anal. Bioanal. Chem.,* **373,** 678–684 (2002).

22. Manger, R. L., Leja, L. S., Lee, S. Y., Hungerford, J. M., Kirkpatrick, M. A., Yasumoto, T. and Wekell, M. M. Detection of paralytic shellfish poison by rapid cell bioassay: antagonism of voltage-gated sodium channel active toxins *in vitro. J. AOAC Int.,* **86,** 540–543 (2003).

23. Jellett, J. F., Roberts, R. L., Laycock, M. V., Quilliam, M. A. and Barrett, R. E. Detection of paralytic shellfish poisoning (PSP) toxins in shellfish tissue using MIST Alert™, a new rapid test, in parallel with the regulatory AOAC® mouse bioassay. *Toxicon,* **40,** 1407–1425 (2002).

24. Llewellyn, L. E., Dodd, M. J., Robertson, A., Ericson, G., de Koning, C. and Negri, A. P. Post-mortem analysis of samples from a human victim of a fatal poisoning caused by the xanthid crab, *Zosimus aeneus, Toxicon,* **40,** 1463–1469 (2002).

25. Usup, G., Leaw, C. P., Cheah, M. Y., Ahmad, A. and Ng, B. K. Analysis of paralytic shellfish poisoning toxin congeners by a sodium channel receptor binding assay. *Toxicon,* **44,** 37–43 (2004).

26. Lagos, N., Onodera, H., Zagatto, P. A., Andrinolo, D., Azevedo, S. M. F. Q. and Oshima, Y. The first evidence of paralytic shellfish toxins in the freshwater cyanobacterium *Cylindrospermopsis raciborskii,* isolated from Brazil. *Toxicon,* **37,** 1359–1373 (1999).

27. García, C., Bravo, M. d. C., Lagos, M. and Lagos, N. Paralytic shellfish poisoning: post-mortem analysis of tissue and body fluid samples from human victims in the Patagonia fjords. *Toxicon,* **43,** 149–158 (2004).

28. Sullivan, J. J., Wekell, M. M. and Kentala, L. L. Application of HPLC for the determination of PSP toxins in shellfish. *J. Food Sci.,* **50,** 26–29 (1985).

29. Oshima, Y. Post-column derivitization liquid chromatographic method for paralytic shellfish toxins. *J. AOAC Int.* **78,** 528–532 (1995).

30. Asp, T. N., Larsen, S. and Aune, T. Analysis of PSP toxins in Norwegian mussels by a post-column derivatization HPLC method. *Toxicon,* **43,** 319–327 (2004).

31. Robertson, A., Stirling, D., Robillot, C., Llewellyn, L. and Negri, A. First report of saxitoxin in octopi. *Toxicon.* **44,** 765–771 (2004).

11

History and Pharmacology of γ-Hydroxybutyric Acid

Laureen Marinetti

Montgomery County Coroners Office and Miami Valley Regional Crime Laboratory, 361 West Third Street, Dayton, OH 45402, USA

11.1 INTRODUCTION

γ-Hydroxybutyric acid (GHB) is an endogenous compound in the mammalian central nervous system and peripheral tissues and a minor metabolite or precursor of γ-aminobutyric acid (GABA), an inhibitory neurotransmitter. Laborit and his associates first synthesized GHB in 1960 as an experimental GABA analog for possible use in the treatment of seizure disorder (Laborit *et al.*, 1960). They hypothesized that if GHB could readily cross the blood–brain barrier then perhaps it could facilitate the synthesis of GABA in the brain. Although GHB did not produce elevated GABA synthesis, the research revealed that GHB had some properties that rendered it useful as an anesthetic adjuvant. The earliest pharmacological use of GHB in humans was in this application. Blumenfeld *et al.* (1962) and Sprince *et al.* (1966) investigated the potential anesthetic properties of GBL and 1,4BD. Their observations that sleep induction time was the shortest with GBL and longest with 1,4BD as compared with GHB was an early clue to the metabolic and structural relationship among these compounds (Figure 11.1). Additional studies demonstrated that the pharmacologically active form of GBL was in fact GHB (Roth *et al.*, 1966). Roth and Giarman (1965) determined that a lactonase enzyme in blood and liver rapidly catalyzed the hydrolysis of GBL to GHB.

A genetic disorder called GHB aciduria occurs when there is a deficiency of succinic semialdehyde dehydrogenase. Persons with this disorder have elevated concentrations of GHB in their blood, spinal fluid and urine (Gibson *et al.*, 1998). The clinical manifestations of the increased GHB concentration can range from mild oculomotor problems and ataxia to severe psychomotor retardation, but it is most commonly characterized by mental, motor and language delay accompanied by hypotonia.

Chromatographic Methods in Clinical Chemistry and Toxicology   Edited by R. L. Bertholf and R. E. Winecker
© 2007 John Wiley & Sons, Ltd

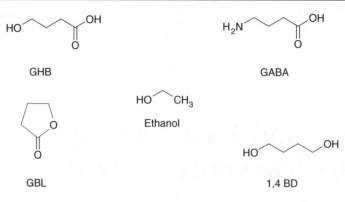

Figure 11.1   Structures of GHB, GBL, 1,4BD, GABA and ethanol

11.2 HISTORY OF ILLICIT USE OF GHB

In 1977, a study was published that would permanently change the relative obscurity of GHB, GBL and 1,4BD. Takahara *et al.* (1977) administered GHB to six healthy adult males and showed an approximately 10-fold increase in plasma growth hormone concentration that peaked at 45 min post-dose. This effect persisted for about 15 min and then the growth hormone concentration declined towards the pretreatment level. The growth hormone concentration at 120 min post-dose was still above baseline but significantly (two-thirds) below the peak concentration. Based on this report, bodybuilders assumed that they could increase growth hormone concentration by using GHB and thereby optimize their muscle-building potential. A more recent study by Van Cauter *et al.* (1997) showed that the increase in growth hormone secretion was correlated with the enhancement of slow wave sleep and growth hormone release did not occur prior to sleep onset. The growth hormone stimulating effect of a 2–3 g dose of GHB was seen during the first 2 h of sleep as an increase in amplitude and duration of the normal growth hormone secretory pulse associated with sleep onset as opposed to an increase in the number of growth hormone release pulses.

The use of GHB by bodybuilders seemed harmless in theory until emergency room reports associated with GHB toxicity started to accumulate (Centers for Disease Control, 1990). Users soon discovered that GHB had a definite mood elevating quality and introduced

Table 11.1   Common street names for GHB

Gamma OH	Liquid ecstasy
Sodium oxybate	Easy Lay
Natural Sleep 500	Salt water
Oxy-Sleep	Vita G
GHBA	Georgia Home Boy
G-caps	Grievous bodily harm
G	Great Hormones at Bedtime
Soap	Liquid X

GHB into the party drug scene, where it remains today. GHB is known by numerous street/ slang names (Table 11.1). Most of these slang names utilize the letters G, H and B, such as 'Georgia Home Boy' or 'Great Hormones at Bedtime'. The slang name 'liquid ecstasy' can be confusing, as 'ecstasy' or XTC is the slang name commonly associated with methylene-dioxymethamphetamine (MDMA), which is not in the same class of compounds as GHB. In November 1990, the US Food and Drug Administration (FDA) warned consumers of the danger of GHB, but the incidents of poisonings continued to rise. GHB and products containing GHB were removed from the market and GHB moved underground. Users soon discovered that GHB could easily be synthesized from readily available precursors. The industrial solvent γ-butyrolactone (GBL), when made basic with lye and heated, would yield GHB. With addition of an acid such as vinegar to adjust the pH, the solution of GHB was ready to consume. This illicit GHB is especially dangerous because its concentration is unknown and can vary greatly from batch to batch. Coupled with the fact that GHB has a very steep dose response curve, it is easy to overdose accidentally. Also, contaminants may be introduced in the clandestine manufacturing process. Toxicity of GHB is characterized by euphoria, dizziness, visual disturbances, decreased level of consciousness, nausea, vomiting, suppression of the gag reflex, bradycardia, hypotension, acute delirium, confusion, agitation, hypothermia, random clonic muscle movements (twitching), coma, respiratory depression and death. Although there have been some reports of seizures associated with GHB intoxication, there is no evidence of true seizure activity as measured by EEG in humans (Entholzner et al., 1995). However, only GHB doses consistent with safe anesthesia have been evaluated in these EEG studies. Clonic muscle movements and severe parasympathomimetic activity, including profuse salvation, defecation and urination, have been documented in dogs treated with large doses (toxic and lethal) of GHB (Lund et al., 1965). The clonic muscle movement was so prominent in the dogs that at anesthetic doses a barbiturate was also administered to effect convenient anesthesia. Another complicating factor is that GHB used outside a clinical setting is frequently used in combination with other drugs. This could affect the pharmacology of GHB in many ways depending upon which additional drug(s) are consumed and their dose. By far the most common drug taken in combination with GHB is ethanol (Louagie et al., 1997; Li et al., 1998; Centers for Disease Control and Prevention, 1999). This combination is especially dangerous because ethanol potentiates GHB's Central Nervous System (CNS) depressant effects, as demonstrated by depression of the startle response (a measure of sensory responsiveness) in rats (Marinetti and Commissaris, 1999). GHB has been implicated in fatalities both when administered alone (Marinetti et al., 2000) and when used in combination with other drug(s) (Ferrara et al., 1995). The use of GHB has been implicated in sexual assault. In fact, a common slang name for GHB is 'date rape drug', although it is not deserving of this title. Actual confirmed cases where GHB has been used in this capacity do exist but they are not common in comparison with other drugs more frequently chosen for this crime, namely ethanol, cannabinoids and benzodiazepines (ElSohly and Salamone, 1999). The most likely negative outcome of chronic GHB use is addiction. A GHB withdrawal syndrome has been documented with chronic GHB use (Hernandez et al., 1998; Craig et al., 2000; Dyer et al., 2001). The clinical presentation of GHB withdrawal ranges from mild clinical anxiety, agitation, tremors and insomnia to profound disorientation, increasing paranoia with auditory and visual hallucinations, tachycardia, elevated blood pressure and extra-ocular motor impairment. Symptoms, which can be severe, generally resolve without sequelae after various withdrawal periods, although one documented death has occurred. Treatment with benzodia-zepines has been successful for symptoms of a mild withdrawal syndrome.

11.3 CLINICAL USE OF GHB IN HUMANS

As discussed at the beginning of this chapter, the first clinical use of GHB was as an anesthetic adjuvant or induction agent. This application is still in use today in Europe (Kleinschmidt *et al.*, 1999). In the USA, the only medically approved use of GHB is as a treatment for cataplexy associated with the disease narcolepsy. The medically formulated GHB approved by the FDA is called Xyrem® and is a controlled substance (Federal Schedule III). This research has been ongoing since the 1970s and subjects participating in the clinical trails have experienced few adverse effects (Broughton and Mamelak, 1979). Another promising area for medically formulated GHB is in the treatment of alcohol withdrawal syndrome. This is not surprising since ethanol and GHB are similar compounds, both in structure and pharmacology (Figure 11.1). Cross-tolerance between ethanol and sub-anesthetic doses of GHB has been observed in rats, which may explain why alcoholics being treated with GHB do not experience sedation at doses that would sedate a non-alcohol-tolerant individual (Colombo *et al.*, 1995). Adverse effects have been mild except for occasional replacement of alcohol addiction with GHB addiction, resulting in some subjects self-medicating with additional GHB to enhance its effects (Addolorato *et al.*, 1997; Beghè and Carpanini, 2000). Treatment with GHB has also been investigated for opiate withdrawal syndrome (Gallimberti *et al.*, 1993), cocaine addiction (Fattore *et al.*, 2000) and fibromyalgia (Scharf *et al.*, 1998). For an in-depth review of GHB use in the treatment of withdrawal, see the April 2000, Volume 20, Issue 3 of the journal *Alcohol*.

11.4 HISTORY OF ILLICIT USE OF GBL AND 1,4BD

On 18 February 2000, GHB was placed in Federal Schedule I of the Controlled Substances Act with GBL as both a list I chemical and a controlled substance analog and 1,4BD falling under the controlled substance analog section (US Congress, 2000). Unfortunately, scheduling has not curbed the illicit use of this trio. With the placement of GHB in Federal Schedule I, there is more interest on the part of the illicit manufacturers in producing a GHB product that will stay in the lactone form and not spontaneously convert to GHB because penalties for GBL are less severe. In fact, a seizure of solid GBL has been reported in California where the liquid GBL was adsorbed on silicon dioxide powder, which was then placed in a clear capsule (DEA Southwest Laboratory, 2001). This adds additional danger because the lactone form, based on its physical characteristics and its increased solubility in lipids, has been shown in animal studies to be absorbed by the gut more efficiently than the GHB acid. The FDA has requested removal of health supplement products containing GBL, but this is a small fraction of the products that contain this compound (Food and Drug Administration, 1999a). The most common use of GBL is as an industrial solvent with domestic production of approximately 80 000 tons per year. In industry, it is widely used and would be difficult to replace based on its excellent properties as a safe, effective and biodegradable degreaser. Some manufacturers of diet aid products containing GBL have masked the presence of this ingredient by using one of the many chemical synonyms for GBL in the list of ingredients on the product label (Table 11.2).

GBL has been detected in alcoholic beverages, tobacco smoke, coffee, tomatoes, cooked meats and several foodstuffs (IARC, 1999). GBL produces the same pharmacological effects as GHB as it is rapidly converted to GHB in the body. At equimolar doses GBL produced a

Table 11.2 Chemical synonyms for GBL

Dihydro-2(3*H*) furanone	4-Hydroxy-γ-lactone	4-Deoxytetronic acid
Butyrolactone-4	Butyrylactone	1,2-Butanolide
4-Butyrolactone	Butyrl lactone	1,4-Butanolide
Butyric acid	Gamma-6480	4-Hydroxybutyric acid γ-lactone
Butyric acid lactone	Gamma BL	γ-Hydroxybutyric acid cyclic ester
Butyrolactone	BLO or BLON	NCI-C55875
α-Butyrolactone	4-Butanolide	Tetrahydro-2-furanone

more prolonged hypnotic effect in rats as compared with GHB (Lettieri and Fung, 1978). To a lesser extent, 1,4BD has followed the same path as GBL, and is gaining in popularity and has recently been associated with adverse events, including death (Kraner *et al.*, 2000; Zvosec *et al.*, 2001). As with GBL, the major use for 1,4BD in the USA is as an industrial compound. Unlike GBL, however, 1,4BD is not used to manufacture GBL or GHB illicitly. This conversion is an industrial process and cannot be accomplished in a household setting. The pharmacological effects of 1,4BD are ultimately those of GHB, which is produced metabolically from 1,4BD (Figure 11.2). With the increased attention on GHB toxicity, the

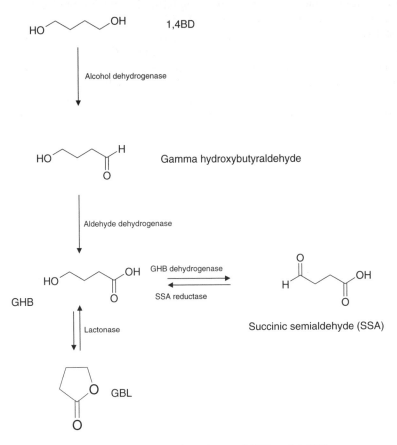

Figure 11.2 Metabolic pathway of GBL and 1,4BD

Table 11.3   Chemical synonyms for 1,4BD

Butane-1,4-diol	Sucol B
1,4-Butylene glycol	1,4-Tetramethylene glycol
1,4-Dihydroxybutane	Butylenes glycol
Diol 14B	Tetramethylene-1,4-diol

FDA has also requested that products containing 1,4BD be removed from the market (Food and Drug Administration, 1999b). As with GBL, this action could also lead manufacturers to replace the name, 1,4-butanediol, with one of its many chemical synonyms to disguise its presence in the product (Table 11.3).

In addition to GBL and 1,4BD, the recent emergence of diet aid products containing γ-valerolactone (GVL) has caused some concern as to the pharmacological effects of this compound and its safety for human consumption. GVL is an FDA-approved food additive and/or adjuvant and it is safe for human consumption at $0.0009322 \, \text{mg kg}^{-1}$ per day per person (FDA data). One currently available diet aid product, Tranquili-G, recommends a dose of GVL around 3 g, which is much greater than the amount that was approved by the FDA for safe human consumption. In rats and rabbits, administration of GVL produced the following effects; marked muscular weakness, mild anesthesia and an increase in the rate of respiration with oral LD_{50} values for GVL at 8.8 and $2.5 \, \text{g kg}^{-1}$ for rats and rabbits respectively (Deichmann et al., 1945). Past research has documented that GVL is quickly metabolized by the lactonase enzyme to 4-Me-GHB in human blood and rat liver microsomes (Fishbein and Bessman, 1966), similar to the way in which GBL is metabolized by this enzyme to GHB (Figures 11.2 and 11.3). 4-Me-GHB is a structural analog of GHB but GVL is not. It has been demonstrated in the rat model that GVL has similar pharmacological properties to GHB mediated through the 4-Me-GHB metabolite (Marinetti, 2003).

11.5 DISTRIBUTION AND PHARMACOKINETICS OF GHB, GBL AND 1,4BD

Normal endogenous concentrations of GHB in CSF in humans are dependent on age and the presence of seizure disorder. Infants had higher concentrations of GHB ($0.26-0.27 \, \mu\text{g mL}^{-1}$) in the CSF than older children ($0.11-0.13 \, \mu\text{g mL}^{-1}$), who, in turn, had higher concentrations than adults ($0.02-0.03 \, \mu\text{g mL}^{-1}$). Children with myoclonic-type seizures had the highest concentrations of GHB in the CSF ($0.78-0.97 \, \mu\text{g mL}^{-1}$), whereas children with other types

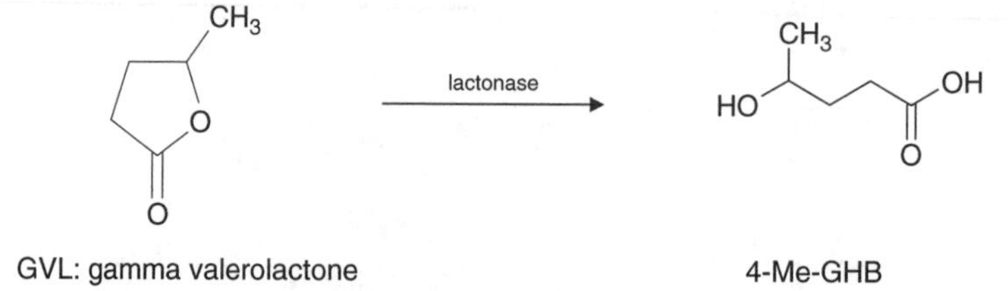

GVL: gamma valerolactone   4-Me-GHB

Figure 11.3   Production of 4-Me-GHB from GVL *in vivo*

of seizures had the next highest concentrations (0.37–0.48 µg mL^{-1}) (Snead et $al.$, 1981). Along with the analysis of the CSF for GHB, the serum of all the subjects was also analyzed. Of the 130 subjects, none had any measurable amount of GHB in the serum, with a limit of sensitivity of the assay of 0.002 µg mL^{-1}.

In another study, GHB concentrations were determined in various tissues and blood of rats (Nelson et $al.$, 1981). The results showed that brown fat, kidney, muscle and heart showed concentrations >1 µg g^{-1}. Liver, lung, blood, brain and white fat had concentrations <0.25 µg g^{-1}. The reason for the variations in GHB content between the specimens analyzed is not known. A study that compared endogenous GHB concentrations in brain from human, monkey and guinea pig showed an elevated concentration compared with that seen in rat brain, 0.2–2, 0.3–1.8 and 0.1–0.6 µg g^{-1}, respectively (Doherty et $al.$, 1978). In this same study, endogenous GHB in CSF and blood of humans was found in only trace amounts (CSF ~0.01 µg mL^{-1}).

The distribution of GHB into the CSF appears to lag behind that in blood or brain. After a 500 mg kg^{-1} intravenous dose of GHB had been administered to dogs, the plasma concentration peaked within 5 min and the brain concentration peaked within 10 min, but it was 170 min before CSF concentrations reached their maximum (Shumate and Snead, 1979). This suggests a passive diffusion of GHB from serum or brain into the CSF. In alcohol-dependent patients GHB did not accumulate in the body on repeated doses nor did it exhibit any protein binding. The mean peak plasma concentrations of therapeutic oral doses of 25 and 50 mg kg^{-1} of GHB per day given to 50 alcohol withdrawal syndrome patients were 55 µg mL^{-1} (range 24–88) and 90 µg mL^{-1} (range 51–158), respectively (Snead et $al.$, 1976; Ferrara et $al.$, 1993).

Absorption of GHB has been shown to be a capacity-limited process with increases in dose resulting in increases in time to peak concentration. The concentration in brain equilibrates with other tissues after approximately 30 min. GHB crosses the placental barrier at a similar rate to that in the blood–brain barrier (Van der Pol et $al.$, 1975). GHB also exhibits first-pass metabolism when given orally with about 65% bioavailability when compared with an equivalent intravenous dose.

GHB exhibits zero-order (constant rate) elimination kinetics after an intravenous dose. Since GHB exhibits zero-order kinetics, it has no true half-life. The time required to eliminate half of a given dose increases as the dose increases. A daily therapeutic dose of 25 mg kg^{-1} has an apparent half-life of about 30 min in humans, as determined in alcohol dependent patients under GHB treatment (Ferrara et $al.$, 1992). In contrast, an apparent half-life of 1–2 h was observed in dogs when they were given high intravenous doses of GHB. In humans it has been documented that there is increased rate of absorption if GHB is administered on an empty stomach, resulting in a reduced time to reach the maximum plasma concentration of GHB (Borgen et $al.$, 2001).

As discussed previously, the absorption of GBL has been documented to occur faster than GHB. It has also been proposed that in addition to the better absorption of GBL, it may also distribute differently than GHB. An early study comparing the distribution of equimolar doses of GHB and GBL in rats found that although peak plasma concentrations were higher with GHB they remained elevated longer with GBL. In addition, concentrations of GBL in the lean muscle mass of the rat were always elevated compared with concentrations of GHB (Roth and Giarman, 1966). This suggests sequestration of GBL into lean muscle prior to its conversion to GHB. Since lean muscle does not contain the lactonase enzyme, it is conceivable that this could occur. The GBL could then redistribute into the blood to be converted to GHB by the blood or liver lactonase enzyme. This could explain the prolonged elevated concentrations of GHB in blood that occur when GBL is given. Neither GHB nor GBL is sequestered in fat.

The absorption and distribution of 1,4BD is similar to that of GHB. It is a lipophobic or polar compound so it does not absorb faster than GHB. After its absorption, it requires a two-step enzymatic conversion to GHB that results in a slightly longer time to peak GHB concentration and also a longer time of elevated GHB concentration. The conversion process of 1,4BD to GHB can be slowed or inhibited by ethanol, pyrazole or disulfiram.

11.6 GHB INTERPRETATION ISSUES AND POST-MORTEM PRODUCTION

The most difficult aspect of GHB analysis is interpretation of the numbers that are generated. Helrich et al. (1964) correlated blood concentrations of GHB with state of consciousness in 16 adult human patients (Table 11.4). This study revealed that GHB blood concentrations as high as 99 $\mu g\,mL^{-1}$ could be achieved with the patient still displaying an 'awake' state, the subject spontaneously coming in and out of consciousness characterized a light sleep state. Subjects in the medium sleep state were clearly asleep but were able to be roused. At the highest concentrations studied, GHB produced a deep sleep characterized by response to stimuli with a reflex movement only. It is clear from these data that blood concentrations of GHB display a large overlap across the four states of consciousness described. For example, a subject with a blood concentration of 250 $\mu g\,mL^{-1}$ could be in a light, medium or deep sleep state. The smallest dose given, 50 $mg\,kg^{-1}$, produced peak plasma concentrations no greater than 182 $\mu g\,mL^{-1}$ and the largest dose given, 165 $mg\,kg^{-1}$, produced peak plasma concentrations $>416\,\mu g\,mL^{-1}$. Fourteen patients received doses of 100 $mg\,kg^{-1}$, resulting in peak blood concentrations ranging from 234 to 520 $\mu g\,mL^{-1}$. Twelve of the 16 patients required intubation, but the need for intubation did not necessarily correlate with those patients who received the higher doses of GHB. Metcalf et al. (1966) observed electroencephalographical (EEG) changes in 20 humans given oral doses of GHB in the range 35–63 $mg\,kg^{-1}$. The EEG pattern was similar to that seen in natural slow wave sleep. Profound coma was observed at approximately 30–40 min post-dose in subjects given oral GHB doses $>50\,mg\,kg^{-1}$.

Animal and human studies have demonstrated that endogenous GHB concentrations can rise post-mortem and under inappropriate specimen storage conditions. Doherty et al. (1978) observed an increase in the GHB concentrations in brain specimens after 6 h with a further increase if the specimens were left at room temperature. Snead et al. (1981) also observed an increase in GHB concentrations in CSF after 12 h of storage at room temperature. It was

Table 11.4 GHB concentrations associated with state of consciousness

State of consciousness	GHB ($\mu g\,mL^{-1}$)
Awake	0–99
Light Sleep	63–265
Medium Sleep	151–293
Deep Sleep	344–395

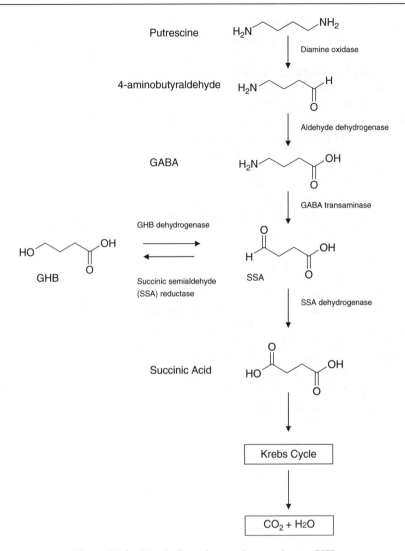

Figure 11.4   Metabolic pathway of putrescine to GHB

subsequently discovered that if animals were killed using microwave irradiation, post-mortem GHB accumulation was blocked (Eli and Cattabeni, 1983). This suggests some type of enzymatic conversion from a GHB precursor. One source of the post-mortem GHB increase is the metabolism of previously sequestered GABA that is being released from storage vesicles as the natural decomposition process occurs. Excess GABA would be exposed to the GABA transaminase enzyme, which could convert it to succinic semialdehyde, which could in turn be converted to GHB in addition to proceeding on to succinic acid (Figure 11.4).

Another source of post-mortem GHB production is putrescine (1,4-butanediamine), a biogenic polyamine initially detected in decaying animal tissues, but now known to be present in all cells, both eukaryotic and prokaryotic, where it is important for cell proliferation and differentiation (Merck, 1996). Research on polyamine metabolism by Seiler (1980) demonstrated the formation of GABA from putrescine both in visceral organs and in the CNS of vertebrates. This is an enzymatic process in the polyamine metabolic pathway that involves diamine oxidase (DAO) and aldehyde dehydrogenase to form GABA (Figure 11.4). In organs that do not contain high activity of DAO, such as brain, an alternative pathway is available for the conversion of putrescine to GABA (Sessa and Perin, 1994). In addition, Snead et al. (1982) observed an 80–100% increase in GHB concentrations in rat brain after intracerebroventricular administration of putrescine. All of these theories are consistent with the observation that microwave irradiation prevents post-mortem accumulation of GHB as the exposure to the microwaves would denature the enzymes. This is also supported by the fact that excessive GHB production is not seen in blood specimens that have an enzyme inhibitor added (Stephens et al., 1999).

Regardless of the source of the increased concentration of GHB post-mortem, it can be a significant problem in determination of a cause of death due to GHB toxicity. Post-mortem production of GHB can result in blood concentrations of GHB that would produce significant effects in a living person. Anderson and Kuwahara (1997) analyzed heart blood, femoral blood and urine from 96 post-mortem cases with no suspected exogenous GHB use and 50 ante-mortem blood specimens also with no evidence of GHB use. The specimens were stored at 4 °C with sodium fluoride added to the blood as a preservative. The following results were obtained for the post-mortem specimens: heart blood 1.6–36, femoral blood 1.7–48, and urine 0–14 µg mL^{-1}, and no detectable amount of GHB in any of the ante-mortem blood specimens with a limit of detection of 0.5 µg mL^{-1}. The upper end of the post-mortem blood range overlaps the range detected in blood during therapeutic application of a 25 mg kg^{-1} per day dose of GHB, 24–88 µg mL^{-1}. If one compares this with the highest concentrations that were discussed earlier in ante-mortem blood and CSF, 1 µg mL^{-1} in the CSF of children with myoclonic-type seizure disorder, it is obvious that these are significant concentrations. Also, the fact that there is very little difference between GHB concentrations in heart and femoral blood suggests that this is not simply a post-mortem redistribution phenomenon. In a similar study by Fieler et al. (1998), a range of 0–168 µg mL^{-1} (average 25 µg mL^{-1}) in post-mortem blood was observed with no detectable GHB in post-mortem urine or ante-mortem blood or urine with a limit of detection of 1 µg mL^{-1}. Although these data showed a larger concentration range than the data of Anderson and Kuwahara (1997), an average of 25 µg mL^{-1} for 20 blood specimens indicates that the majority of the concentrations were at the lower end of the range. Fieler et al. (1998) did not discuss the storage conditions of the specimens prior to analysis, which has been shown to have an effect on the amount of post-mortem GHB produced. Stephens et al. (1999) compared post-mortem GHB concentrations in samples under various storage conditions. They found that GHB concentrations increased by 50% if the specimen did not contain sodium fluoride even when it was stored in a refrigerator, and if the specimen was stored at room temperature without 10 µg mL^{-1} sodium fluoride the concentration could double. The addition of sodium fluoride did not affect the concentration of GHB in urine specimens. Ante-mortem blood buffered with citrate (yellow top tube) has been shown to display an increase in GHB concentration over time (LeBeau et al., 2000a). Ten ante-mortem citrate-buffered whole blood specimens were analyzed for GHB after various storage periods from 6 to 36 months at −20 °C. Although no exogenous GHB use was suspected, all of the specimens had concentrations of GHB ranging from 4 to 13 µg mL^{-1} with a mean of 9 µg mL^{-1}.

The real problem this presents is in the interpretation of exogenous GHB use, GHB toxicity or GHB overdose resulting in a fatality. Since GHB is rapidly cleared from the body, even at elevated doses, if there is any survival time the blood concentrations can easily fall into the range of post-mortem production. Therefore, a urine specimen should be collected in addition to blood in suspected GHB cases. In post-mortem cases, if urine is not available, then eye fluid or CSF is indicated, especially CSF because the normal endogenous concentration of GHB in CSF is well documented. It is advised that all of the specimens collected be preserved with at least 2% sodium fluoride and stored at refrigerator temperature or frozen and that tubes containing citrate be avoided. GHB concentrations in post-mortem eye fluid and urine from decedents with no exogenous GHB exposure have been documented to be $<10\,\mu g\,mL^{-1}$ (Marinetti et al., 2005). Organ specimens may also be helpful, but again there are very few published data on endogenous concentrations of GHB in various organs in humans. The cut-off concentration for reporting exogenous GHB consumption in a specimen must be set above the suspected post-mortem production or ante-mortem endogenous GHB concentration.

Data from analyses of multiple specimens and a good case history can help tremendously, especially if litigation is involved. For example, in a case that involved the death of a young girl and GHB, a post-mortem blood GHB concentration of $15\,\mu g\,mL^{-1}$ was demonstrated. This concentration is in the range of post-mortem GHB production. However, a detailed case history plus a post-mortem urine GHB concentration of $150\,\mu g\,mL^{-1}$, an ante-mortem blood concentration of $510\,\mu g\,mL^{-1}$, an ante-mortem urine concentration of $2300\,\mu g\,mL^{-1}$ and the lack of any other significant autopsy findings made the cause of death an obvious GHB fatality. The decedent had a 14-h survival time in the hospital prior to death, which explains the low post-mortem blood GHB concentration (Marinetti et al., 2000). A second GHB fatality showed a heart blood concentration of $66\,\mu g/mL$, a femoral blood concentration of $77\,\mu g/mL$, an eye fluid concentration of $85\,\mu g/mL$ and a urine concentration of $1260\,\mu g/mL$. This individual had an extensive history of chronic GHB use and no other cause of death (Marinetti et al., 2000).

Ante-mortem blood and urine endogenous levels of GHB have been documented to be $<1\,\mu g\,mL^{-1}$ and $<10\,\mu g\,mL^{-1}$, respectively. Anderson Kuwahara (1997) analyzed 50 ante-mortem blood specimens from individuals with no evidence of GHB use. No detectable amounts of GHB were observed in any of the blood specimens, using a limit of detection of $0.5\,\mu g\,mL^{-1}$. Similarly, endogenous GHB concentrations were measured in 192 blood specimens from living subjects thought to be non-GHB users (Couper and Logan, 1999). All measurable blood GHB concentrations were below $1\,\mu g\,mL^{-1}$, using a detection limit of $0.5\,\mu g\,mL^{-1}$. LeBeau et al. (2000b, 2002) investigated urinary GHB levels to differentiate between endogenous and exogenous concentrations. Every urine void produced by three non-GHB using male subjects over a 1-week period was individually collected and analyzed for the presence of endogenous GHB, using a limit of detection of $0.19\,\mu g\,mL^{-1}$. Overall, 129 urine specimens were analyzed and the mean endogenous GHB concentration detected was $1.58 \pm 1.55\,\mu g\,mL^{-1}$ (range <0.19–$6.62\,\mu g\,mL^{-1}$). Individual mean concentrations in the three subjects were 0.63 ± 0.34, 3.02 ± 1.52 and $0.56 \pm 0.31\,\mu g\,mL^{-1}$, respectively. Urine specimens from females were also analyzed with an endogenous GHB range of 0.00–$1.94\,\mu g\,mL^{-1}$. Although there were significant intra- and inter-individual variations in the urinary levels of endogenous GHB, the concentrations did not fluctuate to levels that were higher than the laboratory's reporting level for urinary GHB of $10\,\mu g\,mL^{-1}$.

11.7 ANALYSIS FOR GHB, GBL AND 1,4BD

GHB does not exist in a static state, even outside the body. In solution, it exists in a dynamic state in equilibrium with its lactone, GBL (Figure 11.2). The ratio of the two forms is dependent on the pH of the matrix or the type of matrix containing the GHB. For example, in blood, the GHB acid form predominates because the lactonase enzyme converts any of the lactone to the acid. However, if the GHB is in a matrix that does not contain this enzyme, such as urine, water or juice, the two forms will reach equilibrium. The pK_as for GHB and 4-Me-GHB are 4.72 and 5.72, respectively, therefore the lactone form predominates at pH below these values and the acidic or salt form predominates above these values. Complete conversion to GBL is favored in dehydrating conditions in a concentrated acid solution (pH < 2). The rate at which the equilibrium is achieved depends on the temperature and the actual pH of the matrix.

There are two basic approaches to GHB and GBL analysis, depending on the matrix being analyzed. If a biological matrix is being analyzed, then conversion of GHB to GBL or derivatization of the GHB without conversion is acceptable, as a biological matrix should not normally contain any GBL with the exception of stomach contents. However, it is possible that a urine specimen (or any other biological specimen that does not contain the lactonase enzyme) with a low pH (5–6) might produce GBL from GHB upon long-term storage. Two studies that demonstrated this were done by Ciolino and Mesmer (2000) and Ciolino et al. (2001). The authors compared solutions of GHB and GBL in various matrices at various pH values and found that the relative amounts of GHB and GBL changed depending on the pH of the matrix and storage conditions, specifically time and temperature. Anderson et al. (2000) reported a fatality that displayed measurable concentrations of GBL in heart and femoral blood, vitreous humor, liver, bile, gastric contents and urine. Perhaps this is evidence of a massive GBL ingestion that caused saturation of the lactonase enzyme since the concentrations of GHB in this case were very high with a heart blood concentration of 1473 $\mu g \, mL^{-1}$. However, if 1,4BD is ingested, because it is not in equilibrium with GHB or GBL, unchanged 1,4BD may be detected in a biological specimen if enough is ingested and/ or the time interval since ingestion is short (McCutcheon et al., 2000). Analysis of illicit or commercial products to determine GHB or GBL require either that two aliquots of sample be analyzed if the GHB to GBL conversion is used, or that a derivatization method is used so that the percentage composition of GHB in the sample can be determined. Knowing the actual percentage of GHB and GBL in the sample may be important in the documentation of a product source in both criminal and civil litigation cases.

Biological specimen extraction can be accomplished by liquid–liquid, solid-phase or solid-phase microextraction with subsequent detection of GHB or GBL by gas chromatography–mass spectrometry (GC-MS) using electron ionization (EI), positive or negative chemical ionization (CI) or gas chromatography with flame ionization detection (GC-FID). LeBeau et al. (1999) describes a method that employs two aliquots of specimen. The first is converted to GBL with concentrated sulfuric acid while the second is extracted without conversion. A simple liquid–liquid methylene chloride extraction was utilized, and the aliquots were then screened by GC-FID without derivatization. Specimens that screened positive by this method were then re-aliquoted and subjected to the same extraction with the addition of the deuterated analog of GBL. The extract was then analyzed by headspace GC-MS in the full-scan mode. Quantitation was performed by comparison of the area of the molecular ion of the parent drug (m/z 86) with that of the hexadeuterated analog (m/z 92). This method displayed linearity in both blood and urine from 5 to 1000 µg mL^{-1} with recoveries that ranged between 75 and 87%.

A liquid–liquid extraction method employed by the toxicology laboratory in the Department of the Coroner of Los Angeles County California (Anderson and Kuwahara, 1997) first converts GHB to the lactone followed by chloroform extraction. Both blood and urine were analyzed using this method with γ-valerolactone as the internal standard. Detection was by GC-MS using the selected-ion monitoring (SIM) mode. The ions monitored were GBL m/z 42, 56 and 86 with m/z 41, 56 and 85 for the internal standard. The method was linear from 5 to 300 µg mL^{-1}, with a lower limit of detection of 5 µg mL^{-1}. However, owing to the introduction of commercial products containing GVL, a different internal standard is recommended.

Couper and Logan (2000) describe a liquid–liquid extraction method that uses ethyl acetate to extract GHB, without conversion to GBL. The extract is derivatized using N,O-bis(trimethylsilyl)trifluoroacetamide (BSTFA) with 1% trimethylchlorosilane (TMCS) and acetonitrile and then analyzed using GC-MS in the EI mode with SIM of m/z 233, 204 and 117 for GHB and m/z 235, 103 and 117 for the diethylene glycol internal standard. The method gave a 55% extraction recovery of GHB in blood with a limit of detection in blood and urine of 0.5 µg mL^{-1} and a linearity of 1–100 µg mL^{-1} in blood and 1–200 µg mL^{-1} in urine. Another liquid–liquid extraction that is very similar to Couper and Logan's method has been described by Elian (2000). The differences between this method and the Couper and Logan method are that it uses GHB-d_6 as the internal standard; it was evaluated only in urine specimens, and gave an 80–85% recovery with linearity from 2 to 50 µg mL^{-1}.

There are three different solid-phase extraction (SPE) methods that differ with respect to the column used, pre-extraction sample treatment, post-extraction sample treatment and internal standard. All three SPE methods extract GHB from urine and blood, derivatize with BSTFA with 1% TMCS, and detect analytes by GC-MS in the EI mode with either SIM or the full-scan mode. The ions monitored for the GHB di-TMS derivative are m/z 233, 234 and 235 with care being taken to avoid the ions that are common between di-TMS GHB and di-TMS urea, at m/z 147, 148 and 149. Urea is a naturally occurring compound in urine and care must be taken that its derivative does not interfere.

The SPE method used by the Miami–Dade County Medical Examiner's Office, Toxicology Laboratory, in Miami, Florida (Andollo and Hearn, 1998) uses Chem-Elute® SPE columns, β-hydroxybutyric acid internal standard, pretreatment of urine with sulfuric acid and pretreatment of blood with sodium tungstate and sulfuric acid. The Dade County method gave an absolute recovery of 30% with a limit of detection of 2 µg mL^{-1} and a limit of quantitation of 10 µg mL^{-1}.

McCusker *et al.* (1999) described a method for urine that uses a United Chemical Technologies GHB SPE column. Urine, GHB-d_6 internal standard and phosphate buffer are combined and extracted. The final eluate is taken to dryness and then reconstituted with hexane and dimethylformamide (DMF). After a simple liquid–liquid extraction, the hexane layer is discarded, the DMF is taken to dryness and the residue is derivatized with BSTFA with 1% TMCS in ethyl acetate. The method gave a linearity range from 5 to 500 µg mL^{-1}. The limit of sensitivity and recovery of the method were not reported.

The United Chemical Technologies GHB SPE column method has been modified to include GHB and 1,4BD analysis in blood, eye fluid and tissue homogenate samples. The sample size is only 200 µL and requires a sample preparation step of extraction with acetone prior to elution on the SPE column. The eluent from the SPE column, which is in 99:1 methanol–ammonium hydroxide, is then taken to dryness and derivatized with BSTFA with 1% TMCS. The ions monitored are at m/z 233, 234 and 235 for GHB di-TMS, m/z 239, 240 and 241 for GHB-d_6 di-TMS, m/z 219, 220 and 221 for 1,4BD di-TMS and m/z 223, 224 and 225 for 1,4BD-d_4 di-TMS (Kraner *et al.*, 2000; Crifasi and Telepchak, 2000).

The third SPE method utilizes a Multi-Prep® Anion Exchange GVSA-200 gravity flow column and is used for urine or blood analysis (Biochemical Diagnostics, 2000). This is a very simple SPE procedure that involves mixing the specimen with pH 9 Tris buffer prior to extraction and passing all solutions through the column by gravity flow. This investigational method utilized GHB-d_6 as the internal standard and displayed linearity from 1 to 500 µg mL^{-1} with a recovery >90%.

Frison *et al.* (1998) describes a method using solid-phase microextraction of GHB in plasma and urine. This is a new approach for GHB analysis that shows promise in that it is simple, sensitive and requires only 0.5 mL of specimen. The linearity range was from 1 to 100 µg mL^{-1} in plasma and from 5 to 150 µg mL^{-1} in urine with a limit of detection of 0.05 and 0.1 µg mL^{-1} for plasma and urine, respectively. The limit of detection was calculated based on aqueous solutions because the blank plasma and urine specimens had endogenous GHB concentrations of 0.1–0.2 and 0.5–1.5 µg mL^{-1}, respectively. The method required conversion of GHB to GBL with GBL-d_6 as the internal standard and detection by headspace GC-MS with spectra from both CI and EI ionization modes. Many methods are available for GHB and GBL analyses depending on the equipment and resources available to the laboratory.

For laboratories interested in measuring concentrations of GHB at or below endogenous levels, a GC-MS method with negative ionization has been described for the measurement of GHB in rat brain (Ehrhardt *et al.*, 1988). It involves a complicated liquid–liquid extraction with subsequent derivatization of GHB with MTBSTFA. This method is very sensitive, with the capability of detecting as little as 2 pg of GHB. The calibration curve showed linearity from 0 to 64 ng of GHB. Another method, described by Shima *et al.* (2005),

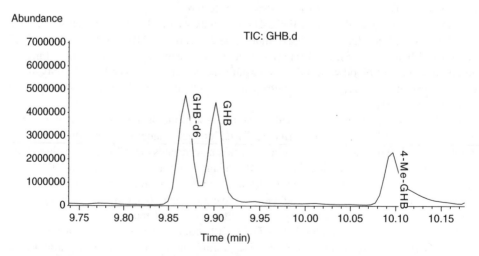

Figure 11.5 Total ion chromatogram of GHB, GHB-d_6 and 4-Me-GHB di-TMS derivatives

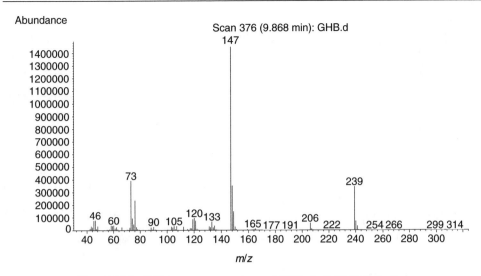

Figure 11.6 Full-scan mass spectrum of GHB-d_6 di-TMS derivative

would also be appropriate with a linear range of 0.003–3.0 μg mL^{-1}. This method involves a simple liquid extraction with subsequent BSTFA derivitization and GC-EI-MS analysis.

However, the same is not true for 1,4BD analysis. Concern over detecting and quantifying 1,4BD is just becoming an issue in the forensic community. As discussed previously, the pharmacologically active form of 1,4BD is GHB. However, determination of the concentration of 1,4BD may be useful in documenting acute intoxication. McCutcheon *et al.* (2000) utilized a simple one-step extraction at physiological pH into

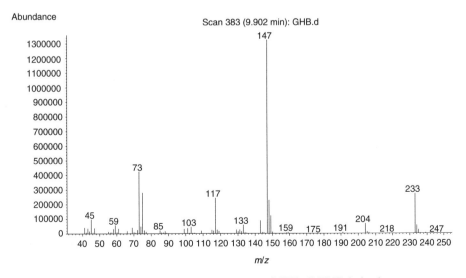

Figure 11.7 Full-scan mass spectrum of GHB di-TMS derivative

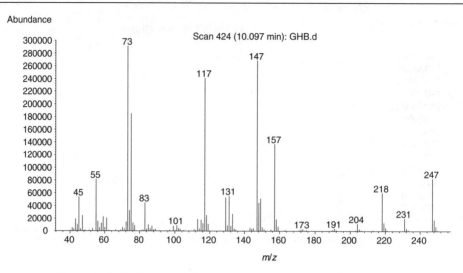

Figure 11.8   Full-scan mass spectrum of 4-Me-GHB di-TMS derivative

n-butyl chloride for the extraction of 1,4BD. The solvent was dried down to about 75 μL and subjected to GC-MS analysis. The 1,4BD eluted prior to GBL with a detection limit between 50 and 100 μg mL^{-1}. Research by the authors is under way to improve this method. A method specific for the detection of 1,4BD in liver and brain tissue was described by Barker *et al.* (1985). This method involved a complicated extraction scheme, which utilized different extraction protocols for the aqueous and lipid fractions of the tissues. Both fractions were then lyophilized, extracted again and subsequently derivatized with heptafluorobutyric anhydride (HFBA). The internal standard used was deuterated 1,4BD with identification and quantitation by GC-EI-MS in the SIM mode. Both rat and human brain tissue were analyzed in addition to rat liver. The method demonstrated a recovery of 74 ± 10% for the aqueous fraction and 88 ± 9% for the lipid fraction. The linearity range was 0.0–1.0 μg g^{-1} wet weight of tissue with a limit of detection of 0.01 μg g^{-1} wet weight of tissue.

Quantitation of 1,4BD by high-performance liquid chromatography (HPLC) with photo-diode-array detection was described by Duer *et al.* (2001). The 1,4BD was extracted from the biological matrix with acetonitrile and derivatized with 3,5-dinitrobenzoyl chloride. This method has a linear range from 3 to 250 mg L^{-1} with a limit of quantitation of 10 mg L^{-1}. Also, as was mentioned previously, the United Chemical Technologies GHB SPE column method has been modified to include 1,4BD analysis (Kraner *et al.*, 2000).

Analysis for 4-Me-GHB can be accomplished along with GHB analysis. An SPE procedure for both analytes has been described by Marinetti *et al.* (2005). This procedure uses 200 μL of specimen with BSTFA derivitization and GC-EI-MS detection using GHB-*d*$_6$ as the internal standard. The ions monitored are at *m/z* 233, 234 and 235 for GHB di-TMS, *m/z* 239, 240 and 241 for GHB-*d*$_6$ di-TMS and *m/z* 231, 247 and 248 for 4-Me-GHB di-TMS (Figures 11.5–11.8). The linearity is from 5 to 150 and from 2 to 200 μg mL^{-1} for GHB and 4-Me-GHB, respectively.

REFERENCES

Addolorato, G., Caputo, F., Stefanini, G.F. and Gasbarrini, G. (1997). Gamma-hydroxybutyric acid in the treatment of alcohol dependence: possible craving development for the drug. *Addiction* **92**:1041–1042.

Anderson, D.T. and Kuwahara, T. (1997) Endogenous gamma hydroxybutyrate (GHB) levels in postmortem specimens. Presented at the combined meeting of CAT/NWAFS/SWAFS/SAT, Las Vegas, NV,7 November.

Anderson, D.T., Muto, J.J. and Andrews, J.M. (2000). Case report: postmortem tissue distribution of gamma hydroxybutyrate (GHB) and gamma butyrolactone (GBL) in a single fatality. *Program and Abstracts of the Society of Forensic Toxicologists Annual Meeting, Milwaukee, WI,* Poster 37.

Andollo, W. and Hearn, W.L. (1998). The characterization of drugs used in sexual battery – the Dade County experience. *American Academy of Forensic Sciences Annual Meeting, San Francisco, CA,* Abstract K52.

Barker, S.A., Snead, O.C., Poldrugo, F., Liu, C., Fish, F.P. and Settine, R.L. (1985). Identification and quantitation of 1,4-butanediol in mammalian tissues: an alternative biosynthetic pathway for gamma-hydroxybutyric acid. *Biochem. Pharmacol.* **34**:1849–1852.

Beghè, F., and Carpanini, M.T. (2000). Safety and tolerability of gamma-hydroxybutyric acid in the treatment of alcohol-dependent patients. *Alcohol* **20**: 223–225.

Biochemical Diagnostics (2000). Method for the measurement of gamma-hydroxybutyrate (GHB) in blood and urine Using GC/MS, personal communication.

Blumenfeld, M., Suntay, R.G. and Harmel, M.H. (1962). Sodium gamma-hydroxybutyric acid: a new anesthetic adjuvant. *Anesthesia Analgesia* **41**:721–726.

Borgen, L.A., Lai, A. and Okerholm, R.A. (2001). Xyrem® (sodium oxybate): the effects of gender and food on plasma kinetics. Orphan Medical, Inc., Minnetonka, MN, personal communication.

Broughton, R. and Mamelak, M. (1979). The treatment of narcolepsy–cataplexy with nocturnal gamma-hydroxybutyrate. *J. Can. Sci. Neurol.* **6**:1–6.

Centers for Disease Control (1990). Multistate outbreak of poisonings associated with illicit use of gamma hydroxy butyrate. *MMWR* **38**:861–863.

Centers For Disease Control and Prevention (1999). Adverse events associated with ingestion of gamma-butyrolactone – Minnesota, New Mexico and Texas, 1998–1999. *MMWR* **48**:137–140.

Ciolino, L.A., and Mesmer, M.Z. (2000). Bridging the gap between GHB and GBL – forensic issues of interconversion. *American Academy of Forensic Science Annual Meeting, Reno, NV,* Abstract B51.

Ciolino, L.A., Mesmer, M.Z., Satzger R.D., Machal C., McCauley H.A. and Mohrhaus A.S. (2001). The chemical interconversion of GHB and GBL: forensic issues and implications. *J Forensic Sci.* **46**:1315–1323.

Colombo, G., Agabio, R., Lobina, C., Reali, R., Fadda, F., and Gessa, G.L. (1995). Cross-tolerance to ethanol and γ-hydroxybutyric acid. *Eur. J. Pharmacol.* **273**: 235–238.

Couper F.J., and Logan B.K. (1999). The determination of GHB in clinical and postmortem specimens. In *Proceedings – Society of Forensic Sciences Annual Meeting, San Juan, PR,* abstract.

Couper, F.J., and Logan, B.K. (2000). Determination of γ-hydroxybutyrate (GHB) in biological specimens by gas chromatography–mass spectrometry. *J. Anal. Toxicol.* **24**: 1–7.

Craig, K., Gomez, H.F., McManus, J.L. and Bania, T.C. (2000). Severe gamma-hydroxybutyrate withdrawal: a case report and literature review. *J. Emerg. Med.* **18**: 65–70.

Crifasi, J.A. and Telepchak, M. (2000). A solid phase method for gamma-hydroxybutyrate (GHB) in blood, urine, vitreous or tissue without conversion to gamma-butyrolactone (GBL) using United Chemical Technologies' ZSGHB020 or CSGHB203 solid phase extraction columns, personal communication.

DEA Southwest Laboratory (2001). Solid GBL in Southern California. DEA Southwest Laboratory in San Diego, CA. *Microgram* **34**: 45.

Deichmann, W.B., Hirose, R. and Witherup, S. (1945). Observations on the effects of gamma-valerolactone upon experimental animals. *J. Ind. Hyg. Toxicol.* **27**: 263–268.

Doherty, J.D., Hattox, S.E., Snead, O.C. and Roth, R.H. (1978). Identification of endogenous γ-hydro-butyrate in human and bovine brain and Its regional distribution in human, guinea pig and rhesus monkey brain. *J. Pharmacol. Exp. Ther.* **207**: 130–139.

Duer, W.C., Byers, K.L. and Martin, J.V. (2001). Application of a convenient extraction procedure to analyze gamma-hydroybutyric acid in fatalities involving gamma-hydroxybutyric acid, gamma-butyrolactone, and 1,4-butanediol. *J. Anal. Toxicol.* **25**: 576–582.

Dyer, J.E., Roth, B. and Hyma, B.A. (2001). Gamma-hydroxybutyrate withdrawal syndrome. *Ann. Emerg. Med.* **37**: 147–153.

Ehrhardt, J.D., Vayer, P.H. and Maitre, M. (1988). A rapid and sensitive method for the determination of γ-hydroxybutyric acid and *Trans*-γ-hydroxycrotonic acid in rat brain tissue by gas chromatography/mass spectrometry with negative ion detection. *Biomed. Environ. Mass Spectrom.* **15**:521–524.

Eli, M. and Cattabeni, F. (1983). Endogenous γ-hydroxybutyrate in rat brain areas: postmortem changes and effects of drugs interfering with γ-aminobutyric acid metabolism. *J. Neurochem.* **41**:524–530.

Elian, A.A. (2000). A novel method for GHB detection in urine and its application in drug-facilitated sexual assaults. *Forensic Sci. Int.* **109**: 83–87.

ElSohly, M.A. and Salmone S.J. (1999). Prevalence of drugs used in cases of alleged sexual assault. *J. Anal. Toxicol.* **23**: 141–146.

Entholzner, E., Mielke, L., Pichlmeier, R., Weber, F. and Schneck, H. (1995). EEG changes during sedation with gamma-hydroxybutyric acid. *Anaesthesist* **44**: 345–350.

Fattore, L., Martellotta, M.C., Cossu, G. and Fratta, W. (2000). Gamma-hydroxybutyric acid an evaluation of its rewarding properties in rats and mice. *Alcohol* **20**: 247–256.

Ferrara, S.D., Zotti, S., Tedeschi, L., Frison, G., Castagna, F., Gallimberti, L., Gessa, G.L. and Palatini, P. (1992). Pharmacokinetics of γ-hydroxybutyric acid in alcohol dependent patients after single and repeated oral doses. *Br. J. Clin. Pharmacol.* **34**: 231–235.

Ferrara, S.D., Tedeschi, L., Frison, G., Castagna, F., Gallimberti, L., Giorgetti, R., Gessa, G.L. and Palatini, P. (1993). Therapeutic gamma-hydroxybutyric acid monitoring in plasma and urine by gas chromatography–mass spectrometry. *J. Pharm. Biomed. Anal.* **11**: 483–487.

Ferrara, S.D., Tedeschi, L., Frison, G. and Rossi, A.(1995). Fatality due to gamma-hydroxybutyric acid (GHB) and heroin intoxication. *J. Forensic Sci.* **40**: 501–504.

Fieler, E.L., Coleman, D.E. and Baselt, R.C. (1998). γ-Hydroxybutyrate concentrations in pre- and postmortem blood and urine. *Clin. Chem.* **44**: 692.

Fishbein, W.N. and Bessman, S.P. (1966). Purification and properties of an enzyme in human blood and rat liver microsomes catalyzing the formation and hydrolysis of gamma-lactones. *J. Biol. Chem.* **241**: 4835–4841.

Food and Drug Administration (1999a). Important message for health professionals – Report of serious adverse events associated with dietary supplements containing GBL, GHB or 1,4BD. FDA website: http://www.fda.gov/medwatch/safety/1999/gblghb.htm.

Food and Drug Administration (1999b). FDA warns about products containing gamma butyrolactone or GBL and asks companies to issue a recall. *FDA Talk Paper*, 21 January, T99-5.

Frison, G.,Tedeschi, L, Maietti, S. and Ferrara, S.D. (1998). Determination of gamma-hydroxybutyric Acid (GHB) in plasma and urine by headspace solid-phase microextraction (SPME) and gas chromatography-positive ion chemical ionization-mass spectrometry, In *Proceedings of the 1998 Joint Society of Forensic Toxicologists and the International Association of Forensic Toxicologists SOFT/TIAFT International Meeting*, Spiehler, V. (Ed.), pp. 394–404.

Gallimberti, L., Cibin, M., Pagin, P., Sabbion, R., Pani, P.P., Pirastu, R., Ferrara, S.D. and Gessa, G.L. (1993). Gamma-hydroxybutyric acid for treatment of opiate withdrawal syndrome. *Neuropsycho-pharmacology* **9**: 77–81.

Gibson, K.M., Hoffman, G.F., Hodson, A.K., Bottiglieri, T. and Jakobs, C. (1998). 4-Hydroxybutyric acid and the clinical phenotype of succinic semialdehyde dehydrogenase deficiency: an inborn error of GABA metabolism. *Neuropediatrics* **29**: 14–22.

Helrich, M., McAslan, T.C., Skolnik, S. and Bessman, S.P. (1964). Correlation of blood levels of 4-hydroxybutyrate with state of consciousness. *Anesthesiology* **25**: 771–775.

Hernandez, M., McDaniel, C.H., Costanza, C.D. and Hernandez, O.J. (1998). GHB-induced delirium: a case report and review of the literature on gamma hydroxybutyric acid. *Am. J. Drug Alcohol Abuse* **24**: 179–183.

IARC (1999). *Gammabutyrolactone. IARC Monographs on the Evaluation of Carcinogenic Risks to Humans*, Vol. 71, Part 2. IARC, Lyon, pp. 367–382.

Kleinschmidt, S., Schellhase, C. and Mertzlufft, F. (1999). Continuous sedation during spinal anaesthesia: gamma-hydroxybutyrate vs. propofol. *Eur. J. Anaesthesiol.* **16**: 23–30.

Kraner, J.C., Plassard, J.W., McCoy, D.J., Rorabeck, J.A., Witeck, M.J., Smith, K.B. and Evans, M.A. (2000). A death from ingestion of 1,4-butanediol, a GHB precursor. In *Program and Abstracts of the Society of Forensic Toxicologists Annual Meeting, Milwaukee, WI*, Poster 39.

Kraner, J.C., Plassard, J.W., McCoy, D.J., Rorabeck, J.A., Witeck, M.J., Smith, K.B. and Evans, M.A. (2000). A death from ingestion of 1,4-butanediol, a GHB precursor. *Program and Abstracts of the Society of Forensic Toxicologists Annual Meeting, Milwaukee, WI*, Poster 39.

Laborit, H., Jouany, J.M., Gerard, J. and Fabiani, F. (1960). Généralites concernant l'étude experimentale de l'emploi clinique du gamma hydroxybutyrate de Na. *Aggressologie* **1**:407.

LeBeau, M.A., Montgomery, M.A., Miller, M.L. and Burmeister, S.G. (1999). Analysis of biofluids for gamma-hydroxybutyrate (GHB) and gamma-butyrolactone (GBL) by headspace GC/FID and GC/MS. Presented in part at the 26th Annual Meeting, Society of Forensic Toxicologists, Rio Mar, PR.

LeBeau, M.A., Montgomery, M.A., Jufer, R.A. and Miller, M.L. (2000a). Elevated GHB in citrate-buffered blood. *J. Anal. Toxicol.* **24**: 383–384.

LeBeau M.A., Darwin W.D. and Huestis M.A. (2000b). Intra- and interindividual variations in urinary levels of endogenous GHB. In *Proceedings – Society of Forensic Toxicologists Annual Meeting, Milwaukee, WI*, p. 5.

LeBeau M.A., Christenson R.H., Levine B., Darwin W.D. and Huestis M.A. (2002). Intra- and interindividual variations in urinary concentrations of endogenous gamma hydroxyl butyrate. *J. Anal. Toxicol.* **26**: 340–346.

Lettieri J. and Fung, H.L. (1978). Improved pharmacological activity via pro-drug modification; comparitive pharmacokinetics of sodium gamma-hydroxybutyrate and gamma-butyrolactone. *Res. Commun. Chem. Pathol. Pharmacol.* **22**: 107–118.

Li, J., Stokes, S.A. and Woeckener, A. (1998). A tale of novel intoxication: seven cases of gamma-hydroxybutyric acid overdose. *Ann. Emerg. Med.* **31**: 723–728.

Louagie, H.K., Verstraete, A.G., DeSoete, C.J., Baetens, D.G. and Calle, P.A. (1997). A sudden awakening from a near coma after combined intake of gamma-hydroxybutyric acid (GHB) and ethanol. *Clin. Toxicol.*, **35**: 591–594.

Lund, L.O., Humphries, J.H. and Virtue, R.W. (1965). Sodium gamma hydroxybutyrate: laboratory and clinical studies. *Can. Anaes. Soc. J.* **12**: 379–385.

Marinetti, L.J. and Commissaris, R.L. (1999). The effects of gammahydroxybutyrate (GHB) administration alone and in combination with ethanol on general CNS arousal in rats as measured using the acoustic startle paradigm. *Society of Forensic Toxicologists Annual Meeting, San Juan, PR*, Abstract 71.

Marinetti, L.J., Isenschmid, D.S., Hepler, B.R., Schmidt, C.J., Somerset, J.S. and Kanluen, S. (2000). Two gamma-hydroxybutyric acid (GHB) fatalities. In *American Academy of Forensic Science Annual Meeting, Reno, NV*, Abstract K16.

Marinetti, L.J. (2003). The pharmacology of gamma valerolactone (GVL) as compared to gamma hydroxybutyrate (GHB), gamma butyrolactone (GBL), 1,4 butanediol (1,4BD), ethanol (EtOH) and baclofen (BAC) in the rat. *PhD Dissertation*, Wayne State University.

Marinetti L.J., Isenschmid D.S., Hepler B.R. and Kanluen S. (2005). Analysis of GHB and 4-methyl-GHB in postmortem matrices after long term storage. *J. Anal. Toxicol.* **29**:41–47.

McCusker, R.R., Paget-Wilkes, H., Chronister, C.W., Goldberger, B.A. and ElSohly, M.A. (1999). Analysis of gamma-hydroxybutyrate (GHB) in urine by gas chromatography–mass spectrometry. *J. Anal. Toxicol.* **23**: 301–305.

McCutcheon, J.R., Hall, B.J., Schroeder, P.M., Peacock, E.A. and Bayardo, R.J. (2000). *American Academy of Forensic Science Annual Meeting, Reno, NV*, Abstract K15.

Merck (1996). *The Merck Index*, 12th edn. Merck, Whitehouse Station, NJ, pp. 1366–1367.

Metcalf, B.R., Emde, R.N. and Stripe, J.T. (1966). An EEG–behavioral study of sodium hydroxybutyrate in humans. *Electroencephalogr. Clin. Neurophysiol.* **20**: 506–512.

Nelson T., Kaufman E., Kline J. and Sokoloff L. (1981). The extraneural distribution of γ-hydroxybutyrate. *J. Neurochem.* **37**: 1345–1348.

Roth, R.H. and Giarman, N.J. (1965). Preliminary report on the metabolism of γ-butyrolactone and γ-hydroxybutyric acid. *Biochem. Pharmacol.* **14**: 177–178.

Roth, R.H. and Giarman, N.J. (1966). γ-Butyrolactone and γ-Hydroxybutyric acid – I. Distribution and metabolism. *Biochem. Pharmacol.* **15**: 1333–1348.

Scharf, M.B., Hauck, M., Stover, R., McDannold, M. and Berkowitz, D. (1998). Effect of gamma-hydroxybutyrate on pain, fatigue and the alpha sleep anomaly in patients with fibromyalgia: preliminary report. *J. Rheumatol.* **25**: 1986–1990.

Seiler, N. (1980). On the role of GABA in vertebrate polyamine metabolism. *Physiol. Chem. Phys.* **12**: 411–429.

Sessa, A. and Perin, A. (1994). Diamine oxidase in relation to diamine and polyamine metabolism. *Agents Actions* **43**: 69–77.

Shima, N., Akihiro, M., Tooru, K., Munehiro, K. and Hitoshi, T. (2005). Endogenous level and *in vitro* production of GHB in blood from healthy humans, and the interpretation of GHB levels detected in antemortem blood samples. *J. Health Sci.* **51**: 147–154.

Shumate, J.S. and Snead, O.C., III (1979). Plasma and central nervous system kinetics of gamma-hydroxybutyrate. *Res. Commun. Chem. Pathol. Pharmacol.* **25**: 241–256.

Snead, O.C., III, Yu, R.K. and Huttenlocher, P.R. (1976). Gamma hydroxybutyrate: correlation of serum and cerebrospinal fluid levels with electroencephalographic and behavioral effects. *Neurology* **26**: 51–56, January.

Snead, O.C., III, Brown, G.B. and Morawetz, R.B. (1981). Concentration of gamma-hydroxybutyric acid in ventricular and lumbar cerebrospinal fluid. *N. Engl. J. Med.* **304**: 93–95.

Snead, O.C., III, Liu, C. and Bearden, L.J. (1982). Studies on the relation of γ-hydroxybutyric acid (GHB) to γ-aminobutyric acid (GABA) – Evidence that GABA is not the sole source for GHB in rat brain. *Biochem. Pharmacol.* **31**: 3917–3923.

Sprince, H., Josephs, J.A. and Wilpizeski, C.R. (1966). Neuropharmacological Effects of 1,4-butanediol and related congeners compared with those of gamma-hydroxybutyrate and gamma-butyrolactone. *Life Sci.* **5**: 2041–2052.

Stephens, B.G., Coleman, D.E. and Baselt, R.C. (1999). *In vitro* stability of endogenous gamma-hydroxybutyrate in postmortem blood. *J. Forensic Sci.* **44**: 231.

Takahara, J., Yunoki, S., Yakushiji, Yamauchi, J., Yamane, Y. and Ofuji, T. (1977). Stimulatory effects of gamma-hydroxybutyric acid on growth hormone and prolactin release in humans. *J. Clin. Endocrinol. Metab.* **44**: 1014–1017.

US Congress (2000). Hillory J. Farias and Samantha Reid Date-Rape Drug Prohibition Act of 2000. *Law Enforcement and Crimes, Public Law 106-172*. 106th Congress, 18 February 2000.

Van Cauter, E., Plat, L., Scharf, M.B., Leproult, R., Cespedes, S., L'Hermite-Baleriaux, M. and Copinschi, G. (1997). *J. Clin. Invest.* **100**: 745–753.

Van der Pol, W., Van der Kleijn, E. and Lauw, M. (1975). Gas chromatographic determination and pharmacokinetics of 4-hydroxybutyrate in dog and mouse. *J. Pharmacokinet. Biopharm.*, **3**:99–113.

Zvosec, D.L., Smith, S.W., McCutcheon, J.R., Spillane, J., Hall, B.J., and Peacock, E.A. (2001). Adverse events, including death, associated with the use of 1,4-butanediol. *N. Engl. J. Med.* **344**: 87–93.

12

Liquid Chromatography with Inductively Coupled Plasma Mass Spectrometric Detection for Element Speciation: Clinical and Toxicological Applications

Katarzyna Wrobel,[1] Kazimierz Wrobel[1] and Joseph A. Caruso[2]

[1]*Instituto de Investigaciones Científicas, Universidad de Guanajauto, L. de Retana 5, 36000 Guanajuato, Gto, Mexico*

[2]*Chemistry Department, University of Cincinnati, P.O. Box 0172, Cincinnati, OH 45221-0172, USA*

12.1 INTRODUCTION

It is common knowledge that the mobility, bioavailability, retention and specific biological functions of an individual element depend on its physicochemical form. However, there are no general rules relating a chemical structure of elemental species to its behavior or biological activity in a given system. Thus, depending on their actual oxidation state, different toxicity is observed for chromium and arsenic: hexavalent chromium [Cr(VI)] presents higher toxicity than the trivalent form [Cr(III)], whereas arsenic in a lower oxidation state [As(III)] is more toxic than pentavalent arsenic [As(V)]. Furthermore, depending on the element, the organic compounds can be more or less harmful than inorganic forms (mercury and arsenic, respectively). It is clear, though, that element speciation information is of great importance in all research areas related to human health. As approved by respective IUPAC commissions, 'chemical species' is a specific and unique molecular, electronic or nuclear structure of an element. The term 'speciation analysis' refers to the measurement of the quantities of one or more individual chemical species in a sample.[1,2] In life sciences, analytical results are helpful

---

Chromatographic Methods in Clinical Chemistry and Toxicology   Edited by R. L. Bertholf and R. E. Winecker

in clarifying element pathways in living organisms with special emphasis on possible beneficial or toxic effects and the development of new metal-based drugs.

Element speciation in biological materials is a difficult analytical task. The challenge is to identify and/or quantify very low concentrations of few to several target species (concentration values far below the total element content) in a complex chemical matrix. Additional difficulties include similar physicochemical properties exhibited by the species of one element and their chemical lability. Very often, not all element forms in the sample are known and, finally, the list of certified reference materials (CRMs) for speciation analysis is still limited.[3,4] The two most important features of an analytical tool suitable for speciation analysis are excellent selectivity and high sensitivity. Special care should be paid to preserve the natural composition and distribution of species in the sample during the entire procedure.

A common speciation scheme after sample preparation involves a fractionation step followed by the element quantification in the fractions obtained. A clear trend exists toward using the techniques that combine separation and detection steps into one operating on-line system. In these coupled techniques, the selectivity is achieved by application of powerful separation modes (different chromatographic or electrophoretic methods), while the use of atomic spectrometric techniques assures high sensitivity of detection. It should be stressed, however, that coupled techniques with element-specific detection do not provide structural information for the species. If the appropriate standards are available, the assignment of chromatographic peaks can be accomplished by spiking experiments. On the other hand, the identification of unknown forms and/or ultimate confirmation of unexpected compounds observed in the sample require the use of complementary techniques (molecular mass spectrometry or NMR).[1,5–8]

In this chapter, the toxicologically and clinically relevant applications of liquid chromatography (LC) with inductively coupled plasma mass spectrometric (ICP-MS) detection are reviewed (a list of abbreviations used is given at the end of this chapter). After brief characterization of the analytical system, the speciation of arsenic, iodine, mercury, selenium and platinum is discussed.

12.2 LIQUID CHROMATOGRAPHY WITH INDUCTIVELY COUPLED PLASMA MASS SPECTROMETRIC DETECTION

The choice of a separation technique is always dictated by the properties of the analyte(s) (volatility, stability, polarity, electrical charge, etc.) and also the sample composition. Since many toxicologically and/or clinically important species are not volatile, LC is often used in speciation analysis. The principal advantage of LC is the extended range of separation mechanisms, based on the convenient selection of column type and mobile phase composition.[9,10] Considering the requirement for high separation power in speciation analysis and, on the other hand, the need to preserve native element speciation, this variety of possible separation conditions is of primary importance. However, in spite of unquestionable versatility of LC, possible interactions of element species with the stationary phase and/or the components of mobile phase (buffers, organic modifiers) may cause unwanted changes in original element distribution among different species.[9,11,12] That is why, in many applications, compromised conditions have to be used, sacrificing chromatographic resolution, but assuring the integrity of species during separation. To improve the separation, a multi-dimensional chromatographic approach has often been explored.[13]

The chromatographic modes typically used in speciation analysis are size-exclusion, ion-exchange, ion-pair reversed-phase, reversed-phase and, to a lesser extent, micellar, vesicular, chiral and affinity LC. Detailed descriptions of their capabilities and limitations can be found in a number of comprehensive reviews.[1,2,9–11,14]

Among element-specific detectors, ICP-MS has become a primary tool in speciation analysis.[15] The unique features observed in total element analysis encompass high sensitivity (detection limits usually in the parts per billion or parts per trillion range), multielement capabilities and practically interference-free linear response over a wide concentration range (typically 4–8 orders of magnitude). Furthermore, ICP-MS detection enables: (i) monitoring effluents for their elemental composition with high sensitivity, (ii) determination of the target element with the selectivity over co-eluting elements, (iii) compensation of incomplete chromatographic resolution from complex matrices, (iv) isotope ratio and isotope dilution capabilities and (v) detection of a number of elements virtually simultaneously.[16] Recent progress in high-resolution instrumentation and collision/reaction technology has assumed high importance, since the modern instruments conveniently eliminate troublesome spectral polyatomic interferences.[7,17–19] The use of stable or radioactive artificial isotopes has expanded the potential of ICP-MS in speciation analysis, with stable isotopes being highly attractive. The major applications are isotope dilution quantification, method validation and studies on possible interconversion of species during the analytical procedure.[1] When the identity of the target species is known and the isotope-labeled species is available, the full possibilities of isotope dilution can be appreciated. In other words, possible loss and/or conversion of substance after spiking the sample with isotope labeled species have no effect on the analytical result.[20]

Finally, it should be noted that, when coupling ICP-MS with LC, factors such as the separation mechanism, composition and flow of the mobile phase and type of nebulizer and sample introduction device must be carefully considered and have consequently been topics of discussion in many papers and reviews.[9,11,16,21–23]

12.3 ANALYTICAL APPLICATIONS OF CLINICAL AND TOXICOLOGICAL RELEVANCE

Within the context of toxicological and clinical importance, speciation studies have been focused on relatively few elements, mainly aluminum, antimony, arsenic, chromium, iodine, lead, mercury, platinum, selenium and tin. However, coupled HPLC-ICP-MS has most often been used for speciation of arsenic, selenium, iodine and, to a lesser extent, mercury. The primary species of these elements include different oxidation states, alkylated metal and/or metalloid compounds, selenoamino acids and selenopeptides.[1,2] In addition, applications in studies on the pharmacokinetics of metal-based drugs (mainly platinum complexes) and metalloproteins should be included.[24,25] In the following sections, the advances in speciation studies of individual elements are reviewed.

12.3.1 Arsenic

The variety of physicochemical forms of arsenic and its well-demonstrated species-dependent toxicity have stimulated progress in speciation analysis of this element. A considerable number of analytical methods for the qualitative and quantitative analysis of several arsenic species have been reported, most of them based on LC separation and element-specific detection.[26,27] In the context of possible health risk, such analyses have been carried out in several environmental, biological and clinical materials under different conditions of exposure. Since arsenic enters the human body in association with food products and drinking water, speciation in aquatic samples, soils, sediments and different food-related products is mandatory. Analysis of clinical samples (mainly urine, bile and liver) helps to clarify the metabolic pathways of this element.

In spite of considerable progress in analytical methodology, full information on all arsenic species contained in complex biological matrices cannot be obtained using a single analytical protocol. The main difficulty is the pretreatment step, in which a complete recovery of all species (characterized by different solubility, polarity, electrical charge, etc.) is not realistic.[27,28] However, of special interest are the species-targeted protocols that focus on (1) more efficient leaching of the species of interest, necessary for its quantification, (2) characterization of the unknown species by mass spectrometric techniques and/or (3) elucidation of the metabolic routes of the element in biological systems.[28,29]

On the other hand, real-world materials usually contain a limited number of arsenic compounds at concentration levels that allow their detection. For example, aquatic, soil and sediment samples mainly contain inorganic arsenic forms with a much lower contribution from methylated pentavalent species that are formed due to the activity of microorganisms.[30–34] Marine organisms are important in global cycling of As, since they accumulate inorganic arsenic from seawater and convert it into organic compounds. Arsenobetaine (AsB) is the main species found in fish tissues[27,35,36] and several arsenosugars have been identified in algae, marine bivalves and crustaceans.[37–42] Food-related products of terrestrial origin usually contain lower concentrations of arsenic than marine food (nanograms per gram versus micrograms per gram). However, certain plants (rice, carrots) are exceptions and speciation analysis is required.[34,43–47] Of special interest for the evaluation of health risk is the analysis of As forms in urine. Two methylated pentavalent element species (MMA and DMA) have often been reported together with smaller amounts of inorganic forms and AsB (after fish consumption). The identification of monomethylarsonous acid [MMA(III)] and dimethylarsinous acid [DMA(III)] provided a new insight into the toxicological aspects related to As biomethylation pathways. Furthermore, the conjugated forms of trivalent As species with glutathione have been identified in bile.[48–51] Worth mentioning is that the variety and quantitative distribution of As species observed in clinical samples depend on the biological species and/or subject examined and the conditions of exposure, but also depend on the analytical methodology applied.

Representative examples of arsenic speciation performed by LC coupled with ICP-MS are presented in Table 12.1. In agreement with the abundance of As species in real-world samples and the health risk involved, most studies have been focused the separation/quantification of As(III), As(V), MMA, DMA and AsB. Depending on the pH conditions, these species can be cationic, anionic or uncharged (pK_{a_1} 9.3, 2.3, 2.6, 6.2 and 2.2, respectively[52]) allowing their separation by anion-exchange, ion-pair and, to a lesser extent, cation-exchange chromatography. Using the anion-exchange mode, good resolution was achieved using phosphate-, carbonate-, hydroxide- or phthalate-based mobile phases (pH in the range 5–11).[26,31,33,34,43,47,53–57] Ammonium salts are preferred in order to minimize the residue formed in the sampler/skimmer cones of the ICP-MS instrument. The most common elution order is AsB (minimum retention), As(III), DMA, MMA, As(V), but it is sometimes different, depending on the pH, gradient conditions, etc. The cation-exchange mode is often

Table 12.1 Applications of LC with ICP-MS detection for arsenic speciation analyses of toxicological and/or relevance

Sample	Analytical procedure	Species of interest	Toxicological and/or clinical relevance	Ref.
Drinking water	IE HPLC-ICP-MS; different mobile phases tested, of which Tris–acetate buffer enabled baseline separation	Arsenate [As(V)], arsenite [As(III)], MMA, DMA	Methodological approach: precision, accuracy, linearity and detection limits evaluated, several samples of drinking water were analyzed	30
Drinking water	Anion-exchange chromatography with a mobile phase containing $2\ mmol\ L^{-1}\ NaH_2PO_4$ and $0.2\ mmol\ L^{-1}$ EDTA, pH 6, coupled to ICP-MS detection	Arsenate [As(V)], arsenite [As(III)], MMA, DMA	Methodological approach: baseline separation achieved within <10 min. The analytical performance characteristics studied (DL < $100\ ng\ As\ L^{-1}$ for all the species evaluated). The feasibility for the analysis of normal As levels in drinking water was demonstrated. However, it was observed that matrix composition and the sample treatment may affect natural As(III)/As(V) distribution	31
Freshwater	Porous graphitic carbon used as the stationary phase for RP separation, elution in a gradient of HCOOH aq.; ICP-MS detection	Arsenate [As(V)], arsenite [As(III)], MMA, DMA	Methodological approach: adsorption of As(V) on the graphite surface observed, efficient separation of four species [except As(V)] within 10 min, DL ranging from 10 to $70\ ng\ As\ L^{-1}$; natural, arsenic-containing freshwater analyzed	32
Soil extracts	Extractions were performed with 0.3 M ammonium oxalate (pH 3), Milli-Q water, 0.3 M sodium carbonate (pH 8) and 0.3 M sodium bicarbonate (pH 11); anion-exchange separation (IonPac AS11 column) with NaOH as a mobile phase and ICP-MS detection	Arsenate [As(V)], arsenite [As(III)], MMA, DMA	Methodological approach focusing mobilization of As species from soils. The results showed that the mobilization of arsenic was pH dependent. Dramatic consequences have to be expected for pH changes in the environment especially in cases where soils contain high amounts of mobile arsenic	33

(continued)

Table 12.1 (*Continued*)

Sample	Analytical procedure	Species of interest	Toxicological and/or clinical relevance	Ref.
Animal feed additives, wine and kelp samples	Ion-pair HPLC on RP microbore column with mobile phase containing 5 mmol L^{-1} TBAH, pH 6.0 [separation time of As(III), As(V), MMA, DMA < 2 min], coupled on-line with ICP-MS	Phenylarsonic acids, arsenate [As(V)], arsenite [As(III)], MMA, DMA, arsenosugars	Development of analytical methodology for fast separation of As species. Application for studies on distribution and possible transformation of phenylarsonic compounds used as poultry and swine feed additives presented; different wine samples analyzed [As(III) the only species found] and arsenosugars determined in kelp samples	62,63
Poultry litter	Three ion-exchange chromatographic separations examined: (1) Dionex AS14 column using a phosphate mobile phase, (2) an AS16 column with hydroxide eluent and (3) an AS7 column with nitric acid mobile phase	ROX, *p*-ASA, arsenate [As(V)], arsenite [As(III)], MMA, DMA	Analytical methodology discriminating between organoarsenic compounds used as animal feed additives and more toxic inorganic species. The values of DL were generally <50 ng As L^{-1}. The major arsenic species in a water extract of a poultry litter sample was identified as ROX and trace concentrations of DMA and As(V) were also detected. A number of unidentified As species were observed at low concentrations, presumably metabolites of ROX	53
Chicken meat	Two anion-exchange LC columns examined, on-line ICP-MS	Arsenite [As(III)], arsenate [As(V)], MMA, DMA, AsB	Evaluation of a candidate reference material; As species found were DMA and AsB. The stability of arsenic species in a chicken meat candidate reference material for at least 12 months was demonstrated	54

Sample	Method	Species	Description	Ref.
Algae	Separation on anion-exchange column (Hamilton PRP-X100) with $20\,mmol\,L^{-1}$ $NH_4H_2PO_4$ (pH 5.6) mobile phase and a column temperature of $40\,^\circ C$ [separation time of seven As species 16 min; As(V) and the ribose with the glycerol aglycone eluted in the dead volume]; on-line ICP-MS detection	Arsenosugars, arsenate [As(V)], arsenite [As(III)], MMA, DMA	Four naturally occurring arsenosugars (AsSug-OH, AsSug-PO$_4$, AsSug-SO$_3$, AsSug-SO$_4$), MMA, DMA and inorganic As species were separated. Extracts obtained from the brown algae *Fucus spiralis* and *Halidrys siliquosa* contained AsSug-SO$_4$ as the major compound ($\sim$55% of total extractable As) together with AsSug-PO$_4$ and AsSug-SO$_3$. Arsenic acid was a significant constituent of *Halidrys siliquosa* ($\sim$6.5%), but was not detected in *Fucus spiralis*	37
Algae	Multidimensional LC: SEC fractionation followed by anion-exchange LC and further purification by RP-HPLC. The ICP-MS detector was coupled on-line and ESI-MS and ESI-MS-MS analyses were additionally carried out for identification of the eluted compounds	Up to 14 standards of arsenic species (including arsenosugars) plus identification of new species	Identification and quantification of As species in algae extracts were undertaken. Several arsenosugars were found, accounting for >99% of the arsenic present in the extract. The identities of all the species, were confirmed and/or assigned by matching the retention times of chromatographically pure (after the 3rd LC dimension) species with standards and by ESI-MS-MS	40,41
Algae and shrimp	Weak anion-exchange column (Dionex AS4A) with elution in a gradient of HNO_3 allowed the separation of 8 species within 12 min; on-line ICP-MS.	arsenate [As(V)], arsenite, [As(III)], MMA, DMA, AsB, TMAO, AsC, TETRA	Methodological approach: very low background in the ICP-MS detection was obtained using nitric acid mobile phase, which enabled for very low DL (30–1600 ng As L^{-1}). Application to real sample analysis shown	38
Lobster tissue	Different extraction procedures tested (Soxhlet, MW-assisted, SFE); anion-exchange column (Hamilton PRP-X100), gradient elution with the two mobile phases containing $(NH_4)_2CO_3$ at different concentrations (separation time 27 min); on-line ICP-MS detection	AsC, AsB, DMA, MMA, arsenite [As(III)], arsenate [As(V)]	Methodological approach, focusing the efficiency of extraction and baseline separation of 6 species. The values of DL were in the range 17–29 ng As kg^{-1} and AsB was a primary species found in the lobster tissue	39

(continued)

Table 12.1 (*Continued*)

Sample	Analytical procedure	Species of interest	Toxicological and/or clinical relevance	Ref.
Fish, crustacean and sediment samples from estuary in the tin mining area	Enzymatic hydrolysis (trypsine) carried out for fauna samples; sediments were treated with 1 mol L^{-1} H$_3$PO$_4$ in an open focused MW system; separation on anion-exchange column (Hamilton PRP-X100); step gradient elution with a phosphate-based mobile phase, pH 6–7.5; on-line ICP-MS	AB, DMA, MMA, arsenite [As(III)], arsenate [As(V)]	In order to assess possible health risk for population living in contaminated area, the important As species were determined in fish and crustaceans which contribute to the local diet. The presence of more toxic inorganic forms of arsenic in both sediments and biota samples has implications for human health, particularly as they are readily 'available'	186
Fish and rice	Water–methanol extraction (for rice α-amylase treatment) followed by IP on an anion-exchange column (IonPac Ag7 + AS7); gradient elution with two mobile phases containing HNO$_3$, benzene-1,2-disulfonic acid dipotassium salt and 0.5% methanol; on-line ICP-MS	Arsenite (As(III)), arsenate (As(V)), DMA, MMA, AsC, TMAO, AsB, AsSug-PO$_4$, AsSug-SO$_3$, AsSug-SO$_4$, AsSug-OH	Health risk associated with dietary arsenic intake: the elevated levels of toxic inorganic As species (up to 90% in rice) found in ~250 food samples. The value of DL in rice was 2 ng As g^{-1}. In marine organisms, inorganic As was mainly present as As (V) whereas in rice As(III) predominated	44
Rice	Water–methanol (1:1) extraction; As(III) and AB separated on cation-exchange column (Hamilton PRP-X200) in pyridine formate 4 mmol L^{-1}, pH 2.8; As(V), MMA and DMA separated on anion-exchange column (Hamilton PRPX-100) with phosphate mobile phase 10 mmol L^{-1}, pH 6; on-line ICP-MS	Arsenite (As(III)), arsenate (As(V)), MMA, DMA, AsB	Study on arsenic species in rice, a candidate reference material representative for terrestrial biological samples. The values of DL expressed as As in dried rice wheat were 2, 3, 3 and 5 ng As g^{-1} for As(III), AsB, As(V), MMA and DMA, respectively. Effect of different processing steps, storage time and temperature conditions on species stability was studied. MMA and As(V) species were not stable under any storage conditions, probably due to microbiological activity	45

Sample	Extraction/Method	Species	Comments	Ref.
Rice	Sample hydrolyzed with 2 mol L^{-1} TFA, 6 h, 100 °C (extraction efficiency 92%); three AE separation conditions applied: (1) Waters IC-Pak Anion HR with carbonate mobile phase at pH 10; (2) Dionex AS7 and AG7 with HNO$_3$ (pH 1.8); and (3) Hamilton PRP-X100 with phosphate-based mobile phase at pH 6.3	Arsenite (As(III)), arsenate (As(V)), MMA, DMA, AsB, AsC	Different procedures examined for the extraction of arsenic species, their separation and quantification. The DL ranged from 6 to 17 ng As g^{-1}. Owing to the reduction of As(V) to As(III) during TFA treatment, the proposed procedure allowed for measuring total inorganic As. Inorganic As accounted for 11–91% of total As in rice. Lower concentrations of DMA were also detected	43
Carrots grown on arsenic-contaminated and non-contaminated soil	Extraction carried out with 1 mmol L^{-1} Ca(NO$_3$)$_2$ aq, As species separated and detected using anion-exchange HPLC coupled with ICP-MS.	Arsenite (As(III)), arsenate (As(V)), MMA, DMA	Study on the soil-to-carrot uptake rate of As (bioavailability). Inorganic arsenic species were prevalent in soil. The ingestion of the potentially toxic inorganic arsenic via consumption of carrots grown in soil contaminated at 30 μg As g^{-1} was conservatively estimated at 37 μg per week	34
Carrots	ASE (extraction yield 80–102%), weak anion-exchange column, isocratic elution with aqueous NH$_4$NO$_3$, on-line ICP-MS	Arsenite [As(III)], arsenate [As(V)], MMA, DMA, AsB	Toxicological consideration: inorganic As(III) and As(V) were the only species found in samples that contained less than 400 ng g^{-1} of total arsenic. MMA and an unidentified arsenic compound were present in some of the samples with higher total arsenic content	46
Apples	Overnight treatment with α-amylase and sonication with acetonitrile–water (40:60), followed by AE separation with phosphate-based mobile phase; on-line ICP-MS	Arsenite [As(III)], arsenate [As(V)], MMA, DMA	Health risk associated with dietary arsenic intake: total arsenic concentrations in the freeze-dried apple samples ranged from 8.2 to 80.9 μg As g^{-1}. The three most abundant species found were As(III), As(V) and DMA	55

(continued)

Table 12.1 *(Continued)*

Sample	Analytical procedure	Species of interest	Toxicological and/or clinical relevance	Ref.
Mushrooms	MW-assisted extraction, ion-exchange HPLC, on-line ICP-MS	Arsenite [(As(III)], arsenite [As(VI)], DMA, AsB, TMAO	DMA was the primary species found in mushrooms from contaminated and non-contaminated soil (68–74% of total extracted As). The results showed that, in contaminated soils, mushrooms or their associated bacteria were able to biosynthesize DMA from inorganic As	187
Nuts	As species were extracted with chloroform–methanol (2:1). After addition of DI, polar species concentrated in aqueous layer and analyzed by AE chromatography (8 min) with ICP-MS detection	Arsenite [As(III)], arsenate [As(V)], MMA, DMA	Health risk associated with dietary arsenic intake: significantly higher contribution of total As (2.6–16.9 ng As g^{-1}) was found in nut oil as compared to defatted material. The primary species found in the oil extracts were As(III) and As(V). Lower concentrations of two methylated species were observed in several nut types	56
Peanut butter	Sonication extraction with 2-butoxyethanol (recovery for spiking experiments 40–122%) followed by ion-exchange HPLC-ICP-MS or HG-ICP-MS	Arsenite [As(III)], arsenate [As(V)], MMA, DMA	Health risk associated with dietary arsenic intake: total As content ranged from 6.5 to 21.4 ng As g^{-1}. Inorganic As was primary element form in the samples analyzed with lower contribution of methylated species	188
Processed infant food products	TFA treatment, 6 h, 100 °C (extraction yield 94–128%), anion-exchange HPLC and ICP-MS detection	Arsenite [As(III)], arsenate [As(V)], MMA, DMA	Health risk associated with dietary arsenic intake: total As levels in the rice-based cereals were in the range 63–320 ng As g^{-1} and in the other food products below 24 ng As g^{-1}. Inorganic As and DMA were the main species found in rice-based and mixed rice/formula cereals. These species were also present in freeze-dried sweet potatoes, carrots, green beans, and peaches. MMA and DMA were detected in minute amounts only in few samples	47

Fish tissues and urine	Arsenite [As(III)], arsenate [As(V)], MMA, DMA, AsB	Sample acidification with H_3PO_4, IP chromatographic separation (Altima C_{18}) with a mobile phase containing citric acid and hexanesulfonic acid at pH 4.5 (separation of 5 species in <4 min); on-line ICP-MS	Methodological approach focusing fast and reliable speciation analysis in environmental and biological samples. The QL for As(V), MMA, As(III), DMA and AsB were 44, 56, 94, 64, 66 ng As L^{-1}, respectively. The procedure was tested using two reference materials (DORM-2 dogfish muscle tissue, NIST SRM 2670 freeze-dried urine, normal level, and then applied to realworld samples	36
Marine organisms, urine	Arsenosugars, AsB, DMA, MMA	Anion-exchange column with bicarbonate (pH 10.3 with aqueous ammonia) mobile phase and on-line ESI-MS	The ability of LC-ESI-MS for both quantitative analysis of As species and structural elucidation of new As compounds in biological samples was demonstrated	42
Urine, hair, fingernail, blood from people exposed to As-contaminated drinking water	AsC, AsB, DMA(III), DMA, MMA(III), MMA, arsenite [As(III)], arsenate [As(V)]	Hot water extraction (TCA treatment for blood plasma), AE separation (polymer-based Shodex Asahipak ES-502N 7C column) with non-buffering mobile phase (15 mmol L^{-1} citric acid adjusted to pH 2 with HNO_3); ICP-MS detection	Monitoring environmental exposure to As: seven As species, including MMA(III) (6.6%) and DMA(III) (13%) were found in urine samples. Primary species in fingernail was As(III) (62.4%) and lower concentrations of As(V), MMA, DMA(III), DMA were also observed. No organic trivalent As species were found in hair and blood. Inorganic arsenic was predominant in hair and DMA in blood plasma. Arsenic in urine, fingernails, and hair correlated positively with water As, suggesting that any of these measurements could be considered as a biomarker to As exposure. Status of urine and exogenous contamination of hair urgently need speciation of As in these samples, but speciation of As in nail is related to its total As content. Therefore, total As concentrations of nails could be considered as biomarker to As exposure in the endemic areas	189–191

(continued)

Table 12.1 (*Continued*)

Sample	Analytical procedure	Species of interest	Toxicological and/or clinical relevance	Ref.
Rat bile	Anion-exchange column (Shodex RSpak JJ50-4D) with 20 mmol L^{-1} oxalic acid (pH 2.3 with aqueous ammonia) and on-line ICP-MS	Arsenite [As(III)], arsenate [As(V)], MMA, MMA(III), DMA, As(GS)$_3$, CH$_3$As(GS)$_3$	Study on the stability of As–glutathione complexes in rat bile and the role of GSH in stabilizing these complexes. The results obtained suggest that GSH plays an important role in preventing hydrolysis of As–GSH complexes and generation of well-known toxic trivalent arsenicals	49
Rat bile and urine after oral or intravenous exposure to As compounds	RP separation (Intersil ODS-3 column, 5 mmol L^{-1} TBAH + 3 mmol L^{-1} malonic acid, 5% methanol mobile phase); ICP-MS detection	Aresenite [As(III)], arsenate [As(V)] DMA, MMA, As(GS)$_3$, CH$_3$As(GS)$_2$	Metabolic speciation studies revealed that, after exposure to inorganic As, the element was excreted into bile as CH$_3$As(GS)$_2$ and DMA. Arsenic in rats exposed to methylated species was mostly excreted into urine in the unchanged chemical forms. It was demonstrated that biliary and urinary arsenic excretion and speciation are affected by the route, dose and chemical forms of arsenical administration, and GSH plays a key role in arsenic metabolism	50
Organs and body fluids of rats after intravenous injection of As(II)	Anion- and cation-exchange separation; on-line ICP-MS	Arsenite [As(III)], MMA, DMA, MMA(III), As conjugates with GSH	The metabolic balance and speciation studies suggested that As(III) is methylated in the liver during its hepato-enteric circulation through the formation of the GSH-conjugated form As(GS)$_3$, and MMA(III) and MMA are partly excreted into the bile, the former being in the conjugated form CH$_{(3)}$As(GS)$_2$. DMA is not excreted into the bile but into the bloodstream, accumulating in RBCs, and then excreted into the urine mostly in the form of DMA in rats	58

Sample	Method	Species	Comments	Ref.
Organs and body fluids of rats after intravenous injection of MMA and DMA	Urine and bile analyzed on AE column (ES-502N 7C) with citric acid (15 mmol L^{-1}, pH 2) as a mobile phase; liver supernatants, plasma and red blood cells lysates on gel filtration column (GS 220 HQ) with ammonium acetate buffer (50 mmol L^{-1}, pH 6.5); on-line ICP-MS	Arsenite [As(III)], DMA, MMA, AsB, possible As conjugates with GSH	Study on species-dependent metabolic pathway of As in rats: DMA and MMA were mostly excreted into urine in the intact forms. Their uptake to organs/tissues was lower with respect to As(III) exposure. The unidentified metabolites were excreted not into the bile but into the bloodstream	51
Urine from four APL (acute promyelocytic leukemia) patients treated with arsenic trioxide	HPLC-ICP-MS, no details given	Arsenite [As(III)], arsenate [As(V)], MMA, DMA, MMA(III), DMA(III)	Pharmacokinetics and metabolism of As(III) in APL patients studied. The intermediate MMA(III) and DMA(III), were detected in most urine samples when diethyldithiocar-bomate, was added to these samples for stabilization of As species. The major urine species As(III), MMA and DMA, accounting for >95% of the total arsenic excreted in urine. On the other hand, these species accounted for 32–65% of the total arsenic, suggesting other pathways of As excretion, such as through the bile, may play an important role in removal of arsenic from the human body when challenged by high levels of As(III)	192
Human urine and blood after ingestion of Chinese seaweed Laminaria	AE column (Hamilton, PRP-X100) with different phosphate-, oxalate- and acetate-based mobile phases containing 3–20% of methanol; ICP-MS detection and ESI-MS-MS for species identification and/or confirmation	Arsenosugars, arsenite [As(III)], DMA, MMA, DMAE	Study on metabolic pathway of As species (mainly arsenosugars) ingested with algae. Total arsenic and speciation analysis revealed no marked increase in arsenic blood, serum and packed cells levels up to 7 h after ingestion. As species identified in urine were DMA, MMA and DMAE. Another 5 species remained unknown. In simulated gastric fluid incubated with algae, a degradation of arsenosugars into a compound with a mass of 254 Da was observed	193

(continued)

Table 12.1 (*Continued*)

Sample	Analytical procedure	Species of interest	Toxicological and/or clinical relevance	Ref.
Urine from children and adults exposed to As in drinking water	HPLC-ICP-MS, no details given	Arsenite [As(III)], arsente [As(V)], DMA, MMA	Health risk assessment: average total urinary arsenic was higher in children than in adults and total arsenic excretion per kg body weight was also higher for children than adults. The values of ratio MMA/(inorganic As) for adults and children were 0.93 and 0.74, respectively, suggesting that the first methylation step could be more active in adults. On the other hand, the values of ratio DMA/MMA in children were higher than in adults (8.15 and 4.11, respectively) indicating that the 2nd methylation step is more active in children than adults. It was concluded that children retain less arsenic in their body than adults, which could explain why children did not show skin lesions compared with adults that had been exposed to this same contaminated water	194
Urine reference material	Samples diluted 1 + 3 and analyzed by gradient elution AE (ICSep ION120 column, carbonate-based mobile phase at pH 10.3) or cation exchange (ChromPack Ionospher 5C with pyridinium ion-based mobile phase) LC with ICP-MS detection	Arsenite [As(III)], arsenate (As(V)), MMA, DMA, AsB, TMAO, TMAs, TMAP, DMAE	Methodological development focusing the separation/quantification of nine As species in urine matrix (reference urine NIES No. 18 and NIST SRM2670a). AE chromatographic mode gave baseline separation of AsB, DMA, As(III), MMA, As(V) in 15 min. Additionally, using a cation-exchange column AsC, TMAs, TMAO and TMAP were separated	59

Sample	Species	Method	Comments	Ref.
Human urine	Arsenite [As(III)], arsenate [As(V)], MMA, DMA, AsB, arsenosugars	Separation on RP column (GL Sciences Inertsil ODS) with TEAH, 4.5 mmol L^{-1} malonic acid, 0.1% methanol mobile phase at pH 6.8 (with HNO_3), on-line ICP-MS detection	Urinary excretion of As species was evaluated during a 3-day period after a meal of blue mussels, *Mytilus edulis*. The effect of cooking on the arsenic speciation was also studied. AsB and DMA were the major arsenic metabolites found in the urine samples. Significant amounts of unknown metabolites were also detected	195
Human urine reference materials (NIST SRM 2670E and 2670N)	Arsenite [As(III)], arsenate [As(V)], MMA, DMA	AE microbore column (150 × 1 mm, Nuceleosil SB 100-5; 5 μm) with phosphate-based mobile phase, pH 5.25 at a flow rate 100 μL min^{-1}. Three nebulizers (DIN, MCN and CFN) were compared for hyphenation of HPLC with ICP-MS detection	A miniaturized speciation method was developed, which is suitable for analysis of the main As metabolites in urine. Enhanced analytical performance was obtained with the low-flow nebulizers (DIN and MCN). The values of DL were 0.2–0.6 ng As mL^{-1}	65
Urine from from the population exposed to As	Arsenate [As(V)], arsenite [As(III)], AsC, AsB, MMA, DMA	Cation- and anion-exchange columns in series, gradient elution with $(NH_4)_2CO_3$ (between 10 and 50 mmol L^{-1}) and ICP-MS detection (separation time <30 min)	As species were determined in human urine ($n = 256$) in a population-based exposure assessment survey	60

(continued)

Table 12.1 (*Continued*)

Sample	Analytical procedure	Species of interest	Toxicological and/or clinical relevance	Ref.
Urine NIST SRM 2670	AE separation (ION-120 InterAction chromatography column) coupled with ICP-MS through DN or HG. For DN: gradient elution using 40 and 70 mmol L^{-1} $(NH_4)_2CO_3$ (pH 10.5, at 60 °C). For HG: isocratic elution with 40 mmol L^{-1} $(NH_4)_2CO_3$. Photo-reactor interface between the column and the HG facilitated the detection of non-hydride active As species	AsB, arsenite [As(III)], DMA, MMA, arsenate [As(V)]	Two procedures for As speciation were proposed that differ in the dilution factor, sample introduction and separation conditions. For DN, resolution among the early eluting species [AsB, DMA, As(III)] was improved at higher column temperatures and lower eluent molarity, but the separation of Cl^- has to be assured. Using HG mode, chloride is eliminated in gas–liquid separator. The high sensitivity of this mode enables 50-fold dilution of urine matrix prior to analysis. The results obtained in the analysis NIST SRM 2670 revealed statistically significant differences, mainly in DMA concentrations	57
Urine in occupational exposure to As_2O_3	As speciation: a weak AE column with the mobile phase containing $NH_4H_2PO_4$, CH_3COONH_4, CH_3COOH and methanol; on-line ICP-MS	Arsenite [As(III)], arsenate [As(V)], DMA, MMA, AsB	The increased urinary excretion of some porphyrin homologues, correlated with As(III) levels, was observed in exposed workers, which confirms the role of inorganic As in the inhibition of URO-decarboxylase in the heme biosynthesis pathway	196

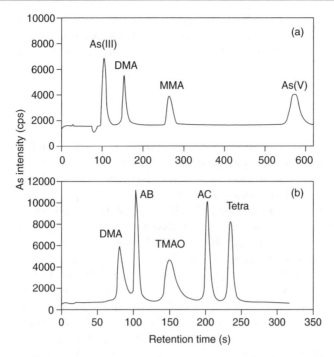

Figure 12.1 HPLC-ICP-SFMS chromatograms of arsenic species ($0.5\,\text{ng mL}^{-1}$), obtained using a conventional concentric nebulizer and low-resolution sector field ($m/\Delta m \approx 300$): (a) anion-exchange chromatography (Hamilton PRP-X100, $20\,\text{mmol L}^{-1}$ NH$_4$H$_2$PO$_4$, pH 5.6); (b) cation-exchange chromatography (Zorbax 300-SCX, $20\,\text{mmol L}^{-1}$ pyridine, pH 2.31). From Zheng, B., Hintelmann, H. Hyphenation of high performance liquid chromatography with sector field inductively coupled plasma mass spectrometry for the determination of ultra-trace level anionic and cationic arsenic compounds in freshwater fish. *J. Anal. At. Spectrom.* 2004, **19**, 191–195. Reproduced by permission of The Royal Society of Chemistry

used to separate AsB, AsC, TMAO and TMAs. These separations are achieved at generally lower pH values (<4) with the mobile phases containing formate or pyridine/formate.[45,48,58-60] In Figure 12.1, the anion- and cation-exchange chromatograms of arsenic species are shown.[61] On the other hand, different ion-pairing agents (mainly tetrabutyl-ammonium ion and various alkylsulfonates) have been used for separation of both charged and neutral species in one chromatographic run within a relatively short time (4–20 min).[26,36,62,63] Typically, separations were achieved on reversed phase columns, but ion-exchange columns were also used.[44,64] Both, isocratic[30,36,43,45,46,53-56,62,63,65] and gradient elution[38,39,44,57,59,60,66] modes were reported. Logically, gradient elution is often preferred, since it offers efficient separation in a shorter time. However, the ICP tolerance to high salt content is limited and the sensitivity of ICP-MS detection can be affected by changes in the composition of the solution entering the plasma (mobile phase). Finally, for the characterization of new As species, a multidimensional chromatographic approach was used prior to electrospray ionization mass spectrometry analysis (ESI-MS).[40,41,67]

The most common interference in As determination by ICP-MS is caused by chloride ($^{40}\text{Ar}^{35}\text{Cl}^+$). Since arsenic is monoisotopic, chromatographic separation of chloride ions or their elimination prior to ICP-MS detection by means of hydride generation is necessary.[36,57,68] On the other hand, high-resolution mass spectrometry and collision cell technology have been shown to be effective in suppressing chloride-based polyatomic interferences.[68–72]

12.3.2 Iodine

Unlike arsenic, iodine is an important human nutrient. It is utilized by the thyroid gland for the biosynthesis of the thyroid hormones thyroxine (T4) and triiodothyronine (T3), necessary for human growth and development.[2] Deficiency of this element leads to iodine deficiency disorders (IDD). Today, about 30% of the world's population is at risk of IDD, 750 million people suffer from goiter, 43 million have IDD-related brain damage and mental retardation and 5.7 million are afflicted by cretinism, the most severe form of IDD.[73] The knowledge about the variety and distribution of different iodine species [mainly mono-iodothyronine (T1), diiodothyronine (T2), T3 and T4] in serum or urine can give information about a malfunction of the thyroid gland and may explain other T4/T3-influenced metabolic abnormalities.[74] Addition of iodine to table salt has become a common practice in order to prevent element deficiency. Several investigations have been performed on iodine speciation in human body fluids and in food-related products, predominantly using chromatographic separations.[10,74–78] Such analyses were also performed in pharmaceutical formulations using ICP-MS as the primary detection tool.[79,80] Even though the determination of halogens by ICP-MS has been somewhat difficult, the relatively low first ionization potential of iodine (10.46 eV) makes it suitable for ICP-MS analysis.[19] As demonstrated by Kannamkumarath *et al.*, the key advantage of this detection system over spectrophotometric detection (225 nm) is its high sensitivity (detection limits for seven iodine species are about 200 times lower with ICP-MS), which permitted the quantification of the degradation products of T4 at trace levels.[79] The chromatograms of seven iodine compounds obtained in the cited work with the two detection systems are shown in Figure 12.2.[79]

The applications of LC coupled to ICP-MS in iodine speciation have been reviewed elsewhere[2,10,19] and some representative examples are presented in Table 12.2.[74–79,81]

12.3.3 Mercury

Where health hazard is concerned, the main requirement for speciation analysis is to distinguish the inorganic forms of mercury from its short-chain alkyl species (MMM, MEM) in environmental, food and clinical matrices. A critical step in any analytical procedure is the extraction of Hg species from the original sample, aiming for a complete recovery of the element without species interconversion. Relatively aggressive treatments have been used, such as acid extraction (mostly combined with solvent extraction), distillation or alkaline extraction.[82] Even though water vapor distillation seems to be a method of choice for complete recovery, the formation of methylated mercury artifacts has been discussed.[83–85] Separation of the Hg species is usually carried out by gas chromatography (GC), after convenient derivatization (hydride generation or alkylation).[86,87] However, the applications of LC have been increasing, since it simplifies sample preparation (no precolumn derivatization needed).[88] As already mentioned, there is a significant risk of mercury

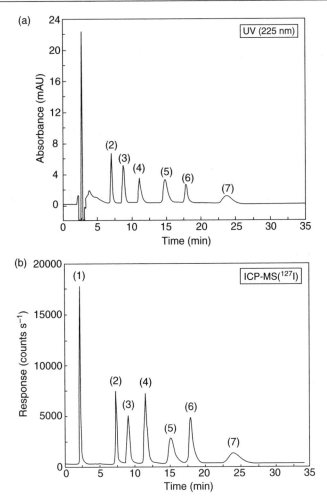

Figure 12.2   Comparison of (a) UV absorbance detection at 225 nm and (b) ICP-MS detection (^{127}I) for the analysis of iodine species in levothyroxine pharmaceutical tablets. Iodine species: (1) inorganic iodine; (2) T2; (3) TTAA2; (4) T3; (5) TTAA3; (6) T4; (7) TTAA4. The concentration of each iodine species was 200 mg L^{-1}. From Kannamkumarath, S. S., Wuilloud, R. G., Stalcup, A., Caruso, J. A., Patel, H., *et al.* Determination of levothyroxine and its degradation products in pharmaceutical tablets by HPLC-UV-ICP-MS. *J. Anal. At. Spectrom.* 2004, **19**, 107–113. Reproduced by permission of The Royal Society of Chemistry

methylation/demethylation during different steps of analytical procedure. Thus, in the particular case of mercury speciation, ICP-MS detection not only provides high selectivity and sensitivity but also offers the opportunity to perform isotope dilution analysis in order to compensate for changes of the original species distribution in the sample. Several representative applications of LC coupled to ICP-MS for mercury speciation are given in Table 12.3.[83–85,89-94] The methodological development has been focused on efficient separation and interference-free, sensitive detection. Typically, reversed-phase columns

Table 12.2 Analytical applications of HPLC-ICP-MS for iodine speciation

Sample	Analytical procedure	Species of interest	Toxicological and/or clinical relevance	Ref
Milk and infant formulas	Milk whey obtained after centrifugation of fresh milk (with SDS) was analyzed by SEC with $30 \, mmol \, L^{-1}$ Tris–HCl buffer as a mobile phase and ICP-MS detection (40 min)	HMW versus LMW iodine species	Iodide was found as the primary iodine species in fresh milk, whereas in the infant formulas more than 50% of element was bound to compounds presenting MW >1000 kDa	77
Milk	Different pretreatment procedures tested, centrifugation finally adopted, separation carried out ion-exchange column (Dionex AS14) with carbonate buffer as the mobile phase	Iodine, iodate	The values of DL were $<2 \, \mu g \, I \, L^{-1}$. Iodide was identified as the main iodine species in milk, but in a few samples also traces of iodate and several unidentified, presumably organoiodine, compounds were observed.	81
Commercial seaweeds	Multidimensional chromatographic approach: alkaline extracts analyzed by SEC (Superdex 75HR, $0.03 \, mol \, L^{-1}$ Tris–HCl, pH 8.0) and by AE chromatography (Ion Pac AS-11, Dionex, $5 \, mmol \, L^{-1}$ NaOH) coupled to ICP-MS; Enzymatic digests (Proteinase K) were analyzed by RP-HPLC (Altima C_{18}, elution with $0.01 \, mol \, L^{-1}$ Tris–HCl, pH 7.3 in a gradient of methanol)	Association of iodine with HMW compounds, iodide, T1, T2,	The association of iodine with both LMW and HMW fractions was observed, which was dependent on the biological species analyzed. Iodine was a primary species found in Kombu, whereas T1 and T2 (bound to proteins) were detected in Wakame. The results obtained suggest that, in different seaweeds, iodine follows different metabolic pathways. Since the bioavailability of iodide is better than any other form of iodine, Kombu seaweed would be preferred as a natural dietary supplement	76

Sample	Method	Species	Comments	Reference
Levothyroxine tablets	RP-HPLC [Cyano-Spherisorb narrow bore column, 22% (v/v) acetonitrile; 0.08% (v/v)trifluoroacetic acid, pH 2.3] with UV or ICP-MS detection. For ICP-MS, post-column, on-line dilution (1:3.3) of the chromatographic eluent with a 2% (v/v) nitric acid (separation time 26 min).	T4, T3, T2, T1, TTAA4, TTAA3, TTAA2	The detection limits obtained with UV detection ranged from 28.9 to 34.5 µg I L^{-1}, whereas those obtained with ICP-MS were about 175–375 times lower. Effect of different common diluents used in the formulation of levothyroxine tablets was studied and iodine species were determined in commercial levothyroxine sodium tablets	79
Human serum and urine	Enzymatic digests (protease) analyzed by RP-HPLC (Phenomenex CAPCELL-C$_{18}$ column, elution with 10 mmol L^{-1} Tris–HCl, pH 7.3, in methanol gradient 1–50% (v/v), at 15 °C)], post-column dilution with dilute HNO$_3$ and ICP-MS detection (separation time 26 min)	T4-TBG, T1, T2, r-T3, T3, T4. iodide	The values of DL were in the range 0.08–1.5 µg I L^{-1}. The concentrations of iodide, T1, T2, r-T3, T3 and T4 found in 'normal' sera were 11, 1.6, 2.1, 3.9, 5.9 and 60 µg L^{-1}, respectively. The method proved to recognize abnormalities in a pathological serum, having reversed -T3 (r-T3) as the predominant species. Iodine was a primary species found in urine	74,75
Homogenates of adult male and female zebrafish	Enzymatic digests (pronase E) were analyzed by RP-HPLC-ICP-MS using the method developed by Michalke et al.[74,75]	T1, T2, r-T3, T3, T4	Male and female zebrafish (D. rerio) and tadpoles of the African clawed frog (X. laevis), at two different development stages, were taken for analysis. The speciation results were different between two biological species and also for different stages of development. Five unknown iodine species were detected	78

Table 12.3 Applications of LC with ICP-MS detection for speciation analysis of mercury

Sample	Analytical procedure	Species of interest	Toxicological and/or clinical relevance	Ref
Seawater, tap water	RP separation (Spherisorb ODS column) with 0.5% cysteine (pH 5) as mobile phase (separation time <6 min), post-column CV (NaBH$_4$) and ICP-MS	Hg(II), MMM, MEM	Methodological approach: the feasibility of post-column CV generation prior to ICP-MS detection was demonstrated. The values of DL were in the range 0.03–0.11 ng Hg mL^{-1}	89
Spiked urine	Separation on C$_{18}$ column with mobile phase acetonitrile–water (20:80) with addition of 5 mmol L^{-1} ammonium pentasulfonate, pH 3.4, flow rate 0.1 mL min^{-1}. Column effluent introduced to ICP-MS through DIN	Hg(II), MMM, MEM, MPhM	Methodological approach: the DLs in urine were 0.018 and 0.016 ng Hg mL^{-1} of MEM and MPhM, respectively	90
Biological tissues	Separation on C$_{18}$ column with an aqueous mobile phase containing 0.08% ammonium acetate and 0.02% L-cysteine, coupled on-line to ICP-MS detection. SSID calibration using enriched isotope standards: ^{198}Hg in MMM and ^{201}Hg in inorganic Hg(II)	MMM, Hg(II)	Methodological approach: abiotic methylation/demethylation process during treatment with TMAH at controlled pH (citric acid) was studied. Depending on the type of sample matrix, up to 11.5% of added Hg(II) was methylated and up to 6.26% MMM was demethylated to Hg(II). The adjustment of pH after samples treatment with TMAH was recommended in order to minimize changes in mercury speciation during analytical procedure	83
Rat tissues after oral intake of HgCl$_2$	A home-made preparative SEC Sephadex G-75 column (30 × 1.9 cm) used for the purification of MT fractions and, after desalting, further separation accomplished by RP-HPLC-UV, ICP-MS and ES-MS detection	Hg binding to MTs	The role of MTs in the detoxification of mercury after its oral intake in mammals was studied. One major and several minor peaks were observed in the HPLC traces of the MT fraction for the kidney sample. Hg binding to MTs observed in kidney but not in liver extracts. Further characterization with ESI-MS allowed the identification of several complexes, containing only Hg and/or Hg and Cu	91

Sample	Procedure	Species	Comments	Ref.
Fish and sediment CRM	RP separation on Hypersil ODS column with acetonitrile–water (65:35) mobile phase adjusted to pH 5.5 with CH_3COONH_4. Column effluent introduced to ICP-MS through HHPN	MA, MMM, MEM, MPhM, Hg(II)	Methodological approach: the DLs were 0.02, 0.015, 0.02, 0.02 and 0.01 ng of Hg as MA, MMM. MEM, MPhM and Hg(II), respectively	92
Fish tissues, DORM-2	An open-vessel MW extraction, RP separation with a mobile phase containing L-cysteine and 2-mercaptoethanol, CV generation and ICP-MS detection	Hg(II), MMM, MEM	Methodological approach: post-column CV generation was used to enhance the detection power (DL in the range 0.05–0.09 ng Hg mL^{-1}).	93
Operationally defined soil extracts	RP-HPLC on Supelcosil LC-18 column with water–methanol (70:30, v/v) mobile phase containing 0.001% (v/v) 2-mercaptoethanol and 0.2 mol L^{-1} CH_3COONH_4; on-line ICP-MS (separation time 11 min)	Hg(II), MMM, MEM	Methodological approach focusing selective determination of toxic mercury species: 'mobile and toxic' fraction was extracted by with acidic ethanol solution (2% HCl + 10% ethanol solution); further SPE fractionation yielded 'soluble inorganic Hg' and 'alkyl-Hg species'	94
Different food products, plants and sediments	SSID (^{200}Hg(II)), steam distillation (NaCl–H_2SO_4), preconcentration on Hypersil in form of SPDC complexes, RP separation (Hypersil ODS) with acetonitrile–water (65:35, v/v), UV irradiation and on-line introduction to ICP-MS through ultrasonic nebulizer	Hg(II), MMM	Methodological approach: application of SSID for the elimination of analytical errors caused by possible interconversion of species. Absolute DL 12 pg of Hg (m/z 202). Methylation of inorganic mercury contained in the sample caused positive errors in the determination of MMM in sediment (54%) and plant materials (0.3–18.7%). No error was detected in the analyses of fish tissues	84,85

have been used with water–acetonitrile or methanol, L-cysteine or 2-mercaptoethanol mobile phases with separation times below 15 min. Size-exclusion chromatography (SEC) was applied for studying mercury–metallothionein complexes.[91] Since the quantification of methylmercury (MMM) in biological samples has become necessary, quality assurance strategies are emerging. There are few CRM materials available for MMM in marine matrices and sediments (fish tissue, CRMs 473 and 464; sediment, CRM 580).[3,4,95] It is worth mentioning that, among other methods, HPLC-ICP-MS was used in interlaboratory studies and certifications.[4]

12.3.4 Platinum

The capability of ICP-MS for the sensitive and specific detection of elements makes this technique suitable for studies on the uptake and metabolism of metal-based drugs.[19] In particular, cisplatin has become one of the most widely used chemotherapy drugs. Its mechanism of action relies on the ability of drug to modify the DNA structure in cancer cells, hence causing their apoptosis. However, it presents severe toxicity and its anticancer activity is limited to a small group of tumor cells. In further development, several other platinum complexes (mainly oxaliplatin, carboplatin, satraplatin, JM-216 and ZD0473) have been synthesized and tested for their possible anticancer activity.[96–100] The challenge is to obtain drugs active against a broad spectrum of tumor cells, to minimize their toxicity and to avoid the development of resistance mechanisms in target cells.[101]

After intravenous or oral administration of drug, it undergoes several biotransformations that activate its toxicity (tumor cells in addition to nephro- and neurotoxicity). A number of speciation studies have been reported that were focused on the characterization and quantification of the parent drug and its possible metabolites *in vitro* and *in vivo*. Typically, LC separation was carried out and platinum was determined in column effluents by atomic absorption sperctrometry[102,103] or by ICP-MS.[96,98,99,104–110] Several applications of mass spectrometry for the structural characterization of platinum compounds have also been reported.[96,101,103,111] The analytical figures of merit for platinum aqua species in plasma were compared for HPLC-ICP-MS system and HPLC with triple-quadrupole MS.[111] Obviously, MS-MS and MS-MS-MS measurements permitted species characterization, but the advantage of HPLC-ICP-MS was its better sensitivity (detection limit lower by an order of magnitude with respect to that achieved by triple-quadrupole instrumentation). The details of analytical procedures, together with toxicological and/or clinical importance of the results obtained, are summarized in Table 12.4.

Since clinical treatment is always based on the administration of a single platinum compound, simultaneous analysis of different drugs has rarely been undertaken. On the other hand, the metal drugs and their metabolites are released into communal waste water. The environmental impact of this discharge becomes an issue and also a challenge for speciation analysis. Recently, the separation of cisplatin, monoaquacisplatin, diaquacisplatin, carbo-platin, and oxaliplatin in environmental and biological samples was studied by Hann *et al.*[98] Owing to different properties of target compounds (neutral and positively charged species), pentafluorophenylpropyl-bonded silica stationary phase was used, providing both reversed-phase and ion-exchange-based retention. The elution was carried out with a mobile phase containing 10 mmol L^{-1} ammonium formate (pH 3.75) and methanol. With increasing concentration of organic modifier, the retention of cisplatin, carboplatin and oxaliplatin

Table 12.4 Summary of HPLC-ICP-MS procedures for platinum speciation in analyses of clinical and/or toxicological relevance

Sample	Analytical procedure	Species of interest	Toxicological and/or clinical relevance	Ref.
Model solutions	IP-HPLC separation on C_{18} column using SDS or sodium heptanesulfonate as ion-pairing agents, on-line ICP-MS detection	Cisplatin, the products of its hydrolysis and reactions with SMet, SCys and GSH	Methodological development: the value of DL evaluated for cisplatin was 1 ng Pt mL^{-1}	104
Dog plasma after ZD0473 administration	HPLC-ICP-MS, HPLC–MS-MS and HPLC–MS-MS-MS	ZD0473 and its 'aqua' compounds	Methodological development: HPLC-ICP-MS procedure was compared with ICP-MS-MS showing extended linear range and superior sensitivity (limit of quantification 0.1 ng Pt mL^{-1} by ICP-MS versus 5 ng Pt mL^{-1} using MS-MS)	111
Platinum drug (JM-216) and blood plasma samples from clinical trial on a new drug	Plasma ultrafiltrates were diluted 1:50 and analyzed by RP-HPLC (column packed with PLRP-S, Polymer Laboratories) with solvent gradient elution (acetonitrile–water from 15:85 to 90:10, 30 min), coupled to ICP-MS through a novel interface, allowing efficient elimination of acetonitrile	JM-216 and its possible metabolites	Methodological development: thanks to the application of the new LC-ICP-MS interface, a gradient of organic solvent could be used and the separation of five metabolites was achieved. The DL was 0.60 ng Pt mL^{-1}. The results obtained in the analysis of clinical samples indicated the complete decomposition of JM-216 in the human body with formation of at least five metabolites	106
Blood plasma after treatment with JM-216	Methanol extracts of clinical samples were analyzed by RP-HPLC (Phenomenex Prodigy C_8 column); elution gradient of methanol (mobile phase A, 25% methanol–0.01% H_3PO_4 at pH 2.5; mobile phase B, 100% methanol, 20 min); ICP-MS detection	JM-216 and its-biotransformation products: JM-118 (reduction product), JM-518 (hydrolysis product), JM-383	Methodological development: an increasing concentration of methanol in the mobile phase caused the suppression of platinum counts for later eluted species by ~70%. However, the quantification of platinum compounds was achieved with good intra- and inter-assay precision (range 1–13% RSD) and the DLs evaluated were in the range 1–2 ng Pt mL^{-1}. The analysis of clinical samples revealed relatively fast degradation of JM-216 with formation of JM-118 and an unidentified Pt species	107

(continued)

Table 12.4 (*Continued*)

Sample	Analytical procedure	Species of interest	Toxicological and/or clinical relevance	Ref.
Red blood cells incubated *in vitro* with satraplatin	Plasma ultrafiltrates and methanol extracts analyzed by RP-HPLC (Phenomenex Prodigy C_8 column) with 0.85% H_3PO_4 mobile phase (pH 2.5), elution in methanol gradient from 20 to 40% (v/v); off-line ICP-MS detection. The fractions were also analyzed by SDS-PAGE ICP-MS	Satraplatin and its possible metabolites	Studies on the biotransformation of satraplatin: the rapid disappearance of drug from human blood *in vitro* was observed (half-life 6.3 min), which was dependent on the presence of RBCs. Two new platinum-containing species observed on HPLC-ICP-MS traces confirmed that satraplatin undergoes rapid biotransformation in whole blood. The unknown metabolite was analyzed by SDS-PAGE and ICPMS and identified as a platinated protein with a similar electrophoretic mobility as serum albumin	99
Pooled serum incubated with Pt or Ru drug	SEC separation (Pharmacia Superdex 75 HR 10/30 or Toso Haas Progel TSK-GEL PWXL column) with 30 m mol L^{-1} Tris–HCl buffer mobile phase (pH 7.0); ICP-MS detection (^{194}Pt, ^{195}Pt, ^{196}Pt, ^{100}Ru, ^{101}Ru, ^{102}Ru, ^{63}Cu, ^{114}Cd and ^{31}P)	2 platinum (incuding cisplatin) and 3 ruthenium drugs	A rapid and simple method was developed, suitable for estimation of (1) the speed of the drug binding to proteins, (2) the efficiency of its transport in the blood stream and (3) possible efficiency of 'releasing agents' (e.g. citrate, EDTA, tartrate)	108
Incubates of cisplatin with guanosine 5-monophosphate	Ion-exchange HPLC (Dionex AS14 column) with mobile phase containing 17.5 mmol L^{-1} Na_2CO_3, 5 mmol L^{-1} $NaHCO_3$ and 5% acetonitrile at pH 10.7 (45 °C), on-line ICP-sector field MS (^{31}P, ^{115}In, ^{195}Pt)	Cisplatin and its GMP adducts	The time-dependent reaction course of the cisplatin–GMP system was followed by monitoring the decrease in the concentration of GMP and the increase in the concentration of adducts formed. The use of high-resolution MS for simultaneous quantification of P and Pt in chromatographic effluent provided unambiguous stoichiometric information about the major GMP adduct.	109

Sample	Method	Species	Comments	Ref.
Incubates of platinum complexes with guanosine 5-monophosphate (GMP)	Anion-exchange HPLC (Nucleosil 5SB column) with 50 mmol L⁻¹ formic acid–triethylamine in methanol–water (5:95) mobile phase at pH 2.6 and 30 °C; cation-exchange separation (Nucleosil 5SA column) with triethylamine in acetonitrile–water–acetic acid (12.5:82.5:5) mobile phase at pH 3.9 and 6 °C; on-line ESI-MS analysis of the column effluents	Cisplatin and four structurally related platinum(II) complexes	The formation of conjugates between Pt complexes and GMP was studied in order to gain a better insight into possible binding of platinum chemotherapy agents with DNA. Three different types of platinum(II)–mono-GMP adducts could be detected via cation-exchange chromatography. Using anion-exchange chromatography, isomers of bis-GMP adducts could be identified	101
Cisplatin incubated with GSH, cysteinylglycine, or N-acetylcysteine	HPLC with off-line atomic absorption spectrometric detection of Pt or MS analysis	Cisplatin and its conjugates	Studies on the nephrotoxicity of cisplatin: the diplatinum and monoplatinum conjugates of the drug were structurally characterized. The toxicity of the preincubated solutions was tested with time and the results obtained showed an increase in toxicity, which correlated with the formation of the monoplatinum conjugate, whereas prolonged preincubation decreased toxicity and correlated with the formation of the diplatinum conjugate	103
Urine from cancer patients	Diluted urine (1:20) was analyzed by adsorption chromatography on a Thermo Hypersil–Keystone Hypercarb column (2.1 mm i.d.) with elution in gradient of NaOH; ICP-sector field MS; IDA analysis (^{196}Pt)	Cisplatin, monoaquacisplatin, diaquacisplatin	The DLs for cisplatin and its metabolites were in the range 0.65–0.74 µg Pt L⁻¹. Diluted urine from a cancer patient contained the parent drug cisplatin and a considerable fraction of highly active monoaquacisplatin, and also several unknown platinum species	110

(continued)

Table 12.4 (*Continued*)

Sample	Analytical procedure	Species of interest	Toxicological and/or clinical relevance	Ref.
Human urine	RP-HPLC separation (Supelco Discovery HS F5 microbore column), step-gradient elution with three mobile phases: (A) 20 mmol L^{-1} ammonium formate–methanol (4%, v/v), (B) water and (C) methanol at 45 °C (separation time 23 min); on-line ICP-MS detection	Cisplatin, mono-aquacisplatin, diaqua-cisplatin, carboplatin, oxaliplatin	For cisplatin, carboplatin and oxaliplatin the DLs were 90.0, 100 and 150 ng Pt L^{-1}, respectively. The urine sample contained more than 17 different reaction products, which demonstrates the extensive biotransformation of the compound	98
Human urine	RP-HPLC separation (YMC ODS-AQ microbore column); gradient elution with two mobile phases: (A) 5 mmol L^{-1} aqueous ammonium acetate with 0.01% (v/v) acetic acid and (B) methanolic 5 mmol L^{-1} ammonium acetate with 0.01% (v/v) acetic acid (separation time 16 min); on-line ESI-MS-MS	ZD047	The assay was developed for IDA–validated RP HPLC-MS quantification of platinum drug in urine. The limit of quantification in urine (100 μL) was 200 ng Pt mL^{-1}. A novel platinum adduct was formed during the storage of ZD0473 in human urine. The addition of 50% (w/v) sodium chloride to the urine prevented the formation of this adduct during storage.	96

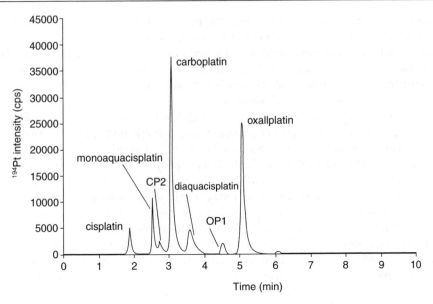

Figure 12.3   Chromatogram obtained from a 24-h mixture of cisplatin ($t_R = 1.9$ min), carboplatin ($t_R = 3.1$ min) and oxaliplatin ($t_R = 5.1$ min) in sub-boiled water. In addition to the parent drugs, the chromatogram shows the major degradation products of cisplatin (monoaquacisplatin, $t_R = 2.5$ min; diaquacisplatin ($t_R = 3.6$ min)), the product of carboplatin (CP2) ($t_R = 2.7$ min) and the product of oxaliplatin (OP1) ($t_R = 4.5$ min). The void volume of the separation system was equivalent to a retention time of 1.5 min. From Hann, S., Stefanka, Z., Lenz, K., Stingeder, G. Novel separation method for highly sensitive speciation of cancerostatic platinum compounds by HPLC-ICP-MS. *Anal. Bioanal. Chem.* 2005, **381**, 405–412, with kind permission of Springer Science and Business Media

decreased (non-polar compounds), whereas the retention of ionic species (monoaquacisplatin, diaquacisplatin) was controlled by the ionic strength. In Figure 12.3, the chromatogram of standards, obtained under optimized conditions (Table 12.4), is presented.[98]

12.3.5 Selenium

Selenium is an element of fundamental importance to human health. Depending on its chemical form and total concentration, it may exhibit essential or toxic effects with a very narrow tolerance band.[112] Thus, the recommended selenium reference nutrient intake (RNI) is 75 and 60 μg day^{-1} for adult males and females, respectively.[113] According to the UK Department of Health, an intake of 750–900 μg is toxic and the maximum safe intake is 450 μg day^{-1} for adult males.[114] The specific biological role of selenium is not fully understood. However, selenocysteine has been characterized as the 21st proteinogenic amino acid, which forms an active center of a number of proteins (selenoenzymes include glutathione peroxidase, iodothyronine deiodinase and thioredoxine reductase and selenoprotein P, among others).[112,115,116] Experimental evidence exists on selenium bioactivity in cancer prevention.[117,118] The ability of the element to reduce toxic effects of heavy metals was also observed.[119] On the other hand, the toxicity of selenium originates in its high affinity to sulfur, which results in non-specific competition of the two elements for sulfur-binding sites in different biomolecules.[120] Consequently, the characterization and selective determination of the particular selenium species in biological and/or clinical materials are necessary in order to understand its metabolism and biological significance in clinical chemistry, biology, toxicology and nutrition.

Several health conditions require supplementation with selenium (element deficiency in many geographical regions, cancer chemoprevention, geriatric treatment). Full characterization of selenium forms in the supplement to be used, their stability and species-dependent bioavailability and pharmacokinetics in the human body are of primary importance. Food products naturally rich in selenium should also be considered. On the other hand, the risk of intoxication due to excessive selenium intake needs to be minimized.[121] Selenium-rich dietary products containing relatively non-reactive, organic element forms are generally preferred, because of their better bioavailability and favorable pharmacokinetics with respect to the inorganic species.[122-124] The biological activity of different selenium species for tumor prevention has been studied in several experimental models.[125-127] The results obtained showed the highest activity for methylselenocysteine followed by selenite, selenocystine and dimethyl selenoxide.[125,128] In addition to inorganic forms and possible selenium metabolites, the activity of synthetic compound [1,4-phenylenebis(methylene)selenocyanate] has been tested.[129,130] A survey of analytical methodology based on HPLC coupled to ICP-MS for selenium speciation in food-related products and in different supplements is presented in Table 12.5. The full list of selenium species is composed of over 30 compounds that comprise inorganic forms, low molecular weight methylated species, Se-amino acids, Se-sugars and Se-containing proteins. Since pure standards of the compounds are not always available, mass spectrometry with soft ionization sources has been used to obtain structural information.[131-134] The feasibility of ion-pairing HPLC coupled to ICP-MS detection for the separation of up to 30 Se compounds was demonstrated.[135,136] However, in applications to real-world samples, a smaller number of species has been found, especially when using one pretreatment and a single chromatographic condition. Logically, the analytical results are determined by the variety of naturally occurring selenium compounds. On the other hand, the design of the pretreatment procedure may enhance the biologically relevant information obtained in speciation analysis.[28] Furthermore, better characterization of species can be achieved by combining different chromatographic approaches.[134,137-142] SEC has typically been used for discrimination between LMW and HMW selenium compounds.[134,137-139,142-145] For LMW inorganic and organic species, ion-exchange[134,139,140,146-149] and ion-pair chromatographic[127,134-136,141-144,150-152] modes have been reported. Anion-exchange columns have been applied for the separation of two inorganic forms [Se(IV) and Se(VI)] and also SeMet, SeCys, SeMC and TMSe in plant and yeast extracts.[2,147,153] A number of organic selenium compounds (selenoamino acids, peptides, TMSe, etc.) are cations, or can be easily converted to cationic species by lowering pH of the solution, thus facilitating separation by a cation-exchange mechanism. Up to 10 selenium compounds (SeMC, allylselenocysteine, propylselenocysteine, SeMet, SeEt, SeCys, dimethylseloniumpropionic acid, SeHcy, TMSe, SeMM) were resolved with different mobile phases at pH 2–5.7 (pyridine, ammonium formate, etc.).[11,35,154] The conjunctive use of anion- and cation-exchange HPLC was shown to provide information on the ionic nature of the retained and separated selenium species with a high degree of selectivity.[140] The majority of selenium speciation analyses in food-related products have been done with ion-pair

Table 12.5 Analytical applications of HPLC-ICP-MS for selenium speciation

Sample	Analytical procedure	Species of interest	Toxicological and/or clinical relevance	Ref
Se-enriched plants and yeast	Hot water and enzymatic extracts were analyzed by IP chromatography on a Waters Symmetry Shield RP8 column using water–methanol (99:1) mobile phases that contained different perfluorinated carboxylic acids as ion-pairing agents. 0.1% HFBA in the mobile phase allowed the separation and ICP-MS detection of >20 Se compounds in 70 min. For ESI-MS analyses, 0.1% TFA was used	23 Se species, among them selenate [Se(VI)], selenite(IV), TMSe, SeMet, SeCys, SeEt, SeHcy, AdoSeHcy, SeMC, γ-glutamyl-SeMC	Methodological development focusing Se speciation in biological materials. Limits of quantification for HPLC-ICP-MS were in the range 2–50 ng Se mL^{-1} and for HPLC–ESI-MS these values were about 100 times higher. Between 60 and 85% of compounds observed in real-world samples were identified by ESI-MS. Primary species found were SeMet, SeMC, γ-glutamyl-SeMC and selenocystathionine, depending on the biological sample and extraction procedure applied	135,150
Se-enriched ramps (leeks)	Hot water and enzymatic (protease XIV) extractions, followed by ion-pairing IP-HPLC-ICP-MS (^{82}Se)	Inorganic Se, organic species with anti-cancer activity	Primary species found in extracts: SeMC (35–50%) and Se(VI) (15–42%), lower levels of Se-cystathionine, γ-glutamyl-SeMC. In rats fed with the selenized ramps, the reduction in chemically induced tumors was observed	127
Se-enriched onion leaves	Aqueous extracts (0.1 mol L^{-1} NaOH) were analyzed by SEC-UV-ICP-MS (Superdex peptide HR 10/30 column, CAPS 10 mmol L^{-1}, pH 10.0); methanol–chloroform–water (12:5:3) extracts or enzymatic digests (proteinase K, protease XIV) analyzed by IP-HPLC-ICP-MS [Altima C$_8$, 5 mmol L^{-1} citric acid, 5 mmol L^{-1} hexanesulfonic acid, pH 4.5)–methanol (95:5)]; ESI-MS and ESI-MS-MS used for species identification and/or confirmation	Selenite [Se(IV)], SeMet, SeCys, SeMC	Uptake, distribution and speciation of selenium in green parts of onion were evaluated. The results obtained indicated that onion leaves could be an alternative dietary source of SeMC	143,144

(continued)

Table 12.5 *(Continued)*

Sample	Analytical procedure	Species of interest	Toxicological and/or clinical relevance	Ref.
Se-yeast and SeMC-based supplements	ASE (water) and enzymatic digestion; SEC performed on a Supelco TSK gel G 3000 PWxL column with 10 mmol L^{-1} CH$_3$COONH$_4$ (pH 8.5) mobile phase; AE separation on a Hamilton PRP-X 100 column with 5 mmol L^{-1} ammonium citrate (pH 5.9) mobile phase containing 2% (v/v) methanol; IP-HPLC on an Agilent Zorbax Rx-C8 column with 0.1% TFA, 2% (v/v) methanol. On-line ICP-MS detection, sample introduction via USN; IP-HPLC coupled on-line to ESI-MS-MS	SeMet, SeMC	The proposed procedure permitted enhanced separation selectivity and very low DL (ng Se L^{-1} levels). The identification and quantification of SeMC (minor yeast species) were reported	134
Se-yeast	Homogenization of fresh cells and precipitation of proteins (0.4 mol L^{-1}, HClO$_4$, 0 °C); IP-HPLC (hexanesulfonic acid) with UV (254 nm) and ICP-MS detection; ESI-MS and MS-MS used for identification/confirmation of Se species	AdoSeMet, AdoSeHcy, SeCis	AdoSeMet and AdoSeHcy were identified in yeasts extract and the presence of SeCis was also observed. A significantly lower ratio between AdoSeMet and AdoSeHcy was observed in yeast extracts compared with the ratio between sulfur analogs. This observation could contribute to further understanding of the selenium role in biomethylation processes	151
Se-yeast	Water extract analyzed by two-dimensional LC (SEC and RP on porous graphitic carbon stationary phase) with ICP-MS detection. Two Se-containing fractions were isolated and analyzed by nanoESI-MS and MS-MS	Se-containing glutathione S-conjugates	Two selenium species identified: selenodiglutathione (GS-Se-SG) and the mixed selenotrisulfide of glutathione and cysteinylglycine (GS-Se-SCG)	137,138

Sample	Species	Method	Results	Ref.
Se-yeast	D- and L-enantiomers of SeMet and SeEt	Different chiral separation techniques investigated: HPLC on a β-cyclodextrin (β-CD) column, CD modified micellar electrokinetic chromatography (MEKC), GC on a Chirasil-L-Val column and HPLC on a Chirobiotic T column	The analytical results for D- and L-SeMet obtained in selenized yeast were presented [HPLC on a Chirobiotic T column, water–methanol (98:2, v/v), separation time 8 min]	156
Se-enriched onion, garlic and yeast	Enantiomers of SeMet, SeMC, SeCys, γ-glutamyl-SeMC, SeEt, SeHcy; Se-lanthionine, Se-cystathionine	Water and a pepsin enzymatic extract analyzed on chiral crown ether column [Daicel Crownpak CR(+)] with $0.1 \, mol \, L^{-1}$ $HClO_4$ as mobile phase; on-line ICP-MS detection	Separation/determination of optical isomers of SeMet and SeCys was studied. Baseline separation of all enantiomers in one chromatographic run was not achieved; however, enhanced resolution was observed at higher temperatures	158
Se-nutritional supplements	D,L- SeCys, SeMet, SeEt	HPLC separation on a chiral crown ether stationary phase with $0.10 \, mol \, L^{-1}$ $HClO_4$ as mobile phase and UV or ICP-MS detection (separation time 60 min, 22 °C)	Absolute DL obtained with UV detection ranged from 34.5 to 47.1 ng of Se, whereas those obtained with the plasma detector were ca 40–400 times better	159
Se-yeast candidate reference material	Inorganic Se, SeMet, SeCys	Sequential extraction yielding watersoluble and non-soluble fractions; 2-dimensional chromatography (SEC on Superdex 75 HiLoad 16/60 column and AE on Dionex AS10 column with gradient elution, pH 8) and on-line UV-ICP-MS (280 nm, ^{78}Se, ^{82}Se); SDS-PAGE-ETV-ICP-MS	The analytical approach focusing characterization of all Se forms/species in Se-yeast	139
^{77}Se-labeled and enriched yeast	LMW Se species biosynthesized by the yeast during fermentation	Two enzymatic digestions examined, separation by cation-exchange HPLC (Ionosphere-5C, Chrompack International) with gradient elution using pyridinium formate mobile phase (separation of 30 Se compounds within 60 min); ICP-DRC-MS detection (^{77}Se, ^{80}Se).	A batch of ^{77}Se-labeled and enriched yeast was characterized with regard to isotopic composition and content of selenium species for later use in a human absorption study based on the method of enriched stable isotopes. SeMet constituted 53% of the total Se content in the yeast. Oxidation of SeMet to selenomethionine-Se-oxide (SeOMet) occurred during sample preparation	146

(continued)

Table 12.5 (*Continued*)

Sample	Analytical procedure	Species of interest	Toxicological and/or clinical relevance	Ref.
Yeast and wheat flour	Enzymatic hydrolysis (protease, lipase) followed by AE separation (Hamilton PRP-X100 column) with 5 mol L^{-1} ammonium citrate in 2% (v/v) methanol and post-column IDA-ICP-MS with octapole reaction system	Selenite [Se(IV)], selenate [Se(VI)], SeCys, SeMet, SeCM	Analytical approach to more reliable quantification of Se in biological materials. Results showed that SeMet was the main Se species in both matrices (59% of total Se for wheat flour and of 68% for yeast). Inorganic Se was detected in yeast samples, but not in wheat flour	147
Se-yeast	Enzymatic hydrolysis (protease XIV) followed by anion-exchange separation (Hamilton PRP-X100 column) in phosphate gradient (from 10 to 100 mmol L^{-1}, pH 7) in the presence of 2% (v/v) of methanol and SSID-ICP-MS with octapole reaction system (separation time 15 min)	SeMet	Biosynthesis of ^{77}Se-enriched SeMet by *Saccharomyces cerevisiae* investigated, SSID method applied for the determination of SeMet in a candidate reference yeast material	148
Selenite-, SeMet- and Se-yeast nutritional supplements	Hot water extracts analyzed by RP-HPLC [Nucleosil 120 A C$_{18}$ column with 30 mmol L^{-1} ammonium formate, pH 3.0, 5% (v/v) methanol] and IP HPLC (10 mM TBAA in the RP mobile phase) with on-line ICP-MS detection. Three nebulizers were tested (Meinhard, HHPN, MCN) and a hexapole collision reaction cell was used	Selenate [Se(VI)], selenite [Se(IV)], SeMet, SeCys, SeEt, TMSe	Performance of different nebulization and chromatographic systems for the speciation of Se investigated. For selenite- and SeMet-based supplements, the analytical results confirmed the information given by the manufacturer (extraction efficiency 80–95%). For the Se-yeast supplements (extraction efficiency 10–25%), more than 20 selenium compounds were detected, among them SeMet, TMSe and SeEt	152

Se-nutritional supplement	IP-HPLC or LiChrosorb RP18 column using a mixed ion-pair reagent (2.5 mmol L^{-1} sodium 1-butanesulfonate and 8 mmol L^{-1} TMAH); on-line introduction of effluent to ICP-MS through USN and pneumatic nebulizer (separation time 18 min)	Selenite [Se(IV)], selenate [Se(VI)], SeCys, selenourea, SeMet, SeEt, SeCM, TMSe	Methodological development focusing fast separation of 30 Se-species in biological samples. SeMet confirmed as the primary Se compound in supplement analyzed	136
Se-yeast, Se-garlic	HPLC-ICP-MS and ESI-MS for the identification/confirmation of species	SeMet, γ-glutamyl-SeMC	The primary Se species in enriched yeast and garlic were characterized (SeMet and γ-glutamyl-SeMC, respectively) and quantified (73 and 85% of total Se) and certain biological activities of these two materials compared. In the rat feeding studies, supplementation with Se-garlic caused a lower total tissue selenium accumulation compared with Se-yeast. On the other hand, Se-garlic was significantly more effective in suppressing the development of premalignant lesions and the formation of adenocarcinomas in the mammary gland of carcinogen-treated rats. The metabolism of SeMet and γ-glutamyl-SeMC was discussed in an attempt to elucidate how their disposition in tissues might account for the differences in cancer chemopreventive activity	126

(continued)

Table 12.5 (Continued)

Sample	Analytical procedure	Species of interest	Toxicological and/or clinical relevance	Ref.
Se-yeast and algae	Enzymatic digestion of yeast (commercial β-glucosidase and peptidases); freeze-drying and TCA extraction (0 °C) for algae samples. Cation-exchange separation (Ionosphere-C, Chrompack column and gradient elution with piridinium formate pH 3.0–3.2) and anion-exchange HPLC (ION-120 interaction chromatography column and isocratic elution with sodium salicylate at pH 8.5); ICP-MS detection with DRC (^{80}Se)	12 Se species comprising Se-amino acids and inorganic Se	Methodological development focusing the optimization of IC-HPLC. SeMet, SeOMet and SeMC were detected in yeast digests (complementary ESI-MS analyses). Algae extracts contained dimethylselon-iumpropionate, allylSeCys and SeEt	140
Se-yeast intervention agents in cancer prevention	Proteolytic hydrolysis, ion-exchange HPLC coupled to ICP-MS detection	SeMet and unidentified species	Speciation and bioavailability of selenium in yeast-based intervention agents: the results revealed different Se contents, depending on the product source, batch and age. Differences in intake, speciation or bioavailability of selenium from the yeast-based supplements could be responsible for different increases in Se in blood plasma observed in the clinical trials	149
In vitro gastric and intestinal digests of Se-yeast supplements	HPLC-ICP-MS and HPLC–ESI-MS-MS	Se-amino acids	The main compound extracted by both gastric and intestinal fluid was SeMet, which is also the primary Se species in yeast proteolytic digest. Two other minor compounds could be identified as SeCis and SeOMet	197

Sample	Method	Species	Comments	Ref.
Nuts	LMW fraction extracted with 0.4 mol L^{-1} HClO$_4$; proteins solubilized with 0.1 mol L^{-1} NaOH; enzymatic digestion (proteinase K). HMW compounds separated by SEC (Superdex Peptide HR 10/30 acetate buffer containing 0.2% SDS at pH 4.5); LMW compounds separated by IP-HPLC [Altima C$_8$, 5 mmol L^{-1} citrate buffer, 5 mmol L^{-1} hexanesulfonic acid, pH 4.5, with 5% (v/v)] methanol. For HMW compounds, on-line UV and ICP-MS, for LMW ICP-Ms and additional ESI-MS analyses	Inorganic Se, SeMet, SeCys, SeEt	Distribution and speciation of Se in different types of nut studied: ∼12% of Se in Brazil nuts weakly bound to proteins, ∼3–5% found in cytosol, the primary species found in enzymatic digests was SeMet (19–25% of total Se for different types of nuts). Peptide structure studied by ESI-MS and MS-MS. Brazil nuts proposed as naturally rich dietary source of SeMet	141,142
Tuna and mussel tissues	Two-step enzymatic hydrolysis with a non-specific protease (subtilisin); separation on a Spherisorb 5 ODS/AMINO column using two different chromatographic conditions, namely phosphate buffers at pH 2.8 and 6.0 as mobile phases; on-line ICP-MS	Selenate [Se(VI)], selenite [Se(IV)], TMSe, SeCis, SeMet, SeEt	Characterization of Se species in marine food: only organic selenium species were found in the samples (TMSe, SeMet and several unknown species). The sum of identified selenium species in the sample was about 30% of the total selenium present in the enzymatic extract	198
Cod muscle	Four different extraction procedures tested; speciation analysis by SEC and RP-HPLC followed by post-column ID-ICP-MS with octapole reaction system	SeMet	The highest Se recoveries were obtained when using enzymatic digestions whereas only 5% of total Se in cod was extracted when a 'soft' extraction procedure (MeOH-HCl) was used. SeMet was the main seleno compound found in enzymatic hydrolyzates. In MeOH-HCl extracts, SeMet was absent, indicating that Se is mainly incorporated into proteins. A number of unidentified Se species were also detected in cod muscle tissue	145

(continued)

Table 12.5 (Continued)

Sample	Analytical procedure	Species of interest	Toxicological and/or clinical relevance	Ref.
Rat organs and body fluids after [82]Se-enriched selenate, selenite or SeMet treatment	Se distribution in organs determined by ICP-MS, speciation studied by SEC HPLC (Asahipak GS 520 or GS 320 column, elution with 50 mmol L^{-1} Tris–HCl buffer at pH 7.4) and ICP-MS detection	Inorganic and organic species relevant to Se metabolism in rats	Metabolic, nutritional and toxicological aspects of Se were studied. The transport of Se within the organism and the urine metabolites observed were different, depending on the form of Se administred. Rapid and efficient incorporation of Se into SeP in the liver and excretion into the plasma followed by the slow and steady incorporation of Se into GPx in the kidneys and excretion into the plasma described. The formation of Se-methyl-N-acetylselenohexosamine in liver and its excretion via urine were reported. The mechanisms underlying the interaction between Se and mercuric ions in the bloodstream were explained by the formation of a ternary complex, [(HgSe)$_{nlm}$-selenoprotein P	160–163
Rat urine after administration of Se(IV) or Se-garlic	Sample homogenization followed by separation by multi-mode capillary gel filtration (CShodex GS320AM5D and GS320A-M8E, 50 mmol L^{-1} CH$_3$COONH$_4$, pH 6.5) and ICP-MS detection (microflow total consumptionnebulizer)	Selenite [Se(IV)], SeMet, SeMC, γ-glutamy l-SeMC, TMSe, SeGal	Methodological development: the advantages of the capillary HPLC over conventional HPLC was demonstrated as follows: (1) lower sample requirement (as low as 100 nL); (2) improved signal-to-noise ratio; and (3) shorter time of analysis. Separation of SeMC and γ-glutamyl-SeMC directly in homogenates of Se-garlic achieved in 15 min. Urine SeGal and TMSe resolved in 16 min	176
Human volunteer urine	Filtered samples analyzed by IP-HPLC on LiChrosorb RP18 column with a mobile phase containing 2.5 mmol L^{-1} sodium 1-butanesulfonate and 8 mmol L^{-1} TMAH (separation time 15 min); on-line ICP-MS	Selenate [Se(VI)], selenourea, SeMet, SeEt, TMSe	The method developed (DL 0.6–1.5 µg Se L^{-1}) was used for the analysis of urine samples: TMSe and two unknown metabolites were detected. On the other hand, SeMet, selenourea and SeEt were not observed even at a total Se concentration higher than 100 µg Se L^{-1}.	169

Sample	Method	Species	Comments	Ref.
Human volunteer urine	Diluted $(1+1)$ or purified (by benzo-15-crown-5-ether extraction) samples were analyzed by cation-exchange LC (Ionpac CS5 column with a mobile phase containing $10\,mmol\,L^{-1}$ oxalic acid and $20\,mmol\,L^{-1}$ K_2SO_4 at pH 3); on-line ICP-MS detection (^{78}Se and ^{82}Se)	Selenite [Se(IV)], selenate [Se(VI)], SeMet, TMSe	Methodological development: for diluted urine, different retention behavior of TMSe in urine versus standard solution and co-elution of SeMet and Se(VI) were observed. Analyzing the purified urine, six chromatographic peaks were observed, two of them assigned as SeMet and TMSe. Crown ether extraction did not improve the separation. Hence, apart from SeMet and TMSe, at least three more unknown selenium-containing species were present in urine	167,168
Rat urine	Methanol extraction; separation by SEC HPLC (Shodex Asahipak GS-320 7G column with $50\,mmol\,L^{-1}$ Tris-HCl at pH 7.4 or GS-220 HQ with $50\,mmol\,L^{-1}$ CH_3COONH_4 at pH 6.5) and on-line detection by ICP-MS detection (^{77}Se and ^{82}Se); ESI-MS and MS-MS analyses for species identification	TMSe, Se-sugars and their possible oxidation products	Identification of major Se metabolite in urine as methylselenium-N-acetylselenohexosamine. Methodological development focusing possible effect of storage and pretreatment conditions on speciation results. It was proposed that monome-thylseleninic acid observed in purified urine could be the oxidation product of Se-sugar	175,199
Human urine before and during supplement-ation with SeMet	Urine purified by SPE (C_{18} cartridge conditioned with HFBA) and analyzed by IP-HPLC [microbore Luna C_8 column with different mobile phases containing perfluorinated carboxylic acids and 20% (v/v) methanol]; ICP-MS detection with DIN	SeMet, SeMM, SeMC, SeGaba, TMSe	Methodological development aiming characterization of Se metabolites in urine. The DLs were in the range 0.8–1.7 µg Se L^{-1}. Several chromatographic peaks were observed and, in some samples, co-elution of species with TMSe, SeMM or SeMet occurred. The major Se compound remained unidentified	171
Human urine	RP-HPLC separation with ICP-MS detection; column fractions were analyzed by ESI-MS and MS-MS for structural information	SeMet, SeCystamine	Six Se-containing fractions were collected from the RP column and the first two fractions were identified specifically as SeMet and selenocystamine, estimated to be present at ∼11 and 40 µg L^{-1}, respectively	178

(continued)

Table 12.5 (Continued)

Sample	Analytical procedure	Species of interest	Toxicological and/or clinical relevance	Ref.
Human urine	Urine purified by SPE (C$_{18}$ cartridges modified with hexanesulfonic acid) and acidified (HClO$_4$, pH 2.0) was analyzed by IP-HPLC [Altima C$_8$ with a mobile phase containing 2 mmol L^{-1} hexanesulfonic acid, 0.4% acetic acid, 0.2% triethanolamine (pH 2.5) and 5% methanol]; ICP-MS detection (^{77}Se and ^{78}Se), ESI-MS and MS-MS analyses for structural information	SeMet, TMSe, AdoSeMet, SeMM, SeEt	Se species identified in urine were: SeMet, TMSe and AdoSeMet. The identification of AdoSeMet seems to confirm that SeMet enters the metabolic pathway of its sulfur analog in the activated methylation cycle	172
Human urine (volunteers supplemented with Se-yeast)	SPE and multidimensional chromatography (IP, RP and SEC) with ICP-MS detection. Purified fractions analyzed by APCI-MS and MS/MS	Main urinary metabolite of Se	The major selenium metabolite from human urine identified as Se-sugar (methylseleno-N-acetylselenohexosamine)	174
Rat urine after feeding with Se(IV)	SEC HPLC (GS 320 column with 50 mmol L^{-1} Tris–HNO$_3$ buffer at pH 7.4) coupled to ICP-MS detection (^{77}Se, ^{82}Se)	SeGal, TMSe	The changes in the urinary SeGal and TMSe were investigated to clarify the relationship between the dose and two main urinary metabolites. It was also examined whether the metabolites are related to age. In young rats, SeGal was always the major urinary metabolite and TMSe increased with a dose higher than 2.0 μg Se mL^{-1} drinking water. In adult rats, TMSe increased only marginally despite the fact that the rats suffered much more greatly from the Se toxicity, suggesting that TMSe cannot be a biomarker of Se toxicity	179

Sample	Procedure	Species	Results	Ref.
Human urine (after a single dose of Se-yeast)	Three pretreatment procedures tested: (1) preconcentration through evaporation in N_2 stream; (2) evaporation and methanol extraction followed by the analysis of two fractions; (3) as (2), but using 100 mL of urine. Two chromatographic systems compared: RP-HPLC [microbore Luna C_{18} (2), 2 mmol L^{-1} CH_3COONH_4 with 5% methanol) and IP-HPLC [microbore Luna C_8 (2), 0.2% HFBA with 20–30% methanol]; on-line ICP-MS with MCN and modified DIN	TMSe, SeMC, SeMM, SeGal, SeGlu, SeEt, SeMet Se-cystamine	Pretreatment procedures did not affect native species distribution in urine. For large urine volumes (preconcentration factor 100), at least 10 selenium compounds were separated. The elution behavior of some species was slightly changed in the presence of a urine matrix, as observed in spiking experiments. Only two species were identified in the analyzed samples: SeGal – primary species and SeGlu – about 2% of SeGal; TMSe, SeMet, SeMM, SeCM and Se cystamine were not detected	177
Serum and prostate tissue from prostate cancer patients	HPLC-ICP-MS	SeMet	The cancer patients had total Se levels in serum and in prostate tissue in a range expected for normal Se intake. There was a significantly higher total Se level found in peripheral versus transitional zone tissues. This is the first time that Se-Met has been detected in prostate tissue	164
Human serum (healthy volunteers and hemodialyzed patients)	Anion-exchange HPLC (Mono Q HR 5/5 column) and affinity chromatography (Hi-Trap Heparin and Hi-Trap Blue-Sepharose columns) and IDA-ICP-MS with an octapole reaction system (^{78}Se/^{77}Se and ^{80}Se/^{77}Se)	Se-containing proteins	Fractionation of serum proteins was achieved by affinity chromatography: three main selenium fractions corresponded to SeP, albumin and GPx. Mass balance performed under different experimental conditions showed quantitative selenium recovery. The distribution of selenium between plasma GPx (~20%), SeP (~55%) and albumin (~20%) was similar in both populations	14
Human serum	Derivatization of the SeCys residues with iodoacetamide, enzymatic digestion (lipase, protease), purification of Seamino acid fraction by SEC-HPLC and separation by capillary HPLC; SSID–ICP-MS with DRC (^{77}Se/^{78}Se)	Se-amino acids	A ^{77}Se-labeled SeMet was added to serum for species quantification. The accurately determined SeMet was used as an internal standard for the SeCys determination from the same chromatogram	165

HPLC.[141,152] Different perfluorinated carboxylic acids have been examined as the ion-pairing agents (pH 2.5–4.5) in the analyses of enriched yeast and vegetable extracts.[135,153] Alkylsulfonic acids have been successfully used in speciation analysis carried out on nuts, onion leaves and yeast.[142,143,151,155] Owing to their different biological activities, optical enantiomers of Se-amino acids were analyzed by several chiral separation techniques.[156–159]

The distribution and speciation of selenium in different organs, tissues and body fluids have been investigated by HPLC-ICP-MS for a better understanding of selenium metabolism[14,160–165] (Table 12.5). In particular, the characterization of selenium metabolites in urine has often been undertaken in the context of element metabolism and evaluation of possible health risk. Depending on the element intake, its chemical form, total urine selenium and the pretreatment procedure applied, different compounds were observed. Inorganic selenium and TMSe have often been detected, confirming the biomethylation route of element elimination.[166–170] In recent studies, significant progress in species characterization has been achieved by application of different clean-up and extraction procedures and the use of collision and reaction cells in ICP-MS.[170–174] Thus, selenosugars were identified as primary selenium species both in a rat model[175] and in human urine.[174,176,177] Less abundant metabolites (SeMet, SeMC, Se-cystamine, AdoSeMet) were also identified and confirmed by ESI-MS.[168,169,171,172,178] Based on the results obtained, a putative pathway and dose-dependent excretion of selenium in human organism have been proposed, which is presented in Figure 12.4.[177] Dose- and age-related changes in the excretion of SeGal and TMSe in rats were studied by Suzuki *et al.*[179] As can be observed in Figure 12.5, in young rats, selenosugar was always the major urinary metabolite and TMSe increased with a dose higher than $2.0\,\mu g$ Se ml^{-1} drinking water.[179] On the other hand, in adult rats, TMSe increased only marginally despite the fact that the rats suffered much more greatly from the Se toxicity, suggesting that TMSe cannot be a biomarker of Se toxicity. The results suggested that sources of the sugar moiety of selenosugar are more abundant in adult

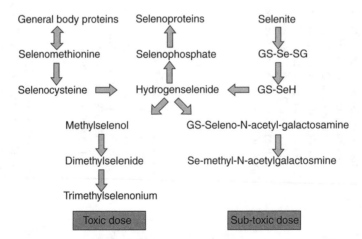

Figure 12.4 The proposed selenium metabolism (GS-: glutathione conjugate). From Gammelgaard, B., Bendahl, L. Selenium speciation in human urine samples by LC- and CE-ICP-MS separation and identification of selenosugars. *J. Anal. At. Spectrom.* 2004, **19**, 135–142. Reproduced by permission of The Royal Society of Chemistry

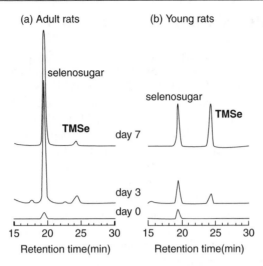

Figure 12.5   Changes in the distribution of Se in the urine of adult (a) and young (b) rats fed selenite at the concentration of 5.0 μg Se mL^{-1} drinking water for 7 days. Male Wistar rats of 36 (a) and 5 (b) weeks of age (three rats/group) were fed selenite at the concentration of 5.0 μg Se mL^{-1} drinking water *ad libitum* for 7 days, and 24-h urine was collected daily for each rat maintained in a plastic cage for metabolism. Urine samples from three rats in each group were combined and applied to a GS 320HQ column, and the column was eluted with 50 mM Tris–HNO$_3$ (pH 7.4, 25.8 °C). Reprinted from *Toxicology and Applied Pharmacology*, **206**, Suzuki, K.T., Kurasaki, K., Okazaki, N., Ogra, Y. Selenosugar and trimethylselenonium among urinary Se metabolites: Dose- and age-related changes, 1–8. Copyright 2005, with permission from Elsevier

than in young rats. Chondroitin 4-sulfate did not affect the ratio of the two urinary metabolites, suggesting that the sugar source is of endogenous origin and that it increases with age.[179]

Regarding element-specific detection in selenium speciation analyses, the conventional ICP-MS instruments pose two important difficulties: (1) low ionization efficiency (about 33% of atoms entering plasma) caused by relatively high ionization potential of selenium [9.8 eV, close to the first ionization potential of argon (15.8 eV)][180] and (2) spectral interference on the most abundant selenium isotope (^{80}Se, 49.61%) caused by argon dimer ion. Hence the sensitivity of selenium speciation is often compromised by incomplete ionization and the necessity of monitoring less abundant isotopes (^{77}Se or ^{78}Se). Furthermore, the effect of the chromatographic mobile phase composition on selenium response in ICP-MS should be mentioned.[151] Consequently, the reported detection limits for selenium species by ICP-MS coupled to chromatographic separation range from 0.5 to 50 μg L^{-1} (as selenium concentration in the solution introduced on to the column).[140,150,181,182] The progress in ICP-MS instrumentation and, in particular, the commercial availability of high-resolution instruments (HR-ICP-MS) and collision/reaction cell technology, have offered improved conditions of selenium determination.[19,140,145,147,148,165,183,184] Isotope dilution analysis has been explored for accurate speciation results.[14,147] In particular, the species-specific ID mode (SSID) permits correction for analyte losses and/or conversion during the entire analytical procedure and reliable quantification.[148,165,185]

12.4 CONCLUSIONS AND FUTURE TRENDS

Element speciation studies are emerging in the life sciences. Such analyses are necessary for assessing pharmacokinetic aspects of elements, for elucidation of their biological pathways and for better understanding of possible cellular mechanisms responsible for the observed health effects. The list of toxicologically and clinically relevant elements/species has gradually grown larger as a result of methodological developments and the progress in our knowledge of element-specific biological functions. Coupling of LC and ICP-MS is well established and considered as a primary speciation tool for non-volatile compounds. The high separation power and versatility of HPLC make this technique particularly suited for the separation of element species. Owing to different molecular size, polarity and/or electrical charge of target compounds, size-exclusion, ion-exchange, reversed-phase and ion-pair reversed-phase separation mechanisms are commonly used. On the other hand, ICP-MS offers outstanding detection conditions in terms of sensitivity and multielement and isotopic capabilities. The use of high-resolution instruments and collision/reaction cell technology eliminates troublesome polyatomic interferences and offers enhanced performance for non-metal determinations. Several elements can be monitored in one chromatographic run, providing information on their possible binding in the sample. However, for the identification and/or confirmation of species, structural analysis is needed. On-line coupling of HPLC with ICP-MS helps to minimize possible analyte loss or sample contamination, to avoid problems related with the storage of fractions, provides higher reproducibility and permits automation. The importance of compatibility between column effluent and plasma and the importance of state of the art interface designs should not be under-emphasized. Within this context, optimization of mobile phase composition, multidimensional separations, the development and applications of both low-flow separation systems and high-consumption nebulizers have been addressed. Finally, the toxicological and clinical importance of speciation results call for more reliable analytical results. Thus, an increase in applications of isotope dilution and, in particular, species-specific isotope dilution can be expected.

12.5 ABBREVIATIONS

AdoSeHcy	selenoadenosylhomocysteine
AdoSeMet	selenoadenosylmethionine
APCI	atmospheric pressure chemical ionization
AsB	arsenobetaine
AsC	arsenocholine
ASE	accelerated solvent extraction
AsSug-OH	arsenosugar: glycerol–ribose
AsSug-PO$_4$	arsenosugar: phosphate–ribose
AsSug-SO$_3$	arsenosugar: sulfonate–ribose
AsSug-SO$_4$	arsenosugar: sulfate–ribose
As(GS)$_3$	arsenic triglutathione
CFN	cross-flow nebulizer
CH$_3$As(GS)$_2$	methylarsenic diglutathione

Cisplatin	*cis*-diamminedichloroplatinum(II)
CRM	certified reference material
CV	cold vapor
DI	deionized water
DIN	direct injection nebulizer
DL	detection limit
DMA	dimethylarsinic acid
DMA(III)	dimethylarsinous acid
DMAE	dimethylarsinoylethanol
DN	direct nebulization
DRC	dynamic reaction cell
ESI-MS	electrospray ionization mass spectrometry
ETV	electrothermal vaporization
GMP	guanosine 5-monophosphate
GPx	glutathione peroxidase
GSH	glutathione
HFBA	heptafluorobutanoic acid
HG	hydride generation
HHPN	hydraulic high-pressure nebulizer
HMW	high molecular weight
HPLC	high-performance liquid chromatography
ICP-MS	inductively coupled plasma mass spectrometry
IDA	isotope dilution analysis
IEC	ion-exchange chromatography
IP	ion-pair
JM-216	bisacetatoamminedichlorocyclohexylamineplatinum(IV)
LC	liquid chromatography
LMW	low molecular weight
MA	mersalylic acid
MCN	micro flow nebulizer
MEM	monoethylmercury
MMA	monomethylarsonic
MMA(III)	monomethylarsonous acid
MMM	monomethylmercury
MPhM	monophenylmercury
MT	metallothionein
Oxaliplatin	(1*R*, 2*R*)-1,2-cyclohexanediamine-*N*,*N'*-oxalato(2−)-O,O'-platinum
p-ASA	*p*-arsanilic acid (4-aminobenzenearsenic acid)
ROX	roxarsone (4-hydroxy-3-nitrobenzenearsenic acid)
RP	reversed-phase
Satraplatin (JM216)	bisacetatoamminedichlorocyclohexylamineplatinum(IV)
SDS	sodium dodecyl sulfate
SDS-PAGE	sodium dodecyl sulfate polyacrylamide gel electrophoresis
SEC	size-exclusion chromatography
SeCis	selenocystine
SeCys	selenocysteine
SeEt	selenoethionine

SeGaba	seleno-γ-aminobutyric acid
SeGal	Se-sugar: methylseleno-N-acetyl-D-galactosamine
SeGlu	Se-sugar: methylseleno-N-acetyl-D-glucosamine
SeHcy	selenohomocysteine
SeMC	methylselenocysteine
SeMet	selenomethionine
SeMM	methylselenomethionine
SeOMet	selenomethionine oxide
SeP	selenoprotein P
SPDC	sodium pyrrolidinedithiocarbamate
SPE	solid-phase extraction
SSID	species-specific isotope dilution
T1	monoiodothyronine
T2	3,5-diiodothyronine
T3	3,3′,5-triiodothyronine
T4	3,3′,5,5′-tetra-iodothyronine
T4–TBG	thyroxin–thyroxin-binding globulin
TBAA	tetrabutylammonium acetate
TBAH	tetrabutylammonium hydroxide
TCA	trichloroacetic acid
TEAH	tetraethylammonium hydroxide
TETRA	tetramethylarsonium ion
TFA	trifluoroacetic acid
TMAH	tetramethylammonium hydroxide
TMAO	trimethylarsine oxide
TMAP	trimethylarsoniopropionate
TMAs	tretramethylarsonium ion
TMSe	trimethylselonium cation
TTAA2	3,5-diiodothyroacetic acid
TTAA3	3,3′,5-triiodothyroacetic acid
TTAA4	3,3′,5,5′-tetraiodothyroacetic acid
USN	ultrasonic nebulizer
ZD0473	cis-amminedichloro(2-methylpyridine)platinum(II)

REFERENCES

1. Caruso, J. A., Klaue, B., Michalke, B., Rocke, D. M. Group assessment: element speciation. *Ecotoxicol Environ Saf* 2003, **56**, 32–44.
2. Michalke, B. Element speciation definitions, analytical methodology, and some examples. *Ecotoxicol Environ Saf* 2003, **56**, 122–139.
3. Cornelis, R., Crews, H., Donard, O. F., Ebdon, L., Quevauviller, P. Trends in certified reference materials for the speciation of trace elements. *Fresenius' J Anal Chem* 2001, **370**, 120–125.
4. Quevauviller, P. Certified reference materials: a tool for quality control of elemental speciation analysis. In Caruso, J. A., Sutton, K. L., Ackley, K. L. (Eds), *Elemental Speciation. New Approaches for Trace Element Analysis*. Elsevier Science: Amsterdam, 2000, pp. 531–569.
5. Ray, S. J., Andrade, F., Gamez, G., McClenathan, D., Rogers, D., *et al*. Plasma-source mass spectrometry for speciation analysis: state-of-the-art. *J Chromatogr A* 2004, **1050**, 3–34.

6. Rosenberg, E. The potential of organic (electrospray- and atmospheric pressure chemical ionisation) mass spectrometric techniques coupled to liquid-phase separation for speciation analysis. *J Chromatogr A* 2003, **1000**, 841–889.

7. Feldmann, I. What can the different current-detection methods offer for element speciation? *Trends Anal Chem* 2005, **24**, 228–242.

8. Gómez-Ariza, J. L., Garcia-Barrera, T., Lorenzo, F., Bernal, V., Villegas, M. J., *et al.* Use of mass spectrometry techniques for the characterization of metal bound to proteins (metallomics) in biological systems. *Anal Chim Acta* 2005, **524**, 15–22.

9. Michalke, B. The coupling of LC to ICP-MS in element speciation: I. General aspects. *Trends Anal Chem* 2002, **21**, 142–153.

10. Michalke, B. The coupling of LC to ICP-MS in element speciation – Part II: recent trends in application. *Trends Anal Chem* 2002, **21**, 154–165.

11. Montes-Bayon, M., DeNicola, K., Caruso, J. A. Liquid chromatography–inductively coupled plasma mass spectrometry. *J Chromatogr A* 2003, **1000**, 457–476.

12. Caruso, J. A., Montes-Bayon, M. Elemental speciation studies – new directions for trace metal analysis. *Ecotoxicol Environ Saf* 2003, **56**, 148–163.

13. Szpunar, J., Lobinski, R. Multidimensional approaches in biochemical speciation analysis. *Anal Bioanal Chem* 2002, **373**, 404–411.

14. Hinojosa Reyes, L., Marchante-Gayón, J. M., Garcia Alonso, J. I., Sanz-Medel, A. Quantitative speciation of selenium in human serum by affinity chromatography coupled to post-column isotope dilution analysis ICP-MS. *J Anal At Spectrom* 2003, **18**, 1210–1216.

15. Szpunar, J. Advances in analytical methodology for bioinorganic speciation analysis: metallomics, metalloproteomics and heteroatom-tagged proteomics and metabolomics. *Analyst* 2005, **130**, 442–465.

16. Sutton, K., Sutton, R. M., Caruso, J. A. Inductively coupled plasma mass spectrometric detection for chromatography and capillary electrophoresis. *J Chromatogr A* 1997, **789**, 85–126.

17. Thomas, R. A Beginner's Guide to ICP-MS: Part IX – Mass analyzers: collision/reaction cell technology. *Spectroscopy* 2002, **17**, 42–48.

18. Bandura, D. R., Baranov, V. I., Tanner, S. D. Reaction chemistry and collisional processes in multiple devices for resolving isobaric interferences in ICP-MS. *Fresenius' J Anal Chem* 2001, **370**, 454–470.

19. Wrobel, K., DeNicola, K., Wrobel, K., Caruso, J. A. ICP-MS: metals and much more. Ashcroft, A. E., Brenton, G., Monaghan, J. J. (Eds), *Advances in Mass Spectrometry*. Elsevier Science: Amsterdam, 2004; Chapter 4.

20. Heumann, K. G. Isotope-dilution ICP-MS for trace element determination and speciation: from a reference method to a routine method? *Anal Bioanal Chem* 2004, **378**, 318–329.

21. Hill, S. J., Bloxham, M. J., Worsfold, P. J. Chromatography coupled with inductively coupled plasma atomic emission spectrometry and inductively coupled plasma mass spectrometry. A review. *J Anal At Spectrom* 1993, **8**, 499–515.

22. Byrdy, F. A., Caruso, J. A. Selective chromatographic detection by plasma mass spectrometry. In *Selective Detectors for Chemical Analysis*. Chemical Analysis Series, Vol. 131. John Wiley & Sons, Inc.: New York, 1995, Chapt. 7, pp. 171–207.

23. Ponce de León, C. A., Montes-Bayón, M., Caruso, J. A. Elemental speciation by chromatographic separation with inductively coupled plasma mass spectrometry detection. *J Chromatogr A* 2002, **974**, 1–21.

24. Prange, A., Schaumloffel, D. Hyphenated techniques for the characterization and quantification of metallothionein isoforms. *Anal Bioanal Chem* 2002, **373**, 441–453.

25. Sanz-Medel, A., Montes-Bayon, M., Luisa Fernandez Sanchez, M. Trace element speciation by ICP-MS in large biomolecules and its potential for proteomics. *Anal Bioanal Chem* 2003, **377**, 236–247.

26. B'Hymer, C., Caruso, J. A. Arsenic and its speciation analysis using high-performance liquid chromatography and inductively coupled plasma mass spectrometry. *J Chromatogr A* 2004, **1045**, 1–13.

27. Francesconi, K. A., Kuehnelt, D. Determination of arsenic species: a critical review of methods and applications, 2000–2003. *Analyst* 2004, **129**, 373–395.

28. Wrobel, K., Wrobel, K., Caruso, J. A. Pretreatment procedures for characterization of arsenic and selenium species in complex samples utilizing coupled techniques with mass spectrometric detection. *Anal Bioanal Chem* 2005, **381**, 317–331.

29. Francesconi, K. A. Complete extraction of arsenic species: a worthwhile goal? *Appl Organomet Chem* 2003, **17**, 682–683.

30. Milstein, L. S., Essader, A., Pellizzari, E. D., Fernando, R. A., Akinbo, O. Selection of a suitable mobile phase for the speciation of four arsenic compounds in drinking water samples using ion-exchange chromatography coupled to inductively coupled plasma mass spectrometry. *Environ Int* 2002, **28**, 277–283.

31. Day, J. A., Montes-Bayón, M., Vonderheide, A. P., Caruso, J. A. A study of method robustness for arsenic speciation in drinking water samples by anion exchange HPLC-ICP-MS. *Anal Bioanal Chem* 2002, **373**, 664–668.

32. Mazan, S., Cretier, G., Gilon, N., Mermet, J. M., Rocca, J. L. Porous graphitic carbon as stationary phase for LC–ICPMS separation of arsenic compounds in water. *Anal Chem* 2002, **74**, 1281–1287.

33. Bissen, M., Frimmel, F. H. Speciation of As(III), As(V), MMA and DMA in contaminated soil extracts by HPLC-ICP/MS. *Fresenius' J Anal Chem* 2000, **367**, 51–55.

34. Helgesen, H., Larsen, E. H. Bioavailability and speciation of arsenic in carrots grown in contaminated soil. *Analyst* 1998, **123**, 791–796.

35. Kuehnelt, D., Schlagenhaufen, C., Slejkovec, Z., Irgolic, K. J. Arsenobetaine and other arsenic compounds in the Natural Research Council of Canada Certified Reference Materials DORM 1 and DORM 2. *J Anal At Spectrom* 1998, **13**, 183–187.

36. Wrobel, K., Wrobel, K., Parker, B., Kannamkumarath, S. S., Caruso, J. A. Determination of As(III), As(V), monomethylarsonic acid, dimethylarsinic acid and arsenobetaine by HPLC-ICP-MS: Analysis of reference materials, fish tissues and urine. *Talanta* 2002, **58**, 899–907.

37. Raber, G., Francesconi, K. A., Irgolic, K. J., Goessler, W. Determination of 'arsenosugars' in algae with anion-exchange chromatography and an inductively coupled plasma mass spectrometer as element-specific detector. *Fresenius' J Anal Chem* 2000, **367**, 181–188.

38. Karthikeyan, S., Hirata, S., Honda, K., Shikino, O. Speciation of arsenic in marine algae and commercial shrimp using ion chromatography with ICP-MS detection. *At Spectrosc* 2003, **24**, 79–88.

39. Brisbin, J. A., B̆Hymer, C., Caruso, J. A. A gradient anion exchange chromatographic method for the speciation of arsenic in lobster tissue extracts. *Talanta* 2002, **58**, 133–145.

40. McSheehy, S., Pohl, P., Lobinski, R., Szpunar, J. Complementarity of multidimensional HPLC-ICP-MS and electrospray MS-MS for speciation analysis of arsenic in algae. *Anal Chim Acta* 2001, **440**, 3–16.

41. McSheehy, S., Pohl, P., Velez, D., Szpunar, J. Multidimensional liquid chromatography with parallel ICP MS and electrospray MS/MS detection as a tool for the characterization of arsenic species in algae. *Anal Bioanal Chem* 2002, **372**, 457–466.

42. Francesconi, K. Applications of liquid chromatography–electropspray ionization–single quadrupole mass spectrometry for determining arsenic compounds in biological samples. *Appl Organomet Chem* 2002, **16**, 437–445.

43. Heitkemper, D. T., Vela, N. P., Stewart, K. R., Westphal, C. S. Determination of total and speciated arsenic in rice by IC and ICP-MS. *J Anal At Spectrom* 2001, **16**, 299–306.

44. Kohlmeyer, U., Jantzen, E., Kuballa, J., Jakubik, S. Benefits of high resolution IC-ICP-MS for the routine analysis of inorganic and organic arsenic species in food products of marine and terrestrial origin. *Anal Bioanal Chem* 2003, **377**, 6–13.

45. Pizarro, I., Gomez, M., Palacios, M. A., Camara, C. Evaluation of stability of arsenic species in rice. *Anal Bioanal Chem* 2003, **376**, 102–109.

46. Vela, N. P., Heitkemper, D. T., Stewart, K. R. Arsenic extraction and speciation in carrots using accelerated solvent extraction, liquid chromatography and plasma mass spectrometry. *Analyst* 2001, **126**, 1011–1017.

47. Vela, N. P., Heitkemper, D. T. Total arsenic determination and speciation in infant food products by ion chromatography–inductively coupled plasma-mass spectrometry. *J AOAC Int* 2004, **87**, 244–252.

48. Suzuki, K. T., Mandal, B. K., Ogra, Y. Speciation of arsenic in body fluids. *Talanta* 2002, **58**, 111–119.

49. Kobayashi, Y., Cui, X., Hirano, S. Stability of arsenic metabolites, arsenic triglutathione [As(GS)$_3$] and methylarsenic diglutathione [CH$_3$As(GS)$_2$], in rat bile. *Toxicology* 2005, **211**, 115–123.

50. Cui, X., Kobayashi, Y., Hayakawa, T., Hirano, S. Arsenic speciation in bile and urine following oral and intravenous exposure to inorganic and organic arsenics in rats. *Toxicol Sci* 2004, **82**, 478–487.

51. Suzuki, K. T., Katagiri, A., Sakuma, Y., Ogra, Y., Ohmichi, M. Distributions and chemical forms of arsenic after intravenous administration of dimethylarsinic and monomethylarsonic acids to rats. *Toxicol Appl Pharmacol* 2004, **198**, 336–344.

52. Vassileva, E., Becker, A., Broekaert, J. A. *Anal Chim Acta* 2001, **441**, 135–146.

53. Jackson, B. P., Bertsch, P. M. Determination of arsenic speciation in poultry wastes by IC-ICP-MS. *Environ Sci Technol* 2001, **35**, 4868–4873.

54. Polatajko, A., Szpunar, J. Speciation of arsenic in chicken meat by anion-exchange liquid chromatography with inductively coupled plasma-mass spectrometry. *J AOAC Int* 2004, **87**, 233–237.

55. Caruso, J. A., Heitkemper, D. T., B'Hymer, C. An evaluation of extraction techniques for arsenic species from freeze-dried apple samples. *Analyst* 2001, **126**, 136–140.

56. Kannamkumarath, S. S., Wrobel, K., Caruso, J. A. Speciation of arsenic in different types of nuts by ion chromatography–inductively coupled plasma mass spectrometry. *J Agric Food Chem* 2004, **52**, 1458–1463.

57. Wei, X., Brockhoff-Schwegel, C. A., Creed, J. T. Comparison of urinary arsenic speciation via direct nebulization and on-line photo-oxidation–hydride generation with IC separation and ICP-MS detection. *J Anal At Spectrom* 2001, **16**, 12–19.

58. Suzuki, K. T., Tomita, T., Ogra, Y., Ohmichi, M. Glutathione-conjugated arsenics in the potential hepato-enteric circulation in rats. *Chem Res Toxicol* 2001, **14**, 1604–1611.

59. Sloth, J. J., Julshamn, K., Larsen, E. H. Selective arsenic speciation analysis of human urine reference materials using gradient elution ion-exchange HPLC-ICP-MS. *J Anal At Spectrom* 2004, **19**, 973–978.

60. Milstein, L. S., Essader, A., Pellizzari, E. D., Fernando, R. A., Raymer, J. H., *et al.* Development and application of a robust speciation method for determination of six arsenic compounds present in human urine. *Environ Health Perspect* 2003, **111**, 293–296.

61. Zheng, B., Hintelmann, H. Hyphenation of high performance liquid chromatography with sector field inductively coupled plasma mass spectrometry for the determination of ultra-trace level anionic and cationic arsenic compounds in freshwater fish. *J Anal At Spectrom* 2004, **19**, 191–195.

62. Pergantis, S. A., Heithmar, E. M., Hinners, T. A. Speciation of arsenic animal feed additives by microbore high-performance liquid chromatography with inductively coupled plasma mass spectrometry. *Analyst* 1997, **122**, 1063–1068.

63. Wangkarn, S., Pergantis, S. A. High-speed separation of arsenic compounds using narrow-bore high-performance liquid chromatography on-line with inductively coupled plasma mass spectrometry. *J Anal At Spectrom* 2000, **15**, 627–633.

64. Londesborough, S., Mattusch, J., Wennrich, R. Separation of organic and inorganic species by HPLC-ICP-MS. *Fresenius' J Anal Chem* 1999, **363**, 577–581.

65. Sun, Y. C., Lee, Y. S., Shiah, T. L., Lee, P. L., Tseng, W. C., *et al.* Comparative study on conventional and low-flow nebulizers for arsenic speciation by means of microbore liquid chromatography with inductively coupled plasma mass spectrometry. *J Chromatogr A* 2003, **1005**, 207–213.

66. Mazan, S., Gilon, N., Crétier, G., Rocca, J. L., Mermet, J. M. Inorganic selenium speciation using HPLC-ICP-hexapole collision/reaction cell-MS. *J Anal At Spectrom* 2002, **17**, 366–370.

67. Miguens-Rodriguez, M., Pickford, R., Thomas-Oates, J. E., Pergantis, S. A. Arsenosugar identification in seaweed extracts using high-performance liquid chromatography/electrospray ion trap mass spectrometry. *Rapid Commun Mass Spectrom* 2002, **16**, 323–331.

68. Klaue, B., Blum, J. D. Trace analyses of arsenic in drinking water by inductively coupled plasma mass spectrometry: high resolution versus hydride generation. *Anal Chem* 1999, **71**, 1408–1414.

69. Polya, D. A., Lythgoe, P. R., Gault, A. G., Brydie, J. R., Abou-Shakra, F., *et al.* IC-ICP-MS and IC-ICP-HEX-MS determination of arsenic speciation in surface and groundwaters: preservation and analytical issues. *Mineral Mag* 2003, **67**, 247–261.

70. Qianli, X., Kerrich, R., Irving, E., Liber, K., Abou-Shakra, F. Determination of five arsenic species in aqueous samples by HPLC coupled with hexapole collision cell ICP-MS. *J Anal At Spectrom* 2002, **17**, 1037–1041.

71. Shinohara, A., Chiba, M., Inaba, Y., Kondo, M., Abou-Ahakra, F. R., *et al.* Speciation of arsenic compounds in human urine by HPLC/hexapole collision cell ICP-MS. *Bunseki Kagaku* 2004, **53**, 589–593.

72. Moldovan, M., Gómez, M. M., Palacios, M. A., Cámara, C. Arsenic speciation in water and human urine by HPLC-ICP-MS and HPLC–MO-HG-AAS. *Microchem J* 1998, **59**, 89–99.

73. Dobson, J. E. The iodine factor in health and evolution. *Geogr Rev* 1998, **88**, 1–28.

74. Michalke, B., Schramel, P., Witte, H. Method developments for iodine speciation by reversed-phase liquid chromatography–ICP-mass spectrometry. *Biol Trace Elem Res* 2000, **78**, 67–79.

75. Michalke, B., Schramel, P., Witte, H. Iodine speciation in human serum by reversed-phase liquid chromatography-ICP-mass spectrometry. *Biol Trace Elem Res* 2000, **78**, 81–91.

76. Shah, M., Kannamkumarath, S. S., Caruso, J. A., Wuilloud, R. G. Iodine speciation studies in commercially available seaweed by coupling different chromatographic techniques with UV and ICP-MS detection. *J Anal At Spectrom* 2005, **20**, 176–182.

77. Fernandez Sanchez, L. F., Szpunar, J. Speciation analysis for iodine in milk by size-exclusion chromatography with inductively coupled plasma mass spectrometric detection (SEC-ICP MS). *J Anal At Spectrom* 1999, **14**, 1697–1702.

78. Simon, R., Tietge, J. E., Michalke, B., Degitz, S., Schramm, K. W. Iodine species and the endocrine system: thyroid hormone levels in adult *Danio rerio* and developing *Xenopus laevis*. *Fresenius' J Anal Chem* 2002, **372**, 481–485.

79. Kannamkumarath, S. S., Wuilloud, R. G., Stalcup, A., Caruso, J. A., Patel, H., *et al.* Determination of levothyroxine and its degradation products in pharmaceutical tablets by HPLC-UV-ICP-MS. *J Anal At Spectrom* 2004, **19**, 107–113.

80. Axelsson, B. O., Jornten-Karlsson, M., Michelsen, P., Abou-Shakra, F. The potential of inductively coupled plasma mass spectrometry detection for high-performance liquid chromatography combined with accurate mass measurement of organic pharmaceutical compounds. *Rapid Commun Mass Spectrom* 2001, **15**, 375–385.

81. Leiterer, M., Truckenbrodt, D., Franke, K. Determination of iodine species in milk using ion chromatographic separation and ICP-MS detection. *Eur Food Res Technol* 2001, **213**, 150–153.

82. Carro, A. M., Mejuto, M. C. Application of chromatographic and electrophoretic methodology to the speciation of organomercury compounds in food analysis. *J Chromatogr A* 2000, **882**, 283–307.

83. Qvarnström, J., Frech, W. Mercury species transformations during sample pre-treatment of biological tissues studied by HPLC-ICP-MS. *J Anal At Spectrom* 2002, **17**, 1486–1491.

84. Hintelmann, H., Falter, R., Ilgen, G., Evans, R. D. Determination of artifactual formation of monomethylmercury (CH_3Hg^+) in environmental samples using stable Hg^{2+} isotopes with ICP-MS detection: calculation of contents applying species specific isotope addition. *Fresenius' J Anal Chem* 1997, **358**, 363–370.

85. Wilken, R. D., Falter, R. Determination of methylmercury by the species-specific isotope addition method using a newly developed HPLC-ICP-MS coupling technique with ultrasonic nebulization. *Appl Organomet Chem* 1998, **12**, 551–557.

86. Leermakers, M., Baeyens, W., Quevauviller, P., Horvat, M. Mercury in environmental samples: speciation, artifacts and validation. *Trends Anal Chem* 2005, **24**, 383–393.

87. Bouyssiere, B., Szpunar, J., Lobinski, R. Gas chromatography with inductively coupled plasma mass spectrometric detection in speciation analysis. *Spectrochim Acta, Part B* 2002, **57**, 805–828.

88. Harrington, C. F. The speciation of mercury and organomercury compounds by using high-performance liquid chromatography. *Trends Anal Chem* 2000, **19**, 167–179.

89. Wan, C.-C., Chen, C.-S., Jiang, S.-J. Determination of mercury compounds in water samples by liquid chromatography–inductively coupled plasma mass spectrometry with an *in situ* nebulizer/vapor generator. *J Anal At Spectrom* 1997, **12**, 683–687.

90. Shum, S. C. K., Pang, H., Houk, R. S. Speciaiton of mercury and lead compounds by microbore column liquid chromatography–inductively coupled plasma mass spectrometry with direct injection nebulization. *Anal Chem* 1992, **64**, 2444–2450.

91. Shen, J.-C., Zhuang, Z.-X., Wang, X.-R., Lee, F. S. C., Huang, Z.-Y. Investigation of mercury metallothionein complexes in tissues of rat after oral intake of $HgCl_2$. *Appl Organomet Chem* 2005, **19**, 140–146.

92. Falter, R., Ilgen, G. Determination of trace amounts of methylmercury in sediments and biological tissue by using water vapor distillation in combination with RP C18 preconcentration and HPLC-HPF/HHPN-ICP-MS. *Fresenius' J Anal Chem* 1997, **358**, 401–406.

93. Chiou, C.-S., Jiang, S.-J., Kumar Danadurai, K. S. Determination of mercury compounds in fish by microwave-assisted extraction and liquid chromatography–vapor generation–inductively coupled plasma mass spectrometry. *Spectrochim Acta, Part B* 2001, **56**, 1133–1142.

94. Han, Y., Kingston, H. M., Boylan, H. M., Rahman, G. M., Shah, S., *et al.* Speciation of mercury in soil and sediment by selective solvent and acid extraction. *Anal Bioanal Chem* 2003, **375**, 428–436.

95. Quevauviller, P. Accuracy and traceability in environmental monitoring–pitfalls in methylmercury determinations as a case study. *J Environ Monit* 2000, **2**, 292–299.

96. Oe, T., Tian, Y., O'Dwyer, P. J., Blair, I. A., Roberts, D. W., *et al.* Determination of the platinum drug *cis*-amminedichloro(2-methylpyridine)platinum(II) in human urine by liquid chromatography–tandem mass spectrometry. *J Chromatogr B* 2003, **792**, 217–227.

97. Falter, R., Wilken, R. D. Determination of carboplatinum and cisplatinum by interfacing HPLC with ICP-MS using ultrasonic nebulisation. *Sci Total Environ* 1999, **225**, 167–176.

98. Hann, S., Stefanka, Z., Lenz, K., Stingeder, G. Novel separation method for highly sensitive speciation of cancerostatic platinum compounds by HPLC-ICP-MS. *Anal Bioanal Chem* 2005, **381**, 405–412.

99. Carr, J. L., Tingle, M. D., McKeage, M. J. Rapid biotransformation of satraplatin by human red blood cells *in vitro*. *Cancer Chem Pharmacol* 2002, **50**, 9–15.

100. Levi, F., Metzger, G., Massari, C., Milano, G. Oxaliplatin: pharmacokinetics and chronopharmacological aspects. *Clin Pharmacokinet* 2000, **38**, 1–21.

101. Zöllner, P., Zenker, A., Galanski, M., Keppler, B. K., Lindner, W. Reaction monitoring of platinum(II) complex–5'-guanosine monophosphate adduct formation by ion-exchange liquid chromatography/electrospray ionization mass spectrometry. *J Mass Spectrom* 2001, **36**, 742–753.

102. Hanada, K., Nishijima, K., Ogata, H., Atagi, S., Kawahara, M. Population pharmacokinetic analysis of cisplatin and its metabolites in cancer patients: possible misinterpretation of covariates for pharmacokinetic parameters calculated from the concentrations of unchanged cisplatin, ultrafiltered platinum and total platinum. *Jpn J Clin Oncol* 2001, **31**, 179–184.

103. Hanigan, M. H., Townsend, D. M., Deng, M., Marto, J. A., MacDonald, T. J. High pressure liquid chromatography and mass spectrometry characterization of the nephrotoxic biotransformation products of cisplatin. *Drug Metab Dispos* 2003, **31**, 705–713.

104. Zhao, Z., Tepperman, K., Dorsey, J. G., Elder, R. C. Determination of cisplatin and some possible metabolites by ion-pairing chromatography with inductively coupled plasma mass spectrometric detection. *J Chromatogr* 1993, **615**, 83–89.

105. Smith, P. F., Booker, B. M., Creaven, P., Perez, R., Pendyala, L. Pharmacokinetics and pharmacodynamics of Mesna-mediated plasma cysteine depletion. *J Clin Pharmacol* 2003, **43**, 1324–1328.

106. Cairns, W. R., Ebdon, L., Hill, S. J. A high performance liquid chromatography–inductively coupled plasma-mass spectrometry interface employing desolvation for speciation studies of platinum in chemotherapy drugs. *Anal Bioanal Chem* 1996, **355**, 202–208.

107. Galettis, P., Carr, J. L., Paxton, J. W., McKeage, M. J. Quantitative determination of platinum complexes in human plasma generated from the oral antitumour drug JM216 using directly coupled high-performance liquid chromatography–inductively coupled plasma mass spectrometry without desolvation. *J Anal At Spectrom* 1999, **14**, 953–956.

108. Szpunar, J., Makarov, A., Lobinski, R., Pieper, T., Keppler, B. K. Investigation of metallodrug–protein interactions by size-exclusion chromatography coupled with inductively coupled plasma mass spectrometry (ICP-MS). *Anal Chim Acta* 1999, **387**, 135–144.

109. Hann, S., Zenker, A., Galanski, M., Bereuter, T. L., Stingeder, G., *et al.* HPIC-UV–ICP-SFMS study of the interaction of cisplatin with guanosine monophosphate. *Fresenius' J Anal Chem* 2001, **370**, 581–586.

110. Hann, S., Koellensperger, G., Stingeder, G., Stefánka, Z., Fürhacker, M., *et al.* Application of HPLC-ICP-MS to speciation of cisplatin and its degradation products in water containing different chloride concentrations and in human urine. *J Anal At Spectrom* 2003, **18**, 1391–1395.

111. Smith, C. J., Wilson, I. D., Payne, R., Parry, T. C., Sinclair, P., *et al.* A comparison of the quantitative methods for the analysis of the platinum-containing anticancer drug {*cis*-[amminedichloro(2-methylpyridine)]platinum(II)} (ZD0473) by HPLC coupled to either a triple quadrupole mass spectrometer or an inductively coupled plasma mass spectrometer. *Anal Chem* 2003, **75**, 1463–1469.

112. Rayman, N. P. The importance of selenium to human health. *Lancet* 2000, **356**, 233–241.

113. Barelay, M. N. I., MacPherson, A., Dixon, J. *J Food Comp Anal* 1995, **8**, 307.

114. Reilly, C. Selenium: a new entrant into the functional food arena. *Trends Food Sci Technol* 1998, **9**, 114–118.

115. Stadtman, T. C. Selenocysteine. *Annu Rev Biochem* 1996, **65**, 83–100.

116. Ganther, H. E. *Carcinogenesis* 1999, **20**, 1657.

117. Clark, L. C., Turnball, B. W., Slate, E. H., Chalker, D. K., Chow, J., *et al.* Effects of selenium supplementation for cancer prevention in patients with carcinoma of the skin: a randomized controlled trial. *J Am Med Assoc* 1996, **276**, 1957–1963.

118. Whanger, P. D. Selenium and its relationship to cancer: an update dagger. *Br J Nutr* 2004, **91**, 11–28.

119. Sasakura, C., Suzuki, K. T. Biological interaction between transition metals (Ag, Cd and Hg), selenide/sulfide and selenoprotein P. *J Inorg Biochem* 1998, **71**, 159–162.

120. Schrauzer, G. N. Selenium. *Elements and Their Compounds in the Environment. Occurence, Analysis and Biological relevance*, 2nd edn. Wiley-VCH: Weinheim, 2002, pp. 1365–1406.

121. Kim, Y. Y., Mahan, D. C. Prolonged feeding of high dietary levels of organic and inorganic selenium to gilts from 25 kg body weight through one parity. *J Anim Sci* 2001, **79**, 956–966.

122. Arthur, J. R. Micronutrient Group Symposium on 'Micronutrient supplementation: when and why?' *Proc Nutr Soc* 2003, **62**, 393–397.

123. Sanz Alaejos, M., Diaz Romero, F. J., Diaz Romero, C. Selenium and cancer: some nutritional aspects. *Nutrition* 2000, **16**, 376–383.

124. Kim, Y. Y., Mahan, D. C. Comparative effects of high dietary levels of organic and inorganic selenium on selenium toxicity of growing-finishing pigs. *J Anim Sci* 2001, **79**, 942–948.

125. Ip, C., Hayes, C., Budnick, R. M., Ganther, H. E. Chemical form of selenium, critical metabolites, and cancer prevention. *Cancer Res* 1991, **51**, 595.

126. Ip, C., Birringer, M., Block, E., Kotrebai, M., Tyson, J. F., *et al*. Chemical speciation influences comparative activity of selenium-enriched garlic and yeast in mammary cancer prevention. *J Agric Food Chem* 2000, **48**, 2062–2070.

127. Whanger, P. D., Ip, C., Polan, C. E., Uden, P. C., Welbaum, G. Tumorgenesis, metabolism, speciation, bioavailability, and tissue deposition of selenium in selenium-enriched ramps (*Allium tricoccum*). *J Agric Food Chem* 2000, **48**, 5723–5730.

128. Block, E., Birringer, M., Jiang, W., Nakahodo, T., Thompson, H. J., *et al*. Allium chemistry: synthesis, natural occurence, biological activity, and chemistry of Se-alk(en)ylselenocysteines and their γ-glutamyl derivatives and oxidation products. *J Agric Food Chem* 2001, **49**, 458–470.

129. Das, A., Desai, D., Pittman, B., Amin, S., El-Bayoumy, K. Comparison of the chemopreventive efficacies of 1,4-phenylenebis(methylene)selenocyanate and selenium-enriched yeast on 4-(methylnitrosamino)-1-(3-pyridyl)-1-butanone induced lung tumorigenesis in A/J mouse. *Nutr Cancer* 2003, **46**, 179–185.

130. Guttenplan, J. B., Spratt, T. E., Khmelnitsky, M., Kosinska, W., Desai, D., *et al*. Effects of 3*H*-1,2-dithiole-3-thione, 1,4-phenylenebis(methylene)selenocyanate, and selenium-enriched yeast individually and in combination on benzo[*a*]pyrene-induced mutagenesis in oral tissue and esophagus in lacZ mice. *Mutat Res* 2004, **559**, 199–210.

131. Kotrebai, M., Birringer, M., Tyson, J. F., Block, E., Uden, P. C. Identification of the principal selenium compounds in selenium-enriched natural sample extracts by ion-pair HPLC with ICP-MS and ESI-MS detection. *Anal Commun* 1999, **36**, 249–252.

132. Kotrebai, M., Bird, S. M., Tyson, J. F., Block, E., Uden, P. C. Characterization of selenium species in biological extracts by enhanced ion-pair liquid chromatography with ICP-MS and by referenced ESI-MS. *Spectrochim Acta, Part B* 1999, **54**, 1573–1591.

133. McSheehy, S., Yang, W., Pannier, F., Szpunar, J., Lobinski, R., *et al*. Speciation analysis of selenium in garlic by two-dimensional high performance LC with parallel ICP-MS and ESI-MS detection. *Anal Chim Acta* 2000, **421**, 147–153.

134. Infante, H. G., O'Connor, G., Wahlen, R., Entwisle, J., Norris, P., *et al*. Selenium speciation analysis of selenium-enriched supplements by HPLC with ultrasonic nebulisation ICP-MS and electrospray MS/MS detection. *J Anal At Spectrom* 2004, **19**, 1529–1538.

135. Kotrebai, M., Tyson, J. F., Block, E., Uden, P. C. High-performance liquid chromatography of selenium compounds utilizing perfluorinated carboxylic acid ion-pairing agents and inductively coupled plasma and electrospray ionization mass spectrometric detection. *J Chromatogr A* 2000, **866**, 51–63.

136. Zheng, J., Ohata, M., Furuta, N., Kosmus, W. Speciation of selenium compounds with ion-pair reversed-phase liquid chromatography using inductively coupled plasma mass spectrometry as element-specific detection. *J Chromatogr A* 2000, **874**, 55–64.

137. Lindemann, T., Hintelmann, H. Identification of selenium-containing glutathione S-conjugates in a yeast extract by two-dimensional liquid chromatography with inductively coupled plasma MS and nanoelectrospray MS/MS detection. *Anal Chem* 2002, **74**, 4602–4610.

138. Lindemann, T., Hintelmann, H. Selenium speciation by HPLC with tandem mass spectrometric detection. *Anal Bioanal Chem* 2002, **372**, 486–490.

139. Chassaigne, H., Chery, C. C., Bordin, G., Rodriguez, A. R. Development of new analytical methods for selenium speciation in selenium-enriched yeast material. *J Chromatogr A* 2002, **976**, 409–422.

140. Larsen, E. H., Hansen, M., Fan, T., Vahl, M. Speciation of selenoamino acids, selenium ions and inorganic selenium by ion exchange HPLC with mass spectrometric detection and its application to yeast and algae. *J Anal At Spectrom* 2001, **16**, 1403–1408.

141. Vonderheide, A. P., Wrobel, K., Kannamkumarath, S. S., B'Hymer, C., Montes-Bayon, M., *et al.* Characterization of selenium species in Brazil nuts by HPLC-ICP-MS and ES-MS. *J Agric Food Chem* 2002, **50**, 5722–5728.

142. Kannamkumarath, S. S., Wrobel, K., Wrobel, K., Vonderheide, A., Caruso, J. A. HPLC-ICP-MS determination of selenium distribution and speciation in different types of nut. *Anal Bioanal Chem* 2002, **373**, 454–460.

143. Wrobel, K., Wrobel, K., Kannamkumarath, S. S., Caruso, J. A., Wysocka, A., *et al.* HPLC-ICP-MS speciation of selenium in enriched onion leaves – a potntial dietary source of Se-methylseleno-cysteine. *Food Chem* 2004, **86**, 617–623.

144. Shah, M., Kannamkumarath, S. S., Wuilloud, J. C. A., Wuilloud, R. G., Caruso, J. A. Identification and characterization of selenium species in enriched green onion (*Allium fistulosum*) by HPLC-ICP-MS and ESI-ITMS. *J Anal At Spectrom* 2004, **19**, 381–386.

145. Huerta, V. D., Sánchez, M. L. F., Sanz-Medel, A. Quantitative selenium speciation in cod muscle by isotope dilution ICP-MS with a reaction cell: comparison of different reported extraction procedures. *J Anal At Spectrom* 2004, **19**, 644–648.

146. Larsen, E. H., Sloth, J., Hansen, M., Moesgaard, S. Selenium speciation and isotope composition in ^{77}Se-enriched yeast using gradient elution HPLC separation and ICP-dynamic reaction cell-MS. *J Anal At Spectrom* 2003, **18**, 310–316.

147. Díaz Huerta, V., Hinojosa Reyes, L., Marchante-Gayón, J. M., Fernández Sánchez, M. L., Sanz-Medel, A. Total determination and quantitative speciation analysis of selenium in yeast and wheat flour by isotope dilution analysis ICP-MS. *J Anal At Spectrom* 2003, **18**, 1243–1247.

148. Hinojosa Reyes, L., Moreno Sanz, F., Herrero Espilez, P., Marchante Gayón, J. M., Garcia Alonso, J. I., *et al.* Biosynthesis of isotopically enriched selenomethionine: application to its accurate determination in selenium-enriched yeast by isotope dilution analysis–HPLC-ICP-MS. *J Anal At Spectrom* 2004, **19**, 1230–1235.

149. Larsen, E. H., Hansen, M., Paulin, H., Moesgaard, S., Reid, M., *et al.* Speciation and bioavailability of selenium in yeast-based intervention agents used in cancer chemoprevention studies. *J AOAC Int* 2004, **87**, 225–232.

150. Kotrebai, M., Tyson, J. F., Uden, P. C., Birringer, M., Block, E. Selenium speciation in enriched and natural samples by HPLC-ICP-MS and HPLC–ESI-MS with perfluorinated carboxylic acid ion-pairing agents. *Analyst* 2000, **125**, 71–78.

151. Wrobel, K., Wrobel, K., Caruso, J. A. Selenium speciation in low molecular weight fraction of Se-enriched yeasts by HPLC-ICP-MS: detection of selenoadenosylmethionine. *J Anal At Spectrom* 2002, **17**, 1048–1054.

152. Marchante Gayón, J. M., Thomas, C., Feldmann, I., Jakubowski, N. Comparison of different nebulizers and chromatographic techniques for the speciation of selenium in nutritional commercial supplements by hexapole collision and reaction cell ICP-MS. *J Anal At Spectrom* 2000, **15**, 1093–1102.

153. Bird, S. M., Ge, H., Uden, P. C., Tyson, J. F., Block, E., *et al.* High-performance liquid chromatography of selenoamino acids and organo selenium compounds. Speciation by inductively coupled plasma mass spectrometry. *J Chromatogr A* 1997, **789**, 349–359.

154. Sloth, J. J., Larsen, E. H. The application of inductively coupled plasma dynamic reaction cell mass spectrometry for measurement of selenium isotopes, isotope ratios and chromatographic detection of selenoamino acids. *J Anal At Spectrom* 2000, **15**, 669–672.

155. Wrobel, K., Kannamkumarath, S. S., Caruso, J. A. Hydrolysis of proteins with methanesulfonic acid for improved HPLC-ICP-MS determination of seleno-methionine in yeast and nuts. *Anal Bioanal Chem* 2003, **375**, 133–138.

156. Pérez Mendez, S., Blanco González, E., Sanz-Medel, A. Hybridation of different chiral separation techniques with ICP-MS detection for the separation and determination of seleno-methionine enantiomers: chiral speciation of selenized yeast. *Biomed Chromatogr* 2001, **15**, 181–188.

157. Méndez, S. P., González, E. B., Medel, A. S. Chiral speciation and determination of seleno-methionine enantiomers in selenized yeast by HPLC-ICP-MS using a teicoplanin-based chiral stationary phase. *J Anal At Spectrom* 2000, **15**, 1109–1114.

158. Ponce De Leon, C. A., Sutton, K. L., Caruso, J. A., Uden, P. C. Chiral speciation of selenoamino acids and selenium enriched samples using HPLC coupled to ICP-MS. *J Anal At Spectrom* 2000, **15**, 1103–1107.

159. Sutton, K. L., Ponce de Leon, C. A., Ackley, K. L., Sutton, R. M., Stalcup, A. M., *et al*. Development of chiral HPLC for selenoamino acids with ICP-MS detection:application to selenium nutritional supplements. *Analyst* 2000, **125**, 281–286.

160. Suzuki, K. T., Ishiwata, K., Ogra, Y. Incorporation of selenium into selenoprotein P and extracellular glutathione peroxidase: HPLC-ICPMS data with enriched selenite. *Analyst* 1999, **124**, 1749–1753.

161. Suzuki, K. T., Ogra, Y. Metabolism of selenium and its interaction with mercury: mechanisms by a speciation study. *Phosphorus, Sulfur Silicon Relat Elem* 2001, **171–172**, 135–169.

162. Suzuki, K. T., Ogra, Y. Metabolic pathway for selenium in the body: speciation by HPLC-ICP MS with enriched Se. *Food Addit Contam* 2002, **19**, 974–983.

163. Shiobara, Y., Ogra, Y., Suzuki, K. T. Speciation of metabolites of selenate in rats by HPLC-ICP-MS. *Analyst* 1999, **124**, 1237–1241.

164. Nyman, D. W., Stratton, M. S., Kopplin, M. J., Dalkin, B. L., Nagle, R. B., *et al*. Selenium and selenomethionine levels in prostate cancer patients. *Cancer Detect Prev* 2004, **28**, 8–16.

165. Encinar, J. R., Schaumlöffel, D., Lobinski, R., Ogra, Y. Determination of selenomethionine and selenocysteine in human serum using speciated isotope dilution–capillary HPLC-inductively coupled plasma collision cell mass spectrometry. *Anal Chem* 2004, **76**, 6635–6642.

166. Quijano, M. A., Gutierrez, A. M., Perez-Conde, M. C., Camara, C. Determination of selenium species in human urine by high performance liquid chromatography and ICP-MS. *Talanta* 1999, **50**, 165–173.

167. Gammelgaard, B., Jessen, K. D., Kristensen, F. H., Jøns, O. Determination of trimethylselenonium ion in urine by ion chromatography and inductively coupled plasma mass spectrometry detection. *Anal Chim Acta* 2000, **404**, 47–54.

168. Gammelgaard, B., Jøns, O., Bendahl, L. Selenium speciation in pretreated human urine by ion-exchange chromatography and ICP-MS detection. *J Anal At Spectrom* 2001, **16**, 339–344.

169. Zheng, J., Ohata, M., Furuta, N. Reversed-phase liquid chromatography with mixed ion-pair reagents coupled with ICP-MS for the direct speciation analysis of selenium compounds in human urine. *J Anal At Spectrom* 2002, **17**, 730–735.

170. Marchante-Gayón, J. M., Feldmann, I., Thomas, C., Jakubowski, N. Speciation of selenium in human urine by HPLC-ICP-MS with a collision and reaction cell. *J Anal At Spectrom* 2001, **16**, 457–463.

171. Gammelgaard, B., Bendahl, L., Sidenius, U., Jøns, O. Selenium speciation in urine by ion-pairing chromatography with perfluorinated carboxylic acids and ICP-MS detection. *J Anal At Spectrom* 2002, **17**, 570–575.

172. Wrobel, K., Kannamkumarath, S. S., Wrobel, K., Caruso, J. A. Identification of selenium species in urine by ion-pairing HPLC-ICP-MS using laboratory-synthesized standards. *Anal Bioanal Chem* 2003, **377**, 670–674.

173. Sloth, J. J., Larsen, E. H., Bügel, S. H., Moesgaard, S. Determination of total selenium and [77]Se in isotopically enriched human samples by ICP–dynamic reaction cell-MS. *J Anal At Spectrom* 2003, **18**, 317–322.

174. Gammelgaard, B., Madsen, A. D., Bjerrum, J., Bendhal, L., Jons, O., *et al*. Separation, purification and identification of the major selenium metabolite from human urine by multi-dimensional HPLC-ICP-MS and APCI-MS. *J Anal At Spectrom* 2003, **18**, 65–70.

175. Ogra, Y., Ishiwata, K., Takayama, H., Aimi, N., Suzuki, K. T. Identification of a novel selenium metabolite, *Se*-methyl-*N*-acetylselenohexosamine, in rat urine by high-performance liquid

chromatography–inductively coupled plasma mass spectrometry and –electrospray ionization tandem mass spectrometry. *J Chromatogr B* 2002, **767**, 301–312.

176. Ogra, Y., Suzuki, K. T. Speciation of selenocompounds by capillary HPLC coupled with ICP-MS using multi-mode gel filtration columns. *J Anal At Spectrom* 2005, **20**, 35–39.

177. Gammelgaard, B., Bendahl, L. Selenium speciation in human urine samples by LC- and CE-ICP-MS separation and identification of selenosugars. *J Anal At Spectrom* 2004, **19**, 135–142.

178. Cao, T. H., Cooney, R. A., Woznichak, M. M., May, S. W., Browner, R. F. Speciation and identification of organoselenium metabolites in human urine using inductively coupled plasma mass spectrometry and tandem mass spectrometry. *Anal Chem* 2001, **73**, 2898–2902.

179. Suzuki, K. T., Kurasaki, K., Okazaki, N., Ogra, Y. Selenosugar and trimethylselenonium among urinary Se metabolites: dose- and age-related changes. *Toxicol Appl Pharmacol* 2005, **206**, 1–8.

180. Houk, R. S. Mass spectrometry of inductively coupled plasmas. *Anal Chem* 1986, **58**, 97A–105A.

181. B'Hymer, C., Caruso, J. A. Evaluation of yeast-based selenium food supplements using high-performance liquid chromatography and inductively coupled plasma mass spectrometry. *J Anal At Spectrom* 2000, **15**, 1531–1539.

182. Montes Bayon, M., B'Hymer, C., Ponce De Leon, C., Caruso, J. A. Resolution of seleno-amino acids optical isomers using chiral derivatization and ICP-MS detection. *J Anal At Spectrom* 2001, **16**, 945–950.

183. Featherstone, A. M., Townsend, A. T., Jacobson, G. A., Peterson, G. M. Comparison of methods for the determination of total selenium in plasma by magnetic sector inductively coupled plasma mass spectrometry. *Anal Chim Acta* 2004, **512**, 319–327.

184. Feldmann, I., Jakubowski, N., Stuewer, D., Thomas, C. *J Anal At Spectrom* 2000, **15**, 371.

185. Heumann, K. G., Rottmann, L., Vogl, J. Elemental speciation with liquid chromatography–inductively coupled plasma isotope dilution mass spectrometry. *J Anal At Spectrom* 1994, **9**, 1351–1355.

186. Rattanachongkiat, S., Millward, G. E., Foulkes, M. E. Determination of arsenic species in fish, crustacean and sediment samples from Thailand using high performance liquid chromatography (HPLC) coupled with inductively coupled plasma mass spectrometry (ICP-MS). *J Environ Monit* 2004, **6**, 254–261.

187. Larsen, E. H., Hansen, M., Goessler, W. Speciation and health risk considerations of arsenic in the edible mushroom *Laccaria Amethystina* collected from contaminated and uncontaminated locations. *Appl Organomet Chem* 1998, **12**, 285–291.

188. Hovanec, B. M. Arsenic speciation in commercially available peanut butter spread by IC-ICP-MS. *J Anal At Spectrom* 2004, **19**, 1141–1144.

189. Mandal, B. K., Ogra, Y., Suzuki, K. T. Identification of dimethylarsinous and monomethylarsonous acids in human urine of the arsenic-affected areas in West Bengal, India. *Chem Res Toxicol* 2001, **14**, 371–378.

190. Mandal, B. K., Ogra, Y., Anzai, K., Suzuki, K. T. Speciation of arsenic in biological samples. *Toxicol Appl Pharmacol* 2004, **198**, 307–318.

191. Mandal, B. K., Ogra, Y., Suzuki, K. T. Speciation of arsenic in human nail and hair from arsenic-affected area by HPLC-inductively coupled argon plasma mass spectrometry. *Toxicol Appl Pharmacol* 2003, **189**, 73–83.

192. Wang, Z., Lu, X., Gong, Z., Le, X. C., Zhou, J. Arsenic speciation in urine from acute promyelocytic leukemia patients undergoing arsenic trioxide treatment. *Chem Res Toxicol* 2004, **17**, 95–103.

193. Van Hulle, M., Schotte, B., Mees, L., Vanhaecke, F., Cornelis, R., *et al.* Identification of some arsenic species in human urine and blood after ingestion of Chinese seaweed *Laminaria*. *J Anal At Spectrom* 2004, **19**, 58–64.

194. Chowdhury, U. K., Rahman, M. M., Sengupta, M. K., Lodh, D., Chanda, C. R., *et al.* Pattern of excretion of arsenic compounds [arsenite, arsenate, MMA(V), DMA(V)] in urine of children

compared to adults from an arsenic exposed area in Bangladesh. *J Environ Sci Health, Part A* 2003, **38**, 87–113.

195. Lai, V. W., Sun, Y., Ting, E., Cullen, W. R., Reimer, K. J. Arsenic speciation in human urine: are we all the same? *Toxicol Appl Pharmacol* 2004, **198**, 297–306.

196. Apostoli, P., Sarnico, M., Bavazzano, P., Bartoli, D. Arsenic and porphyrins. *Am J Ind Med* 2002, **42**, 180–187.

197. Dumont, E., Vanhaecke, F., Cornelis, R. Hyphenated techniques for speciation of Se in *in vitro* gastrointestinal digests of *Saccharomyces cerevisiae*. *Anal Bioanal Chem* 2004, **379**, 504–511.

198. Quijano, M. A., Moreno, P., Gutierrez, A. M., Perez-Conde, M. C., Camara, C. Selenium speciation in animal tissues after enzymatic digestion by high-performance liquid chromatography coupled to inductively coupled plasma mass spectrometry. *J Mass Spectrom* 2000, **35**, 878–884.

199. Ogra, Y., Hatano, T., Ohmichi, M., Suzuki, K. T. Oxidative production of monomethylated selenium from the major urinary selenometabolite, selenosugar. *J Anal At Spectrom* 2003, **18**, 1252–1255.

13

Application of Gas Chromatography-Mass Spectrometry to the Determination of Toxic Metals

Suresh K. Aggarwal,[1] Robert L. Fitzgerald[2] and David A. Herold[2]

[1]*Fuel Chemistry Division Bhabha Atomic Research Centre, Trombay, Mumbai 400 085, India*

[2]*Department of Pathology, University of California, San Diego, and Pathology and Laboratory Medicine, VA San Diego Healthcare System, 3350 La Jolla Village Drive, San Diego, CA 92161, USA*

13.1 INTRODUCTION

Determination of toxic metals such as lead (Pb), cadmium (Cd), mercury (Hg) and arsenic (As) in human samples is necessary to determine the extent of occupational or non-occupational exposure to these metals. Historically, analytical methods were designed to measure the total amount of toxic metal in the biological specimen, but greater under-standing of the mechanisms of toxicity of specific chemical forms of these metals has made it important to know the nature/form of the toxic metal as it exists in biological tissue. As an example, methylmercury is much more toxic to humans than inorganic Hg. Similarly, As(III), As(V) and other organic forms of As such as arsenocholine and arsenobetaine have different toxicities. Therefore, it is essential to obtain information on the type of species present in the sample and their individual concentrations, in addition to determining the total concentration of the toxic metal.

A number of methods have been described for determining the concentrations of toxic metals in biological samples.[1–4] These methods include graphite furnace atomic absorption spectrometry (GF-AAS), electroanalytical techniques such as anodic stripping voltammetry (ASV), neutron activation analysis (NAA), and mass spectrometry (MS). Amongst these techniques, mass spectrometry occupies a unique role due to its potential to measure the

Chromatographic Methods in Clinical Chemistry and Toxicology   Edited by R. L. Bertholf and R. E. Winecker
© 2007 John Wiley & Sons, Ltd

isotopic composition of the metal, which is useful for discriminating among various sources of contamination. Moreover, the use of isotope dilution for quantitation in mass spectrometry makes this technique less susceptible to matrix effects that often limit the accuracy of other analytical techniques. Another advantage of the isotope dilution technique is that the results are not affected by incomplete recovery of the analyte from the matrix. However, isotope dilution depends on the availability of at least two stable isotopes of an element and therefore cannot be used for the determination of monoisotopic elements such as As. In cases where no stable isotope is available, another element may be used as an internal standard. This approach requires the development of a suitable calibration curve to account for the effect of different response factors of the two elements in the mass spectrometer. The use of hyphenated techniques such as gas chromatography–mass spectrometry (GC-MS) and liquid chromatography–mass spectrometry (LC-MS) offers the potential to perform the speciation analysis for toxic elements.[5] High-performance liquid chromatography (HPLC) coupled to inductively coupled plasma mass spectrometry (ICP-MS) has been widely used for determining the concentrations of different species of various toxic metals in biological samples. Recently, capillary electrophoresis (CE) has also been employed in conjunction with ICP-MS. The advent of electrospray ionization mass spectrometry (ESI-MS) has also provided an opportunity to distinguish among the different species of toxic metals. HPLC-ICP-MS requires a dedicated mass spectrometer, which may not be available in many toxicological and biomedical laboratories. The use of a general-purpose instrument, such as GC-MS, may be more practical in many laboratories.

This chapter presents the results of studies carried out for the determination of Pb,[6,7] Cd,[8] Hg[9] and As[10] using GC-MS. Since GC-MS requires a volatile form of the metal, suitable chelating agents were identified, prepared, and examined for the cross-over (memory) effect. Also the validation of isotope dilution GC-MS for the determination of these toxic metals in biological samples was carried out using suitable NIST certified reference materials.

13.2 INSTRUMENTATION

In principle, any mass spectrometer coupled with a gas chromatograph can be used to measure toxic metals by preparing suitable metal chelates. During the early stages of this work at the University of Virginia, we used a Finnigan MAT 8230 reverse-geometry, double-focusing mass spectrometer coupled to a Varian 3700 gas chromatograph. The instrument was calibrated using methyl stearate and measuring the isotope ratios for m/z values 299:298, 300:298, 301:298 and 302:298. These ratios vary by four orders of magnitude, providing a good measure of the overall linearity of the mass spectrometer. A resolution of about 1000 with a sampling frequency of 2 Hz was found to be optimum (resolution is defined as the separation of two masses, M and $M + \Delta M$. The calculated resolution is then $M/\Delta M$.) The instrument was operated in the electron ionization (EI) mode with 70-eV electrons, a source temperature of 200 °C, the conversion dynode at –5000 V and the secondary electron multiplier at 2400 V. The source and the collector slit widths were adjusted to obtain trapezoidal peaks with flat tops. The GC-MS interface was at 280 °C and high-purity He was used as a carrier gas. Data were acquired in the selected-ion monitoring (SIM) mode using voltage peak switching and the quantitation was based on peak areas.

During later development of these methods at the VA Medical Center in San Diego, CA, a Finnigan MAT 4500 mass spectrometer equipped with a quadrupole mass analyzer interfaced to a Hewlett-Packard 5890 gas chromatograph and a Finnigan A200S autosampler was used. In this instrument, electron ionization (EI), positive chemical ionization (PCI) and electron-capture negative chemical ionization (ECNCI) were used as the ionization modes and the electron energy was set at 70 eV. Methane was used as the reagent gas for chemical ionization, at an ionizer pressure of 0.4 ± 0.05 Torr (1 Torr $= 133.3$ Pa) and at an analyzer pressure of 1.4×10^{-5} Torr. A conversion dynode and a secondary electron multiplier were operated at ± 3000 and 1250 V, respectively. The ionizer temperature was 100 °C and the manifold was kept at 106 °C. Perfluorobutylamine (PFTBA) was used for tuning mass and peak shape for EI, PCI and ECNCI. The GC injector was operated in the splitless mode and the data were acquired using SIM.

13.3 EXPERIMENTAL PROCEDURE

One of the main requirements of using GC-MS for the determination of metals is the preparation of a suitable metal chelate, which should have adequate thermal stability and should also be volatile. Further, the metal chelate should not introduce any memory effect or carry-over effect when injecting sequential samples of different isotopic compositions. We have been able to identify suitable metal chelates for the toxic metals Pb, Hg, Cd and As. Another requirement for the determination of toxic metals in biological samples is the development of a suitable digestion procedure, without introducing any contaminants from the reagents employed.

13.3.1 Preparation of internal standard solutions

The concentration of the toxic metal can be determined by isotope dilution mass spectrometry (IDMS), provided that another stable isotope of the element is available to be used as an internal standard. Since IDMS compensates for most matrix effects, it has the potential of being incorporated into a definitive analytical technique. In addition, the IDMS results are mostly unaffected by incomplete recovery of the analyte during specimen digestion and extraction, which are ordinarily required owing to the complex nature of the biological matrix. The internal standard solution is added to the biological sample prior to carrying out any chemical digestion of the sample. An optimal amount of internal standard is required to obtain the most accurate concentration determination for minimizing random errors in the isotope ratio measurements in spiked samples. A 1:1 ratio of internal standard to analyte provides optimal accuracy in the quantitative measurements.

13.3.2 Digestion of biological sample

Urine samples for Pb and Cd analysis were digested with H_2O_2 (50%, stabilized, from Fisher) and HNO_3 (67–70%, double sub-boiling quartz-distilled in Teflon bottles, NIST) to destroy most of the organic material present. Initially, the urine sample (1 mL) was treated with concentrated HNO_3 and was allowed to stand at room temperature for about 1 h. The specimen was subsequently heated at 50 °C on a hot-plate to reduce the volume to ~50 µL. Then 100 µL of 50% H_2O_2 were added and the solution was again heated gently on the hot-plate. The digestion with H_2O_2 was performed 4–5 times until a white residue remained on evaporating the solution. The dry residue was dissolved in 2 mL of deionized water and the solution was extracted with 2 mL of CH_2Cl_2, discarding the organic phase to remove undigested lipids. This solution was then ready for metal chelate formation.

For determination of Pb in whole-blood samples, 1 mL of blood sample was mixed with 2 mL of 4 M HNO_3, which was added to each of the samples dropwise, with constant vortex mixing. The digest was centrifuged for 30 min and the supernatant was further digested with HNO_3 + H_2O_2 and prepared for chelate formation as described for urine samples.[6] A simpler extraction procedure that eliminated the time-consuming hot digestion procedure is described below for blood Pb[7] and Hg.[9]

The simpler method for blood Pb required 200 µL from a whole blood sample drawn into a metal-free tube containing EDTA as an anticoagulant. This sample is mixed with internal standard and vortex mixed briefly. Then, with gentle vortex mixing, 400 µL of 4 M HNO_3 are added dropwise. The sample is centrifuged at $1760\,g$ for 10 min and the supernate transferred to a clean tube. The pH is adjusted to 7–9 with concentrated NH_4OH. This solution is then ready for the chelate preparation step.

Blood Hg required 200 µL from a whole blood sample drawn into a metal-free tube containing EDTA as an anticoagulant. The sample was mixed with 200 µL of 0.6 M HCl and vortex mixed to free any Hg originally bound to protein thiol groups. Then 1 mL of 0.2 M borate buffer was added. This sample could then be chelated.[9]

13.3.3 Preparation of metal chelate

For Pb, lithium bis(trifluoroethyl)dithiocarbamate [Li(FDEDTC)], synthesized in our laboratory, was tested as a derivatizing agent for GC-MS analysis of Pb(FDEDTC)$_2$. However, this approach proved unsuccessful owing to strong memory effect (carryover). The procedure was modified by reacting this derivative with with the Grignard reagent 4-fluorophenylmagnesium bromide (4-FPMgBr), 2.0 mol L^{-1} in diethyl ether, to produce Pb(FC$_6$H$_4$)$_4$.[6] Pb(FC$_6$H$_4$)$_4$ can also be prepared by using inexpensive commercially available sodium diethyldithiocarbamate or ammonium pyrrolidine dithiocarbamate.[7]

The Pb(FDEDTC)$_2$ chelate was prepared at pH 3, using the digested urine or blood samples mixed with 1 mL of acetic acid–sodium acetate buffer and 100 µL of a 20 mmol L^{-1} solution of Li(FDEDTC) in deionized water. The sample was vortex mixed for 2 min before extracting the Pb(FDEDTC)$_2$ chelate into 1 mL of toluene. The organic extract containing the Pb chelate was evaporated to dryness at 60 °C under a stream of argon gas in a laminar-flow hood. The dried residue was dissolved in 200 µL of diethyl ether and approximately 200 µL of 4-FPMgBr were added. Care was taken to add the Grignard reagent quickly, because it is very reactive. The solution in the tube was shaken to allow complete mixing. Excess Grignard reagent was neutralized by adding 100 µL of 100 mL L^{-1} isopropyl alcohol in toluene (i.e. 10%, v/v) and then adding 1 mL of 1 M HNO_3. The Pb(FC$_6$H$_4$)$_4$ chelate was extracted with 1 mL of toluene, evaporated to dryness and reconstituted in 10–50 µL of CH_2Cl_2 for GC-MS analysis.

For Cd and Hg, Cd(FDEDTC)$_2$ and Hg(FDEDTC)$_2$ chelates were prepared with Li(FDEDTC) as the chelating reagent, using the procedure described above.

For As, two different derivatizing agents, N-$tert$-butyldimethylsilyl-N-methyltrifluoroacetamide (MTBSTFA) with 1% $tert$-butyldimethylchlorosilane (t-BDMS) and 4-fluorophenylmagnesium bromide (4-FPMgBr), were used. The chelates of As(III), As(V) and organic arsenic species such as dimethylarsenic acid (DMAA, also known as cacodylic acid) were prepared. For this purpose, an aqueous solution of As(III) as As_2O_3 in 10% HNO_3, a methanolic solution of As(V) as As_2O_5 and a methanolic solution of DMAA were used. First, t-BDMS derivatives were prepared by treating the digest directly with MTBSTFA. Second, the t-BDMS derivatives were prepared by first synthesizing the pyrrolidinedithiocarbamates of As(III), As(V) and DMAA. The toluene solutions containing these derivatives were evaporated to complete dryness at 40 °C under a steady stream of N_2 gas. The dried residues were dissolved in 200 µL of ethyl acetate and 50 µL of MTBSTFA were added to generate the As derivatives.

13.4 GC-MS STUDIES

13.4.1 Memory effect evaluation

Memory effect refers to the carryover of analyte from a previously injected sample, which compromises the reliability of quantitative results by variably contributing to the measured ion intensities. This memory effect can result from poor volatility of a metal derivative in the GC injection port, high reactivity of the metal derivative with active sites on the GC column, or both. One approach that can be used to evaluate the memory effect is sequential analyses of an unaltered specimen, a specimen enriched with isotope (internal standard), followed by the unaltered sample again. Figures 13.1–13.3 show the results of such studies carried out for Pb, Cd and Hg. As Figure 13.1a demonstrates, there is a strong memory effect when sequentially injecting two $Pb(FDEDTC)_2$ chelate preparations with isotope ratios differing by a factor of 10. In contrast, Figure 13.1b shows that when using the $Pb(FC_6H_4)_4$ derivative in the two samples (one unaltered and the other enriched with [204]Pb) with isotope ratios differing by a factor of 300, there was no measurable memory effect. Figure 13.2 shows the results obtained on the evaluation of memory effect for Cd using $Cd(FDEDTC)_2$ chelate. It is clear that a strong memory effect exists across sequential analyses of specimens containing endogenous amounts of Cd and specimens enriched with [106]Cd internal standard. The concentration of Cd in the enriched specimen, however, is much greater than typical Cd concentrations in urine and whole blood specimens. A good design would maintain the isotope ratios in different samples within reasonable limits, making it possible to determine these isotope ratios experimentally without significant contributions from the memory effect. A small memory effect was observed when analyzing two synthetic mixtures with Cd isotope ratios differing by a factor of about 10, demonstrating that the chelating agent Li(FDEDTC) could be used for Cd determination by GC-MS provided that an optimum sample to internal standard ratio is maintained in the spiked samples. Alternatively, the analysis may be carried out over a limited range of isotope ratios, or the analyses may be run in replicate to detect whether significant carryover has occurred. Figure 13.3 shows the results of Hg analyses using $Hg(FDEDTC)_2$ when two mixtures with isotope ratios differing by a factor of about 40 were injected sequentially into the GC-MS system. No measurable memory effect was observed for the $Hg(FDEDTC)_2$ chelate. The memory effect associated with As cannot be assessed using this method, since it is a monoisotopic element.

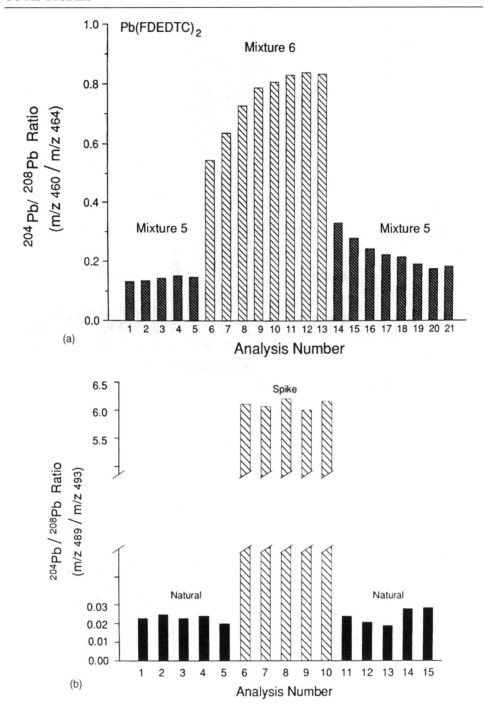

Figure 13.1 (a) Evaluation of cross-contamination in consecutive analyses of two $Pb(FDEDTC)_2$ samples with isotope ratio differing by a factor of ~ 8 (Ref. 6). (b) Evaluation of cross-contamination in consecutive analyses of natural Pb and enriched ^{204}Pb internal standard by using $Pb(FC_6H_4)_4$ with m/z 489:493 ratio differing by a factor of ~ 300 (Ref. 6)

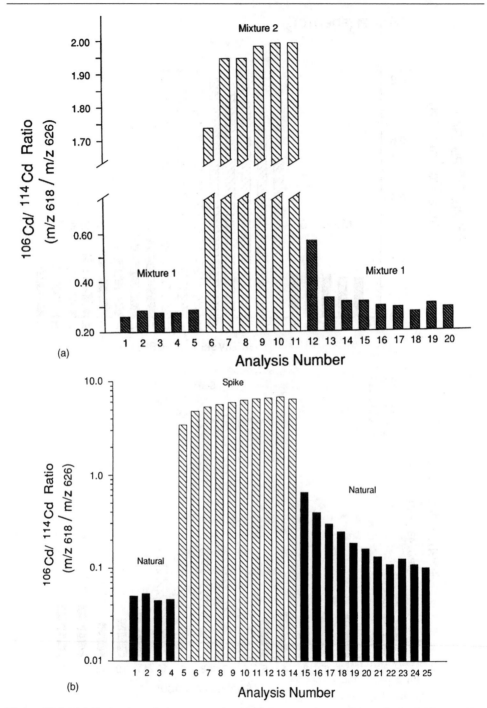

Figure 13.2 (a) Evaluation of cross-contamination in consecutive analyses of two Cd(FDEDTC)$_2$ samples with isotope ratio differing by a factor of ∼10 (Ref. 8). (b) Evaluation of cross-contamination of consecutive analyses of two Cd(FDEDTC)$_2$ samples with isotope ratio differing by a factor of ∼200 (Ref. 8)

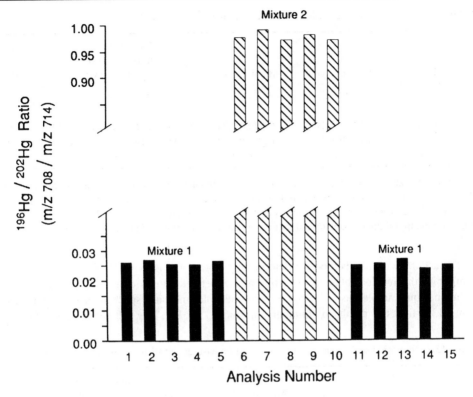

Figure 13.3 Evaluation of cross-contamination of consecutive analyses of two Hg(FDEDTC)$_2$ samples with isotope ratio differing by a factor of 40 (Ref. 4)

13.4.2 Precision and accuracy in measuring isotope ratios

Table 13.1 lists the different metal chelates used for various toxic metals analysis by GC-MS. Table 13.2 summarizes the data on precision and accuracy for isotope dilution measurement of different metals using these metal chelates. These data were generated by injecting 1 μL of the chelate solution containing a few nanograms of the toxic metal. The atom% abundances of different isotopes in an unaltered sample were compared with the calculated abundances of the corresponding ions in the metal chelate, considering the contributions of isotopes of carbon, nitrogen, sulfur, etc., in the molecule. In all cases, there was good agreement between the calculated and experimentally measured abundances of the different isotopes. The data in Table 13.2 are observed values that have not been corrected for any differences in mass discrimination among the isotopes, since any mass discrimination factor would be canceled in isotope dilution. The precision of various isotope ratio measurements was evaluated with chelated endogenous metal over several days. Overall precision was calculated by combining the within-run and between-run precisions. Representative precision data are also included in Table 13.2. It can be seen that typical overall relative precision values of 1–2% are possible using GC-MS of volatile metal chelates when isotope ratios are 0.1 or higher.

Table 13.1 Metal chelates used for different toxic metals in GC-MS[a]

No.	Element	Chelating agent	Metal chelate	Ions containing information on metal in EIMS	Ref.
1	Pb	4-FPMgBr	$Pb(FC_6H_4)_4$	$(M - L)^+$ **(493)**; $(M - 3L)^+$ (308); M^+ (208)	6,7
2	Cd	Li(FDEDTC)	$Cd(FDEDTC)_2$	$M^{+\cdot}$ **(626)**; $(M - L)^+$ (370)	8
3	Hg	Li(FDEDTC)	$Hg(FDEDTC)_2$	$M^{+\cdot}$ **(714)**; $(M - L)^+$ (321)	9
4	As	4-FPMgBr	$As(FC_6H_4)_3$	$M^{+\cdot}$ (360); $(M - 2L)^+$ **(170)**	10

[a]$M^{+\cdot}$ and $(M - zL)^+$ denote the molecular ion and ion obtained after the loss of z ligands, respectively. Ion of the highest intensity is given in bold with m/q value in parentheses, referring to the isotope with the highest percentage abundance in natural element. Li(FDEDTC) and 4-FPMgBr are lithium bis(trifluoroethyl)dithiocarbamate and 4-(fluorophenyl)magnesium bromide, respectively.

The accuracy of this method for determining isotope ratios by GC-MS was evaluated by comparing the results obtained on different synthetic mixtures of Pb with an alternative technique, quadrupole-based inductively coupled plasma mass spectrometry (Q-ICP-MS). Table 13.3 lists the results of comparison studies for Pb. The calculated ratios for m/q

Table 13.2 Precision and accuracy in isotope ratio measurement of toxic metals by GC-MS[a]

No.	Isotope	Atom %	Ion (m/z)	Calculated abundance (%)	Measured abundance (%)	Isotope ratio	Mean of mean values	Within-run precision [CV (%)]	Between-run precision [CV (%)]	Overall precision [CV (%)]
1	Pb-204	1.40	489	1.27	1.31	489/493	0.0256	3.2	8.1	8.7
	Pb-206	24.10	491	21.96	23.26	491/493	0.4528	1.6	0.4	1.7
	Pb-207	22.10	492	24.57	24.04	492/493	0.4679	1.3	0.7	1.5
	Pb-208	52.40	493	52.20	51.38	*	*	*	*	*
2	Cd-106	1.25	618	1.00	1.05	618/626	0.0388	1.8	6.6	6.8
	Cd-108	0.89	620	0.90	0.96	620/626	0.0354	1.3	6.0	6.1
	Cd-110	12.49	622	10.17	10.46	622/626	0.3851	1.2	2.7	3.0
	Cd-111	12.80	623	11.79	12.02	623/626	0.4427	1.2	2.0	2.3
	Cd-112	24.13	624	22.77	22.05	624/626	0.8122	0.6	0.8	1.0
	Cd-113	12.22	625	14.62	15.00	625/626	0.5526	0.9	1.7	2.5
	Cd-114	28.73	626	28.24	27.15	*	*	*	*	*
	Cd-116	7.49	628	10.51	11.30	628/626	0.4161	1.1	2.5	2.7
3	Hg-196	0.14	708	0.11	0.12	708/714	0.0041	10.1	–	10.1
	Hg-198	10.02	710	7.96	8.00	710/714	0.2808	1.6	0.8	1.8
	Hg-199	16.84	711	14.55	14.86	711/714	0.5217	2.1	0.7	2.2
	Hg-200	23.13	712	21.86	21.77	712/714	0.7644	1.5	0.4	1.6
	Hg-201	13.22	713	15.99	16.21	713/714	0.5692	1.5	0.6	1.6
	Hg-202	29.80	714	29.11	28.48	*	*	*	*	*
	Hg-204	6.85	716	10.42	10.56	716/714	0.3709	1.8	1.5	2.3

[a]Asterisks denote the highest abundant isotope in the natural element, used as a reference to represent isotope ratios.

Table 13.3   Comparison of Pb isotope ratios in synthetic mixtures by GC-MS and Q-ICP-MS

Synthetic mixture	$^{204}Pb/^{208}Pb$ atom ratio by Q-ICP-MS	m/z 489:493		
		Calculated from ICP-MS data	Measured by GC-MS	Measured (GC-MS)/ calculated (ICP-MS)
1	0.3177	0.290	0.302	1.042
2	0.4884	0.445	0.452	1.015
3	0.8269	0.752	0.751	0.999
4	1.1511	1.045	1.032	0.988
Mean				1.011

489:493 in $Pb(FC_6H_4)_4$ were obtained by using the experimentally determined Pb isotope ratios by Q-ICP-MS and including the contributions of carbon isotopes. It is apparent from data in Table 13.3 that results obtained using the GC-MS method agree closely with the same measurements by ICP-MS.

13.4.3 Results of concentration determination of toxic metals in biological samples

The IDMS method, using GC-MS, was validated by determining the toxic metals in NIST certified reference materials (SRM 2670 Urine). Prior to formation of the chelate, a suitable wet digestion procedure was used, taking care to eliminate any adventitious contamination from reagents, laboratory ware, environment and personnel. A procedure based on $HNO_3 + H_2O_2$ was used for urine samples. For serum, the reference samples were deproteinated using concentrated nitric acid added dropwise with continuous vortex mixing of the mixture. For Pb and Hg determination, the whole blood specimens were not deproteinated. Table 13.4 summarizes the results obtained for some of the toxic metals by isotope dilution GC-MS; the results obtained were very close to the certified values.

Table 13.5 presents the results of Pb measurements in whole blood specimens obtained from College of American Pathologists (CAP) proficiency testing program and from NIST SRM 955a Lead in Blood. The results demonstrate excellent agreement between GC-MS

Table 13.4   Toxic metals determined by isotope dilution GC-MS in NIST Standard Reference Materials

Element	Matrix	Concentration ($\mu g\,L^{-1}$)	
		Determined	Expected
Pb	Urine	105 ± 4 ($n = 4$)	109 ± 4
	Blood	160 ± 5 ($n = 3$)	150 ± 40
		431 ± 9 ($n = 3$)	430 ± 40
Cd	Urine	94 ± 10 ($n = 6$)	88 ± 3
		9.3 ± 1 ($n = 2$)	8.4 ± 0.3

Table 13.5 Determination of Pb in proficiency and reference standards ($\mu g\,L^{-1}$)

Sample source	Expected value	ID-GC-MS ($n = 3$)	CV (%)	Difference (%)
CAP	105	102	4.9	−2.9
	209	210	0.1	0.5
	316	313	5.4	−0.9
	435	455	1.8	4.6
	483	514	5.1	6.4
NIST SRM 955a	50.1	50.1	11.0	0
	135.3	140.5	4.1	3.8
	306.3	307.3	1.1	0.3
	544.3	547.2	0.6	0.5

and expected values over the range of clinically relevant Pb concentrations. These data were collected using electron-capture negative chemical ionization (ECNCI) and quadrupole mass spectrometry, and involved ambient temperature extraction of Pb from whole blood without hot digestion.

13.5 CONCLUSIONS

GC-MS is a useful analytical technique for the determination of toxic metals in biological specimens. It has the potential to provide relatively high throughput and can be adapted to quadrupole mass spectrometers commonly available in clinical and environmental toxicology laboratories. The isotope dilution GC-MS methodology has been validated for determination of Pb and Cd using NIST certified biological reference materials, e.g. urine and blood. Hg was validated using a standard additions method and ID-MS.

REFERENCES

1. S.K. Aggarwal, M. Kinter, R.L. Fitzgerald and D.A. Herold, Mass spectrometry of trace elements in biological samples. *Crit. Rev. Clin. Lab. Sci.* **31**, 35–87 (1994).
2. S.K. Aggarwal and D.A. Herold, Gas chromatography: separation of metal chelates. In *The Encyclopedia of Analytical Science*, A. Townshend *et al.* (Eds), Academic Press; London, 1995, pp. 1883–1888.
3. D.A. Herold, S.K. Aggarwal and M. Kinter, Trace metal analysis by gas chromatography/mass spectrometry: a less-invasive analytical technique. *Clin. Chem.* **38**, 1647–1649 (1992).
4. D.A. Herold, S. K. Aggarwal and M. Kinter, Analysis by gas chromatography–mass spectrometry. In *Handbook on Metals in Clinical and Analytical Chemistry*, H. G. Seiler, A. Sigel and H. Sigel (Eds). Marcel Dekker, New York, 1994, pp. 149–165.
5. P. Brunmark, G. Skarping and A. Schutz, Determination of methylmercury in human blood using capillary gas chromatography and selected-ion monitoring. *J. Chromatogr. Biomed. Appl.* **573**, 35–41 (1992).
6. S.K. Aggarwal, M. Kinter and D.A. Herold, Determination of lead in urine and whole blood by stable isotope dilution gas chromatography–mass spectrometry, *Clin. Chem.* **40**, 1494–1502 (1994).

7. G.S. Baird, R.L. Fitzgerald, S.K. Aggarwal and D.A. Herold, Determination of blood lead by electron-capture negative chemical ionization gas chromatography–mass spectrometry. *Clin. Chem.* **42**, 286–291 (1996).

8. S.K. Aggarwal, R.G. Orth, J. Wendling, M. Kinter and D.A. Herold, Isotope dilution gas chromatography–mass spectrometry for cadmium determination in urine. *J. Anal. Toxicol.* **17**, 5–10 (1993).

9. S.K. Aggarwal, M. Kinter and D.A. Herold, Mercury determination in blood by gas chromatography–mass spectrometry. *Biol. Trace Elem. Res.* **41**, 89–102 (1994).

10. S.K. Aggarwal, R.L. Fitzgerald and D.A. Herold, Mass spectral studies for the determination and speciation of arsenic by gas chromatography–mass spectrometry (GC-MS). In *Proceedings of Eighth ISMAS Symposium on Mass Spectrometry, 7–9 December 1999, Hyderabad*, Vol. II, S.K. Aggarwal (Ed.). ISMAS: Mumbai, 1999, pp. 553–556.

Index

Note: page numbers in **bold** refer to tables and those in *italic* refer to figures.

Chromatographic Methods in Clinical Chemistry and Toxicology   Edited by R. L. Bertholf and R. E. Winecker
© 2007 John Wiley & Sons, Ltd